Amorphous Semiconductors

Wiley Series in Materials for Electronic and Optoelectronic Applications

www.wiley.com/go/meoa

Series Editors

Professor Arthur Willoughby, *University of Southampton, Southampton, UK*
Dr Peter Capper, Formerly of *SELEX Galileo Infrared Ltd, Southampton, UK*
Professor Safa Kasap, *University of Saskatchewan, Saskatoon, Canada*

Published Titles

Bulk Crystal Growth of Electronic, Optical and Optoelectronic Materials, Edited by P. Capper
Properties of Group-IV, III–V and II–VI Semiconductors, S. Adachi
Charge Transport in Disordered Solids with Applications in Electronics, Edited by S. Baranovski
Optical Properties of Condensed Matter and Applications, Edited by J. Singh
Thin Film Solar Cells: Fabrication, Characterization and Applications, Edited by J. Poortmans and V. Arkhipov
Dielectric Films for Advanced Microelectronics, Edited by M. R. Baklanov, M. Green and K. Maex
Liquid Phase Epitaxy of Electronic, Optical and Optoelectronic Materials, Edited by P. Capper and M. Mauk
Molecular Electronics: From Principles to Practice, M. Petty
CVD Diamond for Electronic Devices and Sensors, Edited by R. S. Sussmann
Properties of Semiconductor Alloys: Group-IV, III–V and II–VI Semiconductors, S. Adachi
Mercury Cadmium Telluride, Edited by P. Capper and J. Garland
Zinc Oxide Materials for Electronic and Optoelectronic Device Applications, Edited by C. Litton, D. C. Reynolds and T. C. Collins
Lead-Free Solders: Materials Reliability for Electronics, Edited by K. N. Subramanian
Silicon Photonics: Fundamentals and Devices, M. Jamal Deen and P. K. Basu
Nanostructured and Subwavelength Waveguides: Fundamentals and Applications, M. Skorobogatiy
Photovoltaic Materials: From Crystalline Silicon to Third-Generation Approaches, G. Conibeer and A. Willoughby
Glancing Angle Deposition of Thin Films: Engineering the Nanoscale, Matthew M. Hawkeye, Michael T. Taschuk and Michael J. Brett
Spintronics for Next Generation Innovative Devices, Edited by Katsuaki Sato and Eiji Saitoh
Physical Properties of High-Temperature Superconductors, Rainer Wesche
Inorganic Glasses for Photonics - Fundamentals, Engineering, and Applications, Animesh Jha

Forthcoming

Materials for Solid State Lighting and Displays, Adrian Kitai

Amorphous Semiconductors

Structural, Optical, and Electronic Properties

Kazuo Morigaki,
Emeritus, University of Tokyo, Japan

Sándor Kugler,
Department of Theoretical Physics,
Budapest University of Technology and Economics,
Hungary

Koichi Shimakawa
Centre of Innovative Photovoltaic Systems,
Gifu University, Japan

This edition first published 2017

Registered Office
John Wiley & Sons Ltd, The Atrium, Southern Gate, Chichester, West Sussex, PO19 8SQ, United Kingdom

For details of our global editorial offices, for customer services and for information about how to apply for permission to reuse the copyright material in this book please see our website at www.wiley.com.

Library of Congress Cataloging-in-Publication Data

Names: Morigaki, Kazuo, author. | Kugler, Sándor, 1950– author. | Shimakawa, Koichi, author.
Title: Amorphous semiconductors : structural, optical, and electronic properties / Kazuo Morigaki, Sándor Kugler and Koichi Shimakawa.
Description: Chichester, West Sussex, United Kingdom : John Wiley & Sons, Ltd, [2017] | Includes bibliographical references and index.
Identifiers: LCCN 2016040352| ISBN 9781118757925 (cloth) | ISBN 9781118758205 (epub) | ISBN 9781118757949 (Adobe PDF)
Subjects: LCSH: Amorphous semiconductors.
Classification: LCC QC611.8.A5 M 2017 | DDC 537.6/226–dc23
LC record available at https://lccn.loc.gov/2016040352

A catalogue record for this book is available from the British Library.

Set in 10/12pt Warnock by SPi Global, Pondicherry, India

Printed and bound in Malaysia by Vivar Printing Sdn Bhd

10 9 8 7 6 5 4 3 2 1

Contents

Series Preface xi
Preface xiii

1 Introduction 1
1.1 General Aspects of Amorphous Semiconductors 1
1.2 Chalcogenide Glasses 3
1.3 Applications of Amorphous Semiconductors 3
References 3

2 Preparation Techniques 5
2.1 Growth of a-Si:H Films 5
2.1.1 PECVD Technique 5
2.1.2 HWCVD Technique 6
2.2 Growth of Amorphous Chalcogenides 6
References 8

3 Structural Properties of Amorphous Silicon and Amorphous Chalcogenides 11
3.1 General Aspects 11
3.1.1 Definitions of Crystalline and Noncrystalline 11
3.2 Optical Spectroscopy 12
3.2.1 Raman Scattering 12
3.2.2 Infrared Absorption 13
3.3 Neutron Diffraction 15
3.3.1 Diffraction Measurements on Amorphous Silicon 17
3.3.2 Diffraction Measurements on Hydrogenated Amorphous Silicon 18
3.3.3 Diffraction Measurements on Amorphous Germanium 19
3.3.4 Diffraction Measurements on Amorphous Selenium 19

3.4 Computer Simulations 20
3.4.1 Monte Carlo-Type Methods for Structure Derivation 20
3.4.2 Atomic Interactions 21
3.4.3 a-Si Models Constructed by Monte Carlo Simulation 25
3.4.4 Reverse Monte Carlo Methods 26
3.4.5 a-Si Model Constructed by RMC Simulation 28
3.4.6 a-Se Model Constructed by RMC Simulation 30
3.4.7 Molecular Dynamics Simulation 32
3.4.8 a-Si Model Construction by Molecular Dynamics Simulation 34
3.4.9 a-Si:H Model Construction by Molecular Dynamics Simulation 34
3.4.10 a-Se Model Construction by Molecular Dynamics Simulation 35
3.4.11 Car and Parrinello Method 38
References 38

4 Electronic Structure of Amorphous Semiconductors 43
4.1 Bonding Structures 43
4.1.1 Bonding Structures in Column IV Elements 44
4.1.2 Bonding Structures in Column VI Elements 45
4.2 Electronic Structure of Amorphous Semiconductors 46
4.3 Fermi Energy of Amorphous Semiconductors 47
4.4 Differences between Amorphous and Crystalline Semiconductors 49
4.5 Charge Distribution in Pure Amorphous Semiconductors 49
4.6 Density of States in Pure Amorphous Semiconductors 52
4.7 Dangling Bonds 54
4.8 Doping 57
References 58

5 Electronic and Optical Properties of Amorphous Silicon 61
5.1 Introduction 61
5.2 Band Tails and Structural Defects 62
5.2.1 Introduction 62
5.2.2 Band Tails 62
5.2.3 Structural Defects 66
5.3 Recombination Processes 68
5.3.1 Introduction 68
5.3.2 Radiative Recombination 68

5.3.3 Nonradiative Recombination 70
5.3.4 Recombination Processes and Recombination Centers in a-Si:H 72
5.3.5 Spin-Dependent Recombination 73
5.4 Electrical Properties 74
5.4.1 DC Conduction 74
5.4.2 AC Conduction 80
5.4.3 Hall Effect 87
5.4.4 Thermoelectric Power 88
5.4.5 Doping Effect 89
5.5 Optical Properties 92
5.5.1 Fundamental Optical Absorption 92
5.5.2 Weak Absorption 94
5.5.3 Photoluminescence 96
5.5.4 Frequency-Resolved Spectroscopy (FRS) 96
5.5.5 Photoconductivity 101
5.5.6 Dispersive Photoconduction 109
5.6 Electron Magnetic Resonance and Spin-Dependent Properties 112
5.6.1 Introduction 112
5.6.2 Electron Magnetic Resonance 112
5.6.3 Spin-Dependent Properties 128
5.7 Light-Induced Phenomena and Light-Induced Defect Creation 131
5.7.1 Introduction 131
5.7.2 Light-Induced Phenomena 132
5.7.3 Light-Induced Defect Creation 134
References 145

6 Electronic and Optical Properties of Amorphous Chalcogenides 157
6.1 Historical Overview of Chalcogenide Glasses 157
6.1.1 Applications 157
6.1.2 Science 158
6.2 Basic Glass Science 159
6.2.1 Glass Formation 159
6.2.2 Glass Transition Temperature 160
6.2.3 Crystallization of Glasses 162
6.3 Electrical Properties 165
6.3.1 Electronic Transport 165
6.3.2 Ionic Transport 170
6.4 Optical Properties 175
6.4.1 Fundamental Optical Absorption 175
6.4.2 Urbach and Weak Absorption Tails 178

6.4.3 Photoluminescence 179
6.4.4 Photoconduction 183
6.5 The Nature of Defects, and Defect Spectroscopy 191
6.5.1 Electron Spin Resonance 196
6.5.2 Optical Absorption 197
6.5.3 Primary Photoconductivity 197
6.5.4 Secondary Photoconductivity 197
6.5.5 Electrophotography 199
6.5.6 Electronic Transport 199
6.6 Light-Induced Effects in Chalcogenides 200
6.6.1 Electron Spin Resonance 200
6.6.2 Optical Absorption 202
6.6.3 Photoluminescence 203
6.6.4 Photoconductivity 205
6.6.5 Electronic Transport 206
6.6.6 Defect Creation Kinetics 207
6.6.7 Structure-Related Properties 210
References 218

7 Other Amorphous Material Systems 231
7.1 Amorphous Carbon and Related Materials 231
7.1.1 Basic Structure of a-C (sp^2 Hybrids) 232
7.1.2 Preparation Techniques 233
7.1.3 Brief Review of Structural Studies on Amorphous Carbon 233
7.1.4 Applications 234
7.2 Amorphous Oxide Semiconductors 235
7.2.1 Preparation Techniques 235
7.2.2 Optical Properties 236
7.2.3 Electronic Properties 237
7.2.4 Applications 239
7.3 Metal-Containing Amorphous Chalcogenides 239
7.3.1 Preparation Techniques 240
7.3.2 Structure of Ag-Chs and Related Physical Properties 240
7.3.3 Photodoping 241
7.3.4 Applications 242
References 242

8 Applications 247
8.1 Devices Using a-Si:H 247
8.1.1 Photovoltaics 247
8.1.2 Thin-Film Transistors 248

8.2 Devices Using a-Chs 249
8.2.1 Phase-Change Materials 249
8.2.2 Direct X-ray Image Sensors for Medical Use 257
8.2.3 High-Gain Avalanche Rushing Amorphous Semiconductor Vidicon 258
8.2.4 Optical Fibers and Waveguides 260
References 261

Index **265**

Series Preface

Wiley Series in Materials for Electronic and Optoelectronic Applications

This book series is devoted to the rapidly developing class of materials used for electronic and optoelectronic applications. It is designed to provide much-needed information on the fundamental scientific principles of these materials, together with how these are employed in technological applications. The books are aimed at (postgraduate) students, researchers and technologists, engaged in research, development and the study of materials in electronics and photonics, and industrial scientists developing new materials, devices and circuits for the electronic, optoelectronic and communications industries.

The development of new electronic and optoelectronic materials depends not only on materials engineering at a practical level, but also on a clear understanding of the properties of materials, and the fundamental science behind these properties. It is the properties of a material that eventually determine its usefulness in an application. The series therefore also includes such titles as electrical conduction in solids, optical properties, thermal properties, and so on, all with applications and examples of materials in electronics and optoelectronics.

The characterization of materials is also covered within the series in as much as it is impossible to develop new materials without the proper characterization of their structure and properties. Structure–property relationships have always been fundamentally and intrinsically important to materials science and engineering.

Materials science is well known for being one of the most interdisciplinary sciences. It is the interdisciplinary aspect of materials science that has led to many exciting discoveries, new materials and new applications. It is not unusual to find scientists with a chemical engineering background working on materials projects with applications in electronics. In selecting titles for the series, we have tried to maintain the interdisciplinary aspect of the field, and hence its excitement to researchers in this field.

Arthur Willoughby
Peter Capper
Safa Kasap

Preface

This book deals with amorphous semiconductors, which are typical disordered systems. Many textbooks and monographs concerning amorphous semiconductors have already been published. In this book, we treat their structural, optical, and electronic properties; in particular, recent developments in these areas are included and are emphasized as features of the book, because since amorphous semiconductors exhibit a disordered nature, there exist difficult problems to be resolved in the interpretation of experimental results. So, we attempt to present recent experimental data and interpretations in order to help in obtaining a deeper understanding of these properties.

The book is composed of eight chapters. The basic ideas for understanding the properties of amorphous semiconductors mentioned above are presented in these chapters, particularly preparation techniques in Chapter 2 and theoretical fundamentals in Chapter 4. In Chapter 3, structural properties are treated, with particular emphasis on neutron diffraction measurements and structural modeling of computer simulations. In Chapter 5, hydrogenated amorphous silicon is treated and emphasis is put on recent developments in the nature of shallow and deep states and of light-induced defects, particularly as investigated by magnetic resonance measurements. In Chapter 6, amorphous chalcogenides are treated, with particular emphasis on their basic glass science, their unique properties, and light-induced effects in them. In Chapter 7, other amorphous materials are treated, particularly amorphous carbon and amorphous semiconducting oxides. In Chapter 8, applications of amorphous semiconductors are mentioned, particularly those of hydrogenated amorphous silicon and amorphous chalcogenides. The eight chapters were written by the authors as follows: Kazuo Morigaki (KM) wrote Chapters 1 and 5 (except for Section 5.4.2, by (KS)), Sándor Kugler (SK) wrote Chapters 3 and 4, and Koichi Shimakawa (KS) wrote Chapters 2, 6, 7 (with contributions from SK), and 8. We hope that the book will be useful for students in physics, chemistry, and materials science as well as for researchers and engineers interested in amorphous semiconductors.

KS acknowledges Professors Keiji Tanaka (Hokkaido University, Japan), Safa Kasap (University of Saskatchewan, Canada), Jai Singh (Charles Darwin University, Australia), Takeshi Aoki (Tokyo Polytechnic University, Japan), Tomas Wagner, Miloslav Frumar (University of Pardubice, Czech Republic), and Noboru Yamada (Kyoto University, Japan), and Dr. Alexander Kolobov (Advanced Institute of Science and Technology, Tsukuba, Japan) for many fruitful discussions. KS also wishes to thank the Grant Project ReAdMat funded by the ESF for financial support. SK expresses his thanks to Krisztian Kohary (University of Exeter, UK) for helpful discussions. SK must also thank Tokyo Polytechnic University for providing him with computer facilities for his large-scale computer simulations.

KM acknowledges Professor Harumi Hikita (Meikai University, Urayasu, Japan) for preparing some of the figures in Chapter 5.

Kazuo Morigaki
Sándor Kugler
Koichi Shimakawa

1 Introduction

1.1 General Aspects of Amorphous Semiconductors

Amorphous solids are typical disordered systems. Two classes of disorder can be defined, namely, compositional disorder as seen in crystalline binary alloys and topological disorder as seen in liquids. Amorphous solids have topological disorder. However, short-range order, that is, chemical bonding of constituent atoms, exists in amorphous covalent semiconductors. Spatial fluctuations in the bond lengths, bond angles, and dihedral angles (Figure 1.1) give rise to tail states in the band gap region, that is, below the edge of the conduction band and above the edge of the valence band. The edges of the conduction and valence bands are called mobility edges and the band gap is called a mobility gap. These edges are the boundaries between delocalized and localized states. This is illustrated in Figure 1.2. Such boundaries are caused by disorder; this is called Anderson localization. In an amorphous network, translational order does not exist, so the Bloch theory of crystalline solids is not applicable, but the tight-binding model, the Hartree–Fock approximation or the density functional method can be applied for understanding the electronic properties of amorphous semiconductors, as shown in Chapter 4.

Although the above considerations are based on a continuous random network, actual samples have a structure deviating from an ideal random network, that is, the coordination of the constituent atoms deviates from the normal coordination following the $8-N$ rule [1], where N denotes the relevant column number in the periodic table. Here, we consider elements only in columns IV–VI of the periodic table. An additional rule can be given as follows: Z (the valency) $= N$ if $N < 4$. For instance, the normal coordination of amorphous silicon is fourfold, but threefold-coordinated silicon atoms are also present. These are called structural defects. For amorphous selenium, since the normal coordination is twofold, the structural defects are onefold- and threefold-coordinated selenium atoms. The band tails and structural defects affect the optical and

Amorphous Semiconductors: Structural, Optical, and Electronic Properties, First Edition.
Kazuo Morigaki, Sándor Kugler, and Koichi Shimakawa.

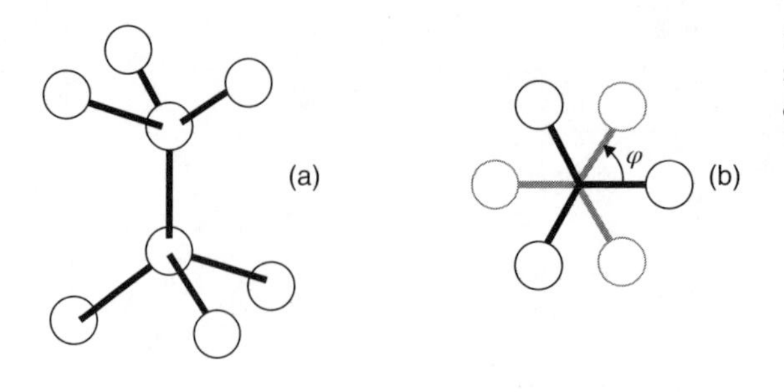

Figure 1.1 Short-range order structure and dihedral angle φ in amorphous silicon.

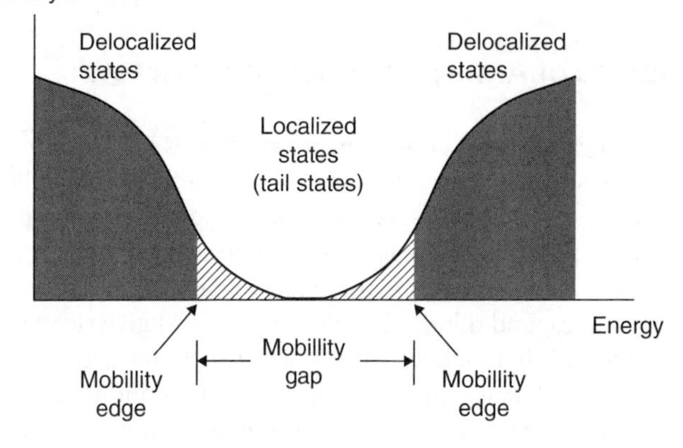

Figure 1.2 Schematic illustration of density of states in an amorphous semiconductor. See text for details.

electronic properties of amorphous semiconductors. Thus it is very important to elucidate the electronic structures of these states in order to understand these properties. An important experimental means of doing this is magnetic resonance. For instance, the electronic structures of these localized states can be elucidated from electron spin resonance (ESR); that is, their symmetry can be determined from a g-value measurement. However, the principal axes of symmetry are randomly oriented in an amorphous network, and that it makes more difficult to identify defects in amorphous semiconductors than in crystals. This identification is normally performed by comparison between observed ESR spectra and computer-simulated spectra. In addition, electron–nuclear double resonance (ENDOR) provides us with a powerful means for identification of defects, as shown in Chapter 5, in which pulsed electron magnetic resonance measurements in particular are presented in detail.

In this book, we consider two types of material as examples of amorphous semiconductors, namely, amorphous silicon and other column IV elemental semiconductors, and amorphous chalcogenides, including amorphous metal chalcogenides. Their preparation, structure, and optical and electronic properties are presented in Chapters 2, 3, 5, 6, and 7. In Chapter 3, definitions of crystalline and noncrystalline structures are given, and the structures of amorphous silicon, hydrogenated amorphous silicon (a-Si:H), and amorphous selenium are treated theoretically and experimentally in more detail.

1.2 Chalcogenide Glasses

Amorphous chalcogenides are also known as chalcogenide glasses, because they exhibit a glass transition. The details of the glass transition and the structural, optical, and electronic properties of these material are dealt with in Chapter 6.

1.3 Applications of Amorphous Semiconductors

Amorphous semiconductors are widely today used as device materials. Devices using a-Si:H include, for example, solar cells and thin-film transistors. Devices using amorphous chalcogenides include, for example, phase-change memories, direct x-ray image sensors for medical use, high-gain avalanche rushing amorphous semiconductor vidicons, and optical fibers and waveguides. These are treated in Chapter 8.

There are several comprehensive books about amorphous semiconductors [1–5], as well as books about hydrogenated amorphous silicon [6] and amorphous chalcogenides [7].

References

1 Mott, N.F. and Davis, E.A. (1979) *Electronic Processes in Non-Crystalline Materials*, 2nd edn, Clarendon Press, Oxford.
2 Elliott, S.R. (1990) *Physics of Amorphous Materials*, Longman Scientific and Technical, Harlow.
3 Morigaki, K. (1999) *Physics of Amorphous Semiconductors*, World Scientific, Singapore and Imperial College Press, London.
4 Singh, J. and Shimakawa, K. (2003) *Advances in Amorphous Semiconductors*, Taylor & Francis, London and New York.
5 Kugler, S. and Shimakawa, K. (2015) *Amorphous Semiconductors*, Cambridge University Press, Cambridge.
6 Street, R.A. (1991) *Hydrogenated Amorphous Silicon*, Cambridge University Press, Cambridge.
7 Tanaka, K. and Shimakawa, K. (2011) *Amorphous Chalcogenide Semiconductors and Related Materials*, Springer, New York.

2

Preparation Techniques

There are many preparation techniques, depending on what kind of material is needed. Quenching from the liquid state, also termed melt-quenching (MQ), is a popular technique for so-called glasses. To meet the requirements of thin-film forms, a variety of techniques such as evaporation, sputtering, and chemical vapor deposition (CVD) are adopted. Ion bombardment and the use of powerful light on crystalline solids can also produce amorphous materials. Here we briefly summarize the principal methods for preparing hydrogenated amorphous silicon (a-Si:H) films and amorphous chalcogenides.

2.1 Growth of a-Si:H Films

a-Si:H films, as well as a-Ge:H, a-C:H, and related compound films, are prepared only by condensation from the gas phase. The most popular preparation techniques for these films belong to the class of CVD techniques. Two CVD techniques are known: one is *glow-discharge deposition*, which is now also called *plasma-enhanced chemical vapor deposition* (PECVD), and the other is *hot-wire chemical vapor deposition* (HWCVD), as described in the following.

2.1.1 PECVD Technique

This was previously called glow-discharge deposition, when a-Si:H films were deposited by decomposition of SiH_4 gas with the help of a glow discharge [1]. It is now termed plasma-enhanced chemical vapor deposition [2]. The application of an RF field (usually at 13.6 MHz; RF PECVD) generates a plasma in a reaction chamber. By introducing PH_3 or B_2H_6 gas into the SiH_4, n- or p-type Si:H films can be prepared [1]. RF PECVD provides high-quality uniform films, but the deposition rate can be slow as 0.3 nm/s. The development of

Amorphous Semiconductors: Structural, Optical, and Electronic Properties, First Edition.
Kazuo Morigaki, Sándor Kugler, and Koichi Shimakawa.

this technique, with control of the gas pressure (0.05–2 Torr), RF power (10–100 mW/cm^2), substrate temperature (150–350 °C), and so on, has led to a huge range of commercial applications, such as large-area photovoltaics and thin-film transistors.

To improve the deposition rate, PECVD with a very high-frequency field (40–100 MHz), called VHF PECVD, has been employed, and the deposition rate has reached 2 nm/s [3]. VHF PECVD also provides high-quality films. However, its poor uniformity is still a problem. Microwave PECVD (MW PECVD) with a 2.45 GHz field provides a significantly enhanced deposition rate, reaching 10 nm/s, but device quality films are not easy to obtain [4]. It is known, however, that VHF PECVD and MW PECVD are very useful means for preparing microcrystalline Si films (μc-Si:H), as well as a-Ge:H films [5], whereas the deposition rate of RF PECVD for μc-Si:H is very low. A high deposition rate is a necessary condition for cost-effective performance in large-scale production.

2.1.2 HWCVD Technique

When we are interested in a very high deposition rate of a-Si:H and related films, HWCVD can be a promising technique, and the deposition rate for this technique has reached 15–30 nm/s [6]. The setup for an HWCVD system is similar to that for RF PECVD except that the RF electrode is replaced with a heated filament (Pt, W, Ta, etc.). This technique is thus *thermal deposition*, in principle. The gas introduced into a chamber is catalytically excited or decomposed into radicals or ions by a metal filament heated to around 1800–2000 °C. Then Si radicals diffuse inside the chamber and are deposited onto a substrate, kept at a relatively high temperature (150–450 °C). Although device quality films can be obtained with a high deposition rate by this technique, a drawback is poor uniformity of the deposited films, and great efforts are being made to overcome this.

2.2 Growth of Amorphous Chalcogenides

Strictly speaking, the term *glass* usually refers to materials prepared from the liquid state. On the other hand, the term *amorphous* is used for materials made by quenching from the gas phase. However, the actual structural and electronic properties are not very different, and hence these two terms are frequently used without distinguishing between them. We shall thus simply use the terms "amorphous chalcogenides" and "chalcogenide glasses" interchangeably. Although many preparation techniques are possible for these materials, depending on what kinds of material are required, the most popular one is quenching from the liquid state (melt-quenching, MQ). So-called glasses and glass fibers are prepared by this method. When thin films are required, techniques such as evaporation (EV), sputtering (SP), and pulsed laser deposition (PLD) are adopted. In the following, we briefly explain the above-mentioned techniques:

1) *Quenching from the liquid state.* Most chalcogenide, as well as oxide, glasses are prepared by the MQ method. When the temperature of the melt decreases, the melt becomes a *supercooled liquid.* A further decrease in temperature below the glass transition temperature T_g produces a glassy material. Thus, glasses prepared by the MQ method undergo a glass transition. It is of interest to note that the empirical relation $T_g \approx 2T_m/3$ holds for most glasses, where T_m is the melting temperature. Chalcogenide materials are sealed in quartz ampoules at 10^{-6} Torr and heated beyond their melting temperature. The samples are agitated by rotation or vibration and then cooled quickly by immersion in cold water or air at ambient temperature. Glasses are formed owing to the relatively high cooling rate. Details will be discussed in Chapter 6, and Figure 6.1 may help in understanding glass formation.
2) *Evaporation.* This is the most conventional method of preparing films. The films are simply condensed at an arbitrary temperature onto a substrate, directly from the gas phase. Thus the T_g used for MQ materials does not have any physical meaning for EV materials. The starting material comprises ingots or powder mounted in a boat and is heated above the melting temperature in a vacuum chamber at around 10^{-6} Torr ($\sim 10^{-4}$ Pa). The deposition rate (~ 10 nm/s) is controllable and depends on the boat temperature. The thickness of the film is usually controlled by the deposition time. Some well-known amorphous chalcogenides, for example Se, $As_2Se(S)_3$, and $GeSe(S)_2$, are prepared using this technique. Evaporation of multiple compounds is not easy and hence so-called *flash evaporation* techniques are often employed, in which powdered components are poured into a heated boat. The main disadvantage of EV films is that the composition of the resulting film is not always the same as that of the starting material. Device quality a-Se thick films (more than 1 mm thick), which are utilized commercially in direct X-ray image detectors [7] and high-gain avalanche rushing amorphous photoconductor (HARP) vidicon TV cameras [8], can be prepared by the EV technique.
3) *Sputtering.* In the sputtering process, a strong RF electric field (a frequency of 13 MHz is usually used) is applied between a target made from the source material and electrodes made from the substrate. A sputtering gas, for example Ar or N, is introduced into the chamber and kept at a pressure of 0.13–2.7 Pa. Positively ionized gas produced by the RF field supplies kinetic energy to the surface of the source material target. Atoms or molecules dissociated from the target are deposited onto the substrate. The deposition

rate (~1 nm/s) is lower than that for the EV method (~10 nm/s), although it depends strongly on the sputtering conditions. SP is superior to EV for multicomponent systems; the composition of the resulting films is almost the same as that of the target material because the sputtering rates of different elements are of the same order. When a magnetic field is applied, the deposition rate is significantly enhanced. This is due to confinement of the ionized gas in a spiral motion by the Lorentz force produced by the magnetic field. This method is called *magnetron sputtering* (MSP). The MSP method is suitable for large-scale production. So-called phase-change materials, such as the $Ge_2Sb_2Te_5$ films used in optical memory devices (DVDs) and electrical memory devices (phase-change random access memory, PRAM) are usually prepared using the MSP technique (Section 6.6).

4) *Pulsed laser deposition.* PLD is very simple technique compared with SP. A pulsed high-power laser beam is focused onto a target material in a vacuum chamber, causing vaporization of the source material. This technique resembles the EV method. In PLD, however, a directional plasma plume is created by the absorption of photons, and hence this method is distinct from simple thermal evaporation [9]. The pulse energy, 300–600 mJ/pulse, induces an instantaneous power density of around 10^9 W/cm^2. Excimer gas (KrF or ArF) lasers are used for this purpose. Stoichiometric transfer of atoms from multicomponent source materials to the substrate can be performed by the PLD technique, and hence this technique can be useful for preparing device quality phase-change chalcogenides such as $Ge_2Sb_2Te_5$ films.

We have discussed only *physical* deposition techniques for amorphous chalcogenides here. However, there are other techniques, for example chemical deposition techniques such as CVD, that are suitable for a-Si:H. The CVD technique is not popular for amorphous chalcogenides.

References

1 Spear, W.E. and Le Comber, P.G. (1975) Substitutional doping of amorphous silicon. *Solid State Commun.*, **17**, 1193–1196.

2 Shiff, E.A., Hegedus, S., and Deng. X. (2010) Amorphous silicon-based solar cells, in *Handbook of Photovoltaic Science and Engineering*, 2nd edn (eds A. Luque, S. Hegedus, and W. Shire), John Wiley & Sons, Inc., Hoboken, NJ, pp. 487–545.

3 Chen, Y., Wang, J., Liu, J., Zheng, W., Gu, J., Yang, S, and Gao, X. (2008) Microcrystalline silicon grown by VHF PECVD and the fabrication of solar cells. *Solar Energy*, **82**, 1083–1087.

4 Mejia, S.R., McLeod, R.D., Kao, K.C., and Card, H.C. (1983) The effects of deposition parameters on a-Si:H films fabricated by microwave glow discharge techniques. *J. Non-Cryst. Solids*, **59&60**, 727–730.

5 Aoki, T., Kato, S., Hirose, M., and Nishikawa, Y. (1989) DC bias effects on growth of a-Ge:H in coaxial type ECR plasma. *Japan. J. Appl. Phys.*, **28**, 849–855.

6 Mahan, A., Xu, Y., Nelson, B., Crandal, R., Cohen, J.D., Palinginis, K., and Gallagher, A. (2001) Saturated defect densities of hydrogenated amorphous silicon grown by hot-wire chemical vapour deposition at rates up to 150 Å. *Appl. Phys. Lett.*, **78**, 3788–3790.

7 Rawlands, J.A. and Kasap, S.O. (1997) Amorphous semiconductors usher in digital X-ray imaging. *Phys. Today*, **50**, 24–31.

8 Tanioka, K. (2007) The ultra sensitive TV pickup tube from conception to recent development. *J. Mater. Sci.*, **18**, S321–S325.

9 Delahoy, A.E. and Guo, S. (2010) Transparent conducting oxides for photovoltaics, in *Handbook of Photovoltaic Science and Engineering*, 2nd edn (eds A. Luque, S. Hegedus, and W. Shire), John Wiley & Sons, Inc., Hoboken, NJ, pp. 716–796.

3

Structural Properties of Amorphous Silicon and Amorphous Chalcogenides

3.1 General Aspects

Amorphous materials are not new. People have been preparing glassy materials (i.e., SiO_2) for thousands of years. The earliest glaze known is one on stone beads from the Badarian age in Egypt, from about 12 000 BC. Green glaze was applied to powdered quartz for making small figures in about 9000 BC. The oldest pure glass is an amulet of deep lapis lazuli color, from about 7000 BC. In the last century, iron-rich siliceous glassy materials estimated to be a billion years old were recovered from the Moon! Amorphous materials have been applied in electrophotography, or xerography (from the Greek for "dry writing"). In the first machines, in the 1930s, a metal cylinder called the drum was coated with amorphous selenium.

The start of a period of very intensive research on amorphous semiconductors, including the structure of these materials, can be traced to work carried out by Chittick, Alexander, and Sterling in 1969 [1] at Standard Telephone Laboratories, Harlow, UK, and by Spear and Le Comber in 1975 [2] in Dundee, UK. These authors demonstrated that amorphous silicon can be doped. From that point on, amorphous silicon became the basis of a multibillion-dollar market, with diverse applications including solar cells, electrophotography, and liquid crystal TV displays.

3.1.1 Definitions of Crystalline and Noncrystalline

There is a considerable amount of confusion in the scientific literature concerning the terms "noncrystalline," "amorphous," "glassy," "vitreous," "randomness," "disorder," "liquid," and even "crystalline." The first question we put is whether an atomic structure is crystalline or noncrystalline. A perfect crystal is one in which atoms or groups of atoms are arranged periodically in three dimensions and has an infinite extent. This model of the atomic configuration provides us with simple methods for calculating various properties of such

Amorphous Semiconductors: Structural, Optical, and Electronic Properties, First Edition.
Kazuo Morigaki, Sándor Kugler, and Koichi Shimakawa.

condensed matter. Unfortunately, these methods are useless in the amorphous case. A more realistic arrangement, an imperfect crystal, is one where the atoms are found in a pattern that repeats periodically but only has a finite extent. Real crystals are not only finite in size but also contain imperfections such as vacancies, interstitial (foreign or self-) atoms, dislocations, impurities, and distortions associated with the surface. Furthermore, at finite temperature the random motion of the atoms about their equilibrium position destroys the perfect periodicity in a snapshot of the crystal. These defects cause distortion in the crystal lattice, but we do not consider such crystals to be amorphous solids.

Until 1992, a crystal was defined by the International Union of Crystallography as "a substance, in which the constituent atoms, molecules, or ions are packed in a regularly ordered, repeating three-dimensional pattern." In 1984 Shechtman *et al.* published a paper [3] on rapidly solidified alloys of Al with 10–14% Mn that possessed icosahedral symmetry in combination with long-range order; these were named quasicrystals. Since this discovery, hundreds of materials with similar atomic structures have been synthesized in laboratories. The International Union of Crystallography had to modify their statement, and the new and broader definition of a crystal became "any solid having an essentially discrete diffraction diagram." Now a possible definition of amorphous materials can be given as follows: *Amorphous materials are in the condensed phase and do not possess long-range translational order (periodicity) in the atomic sites. They have no essentially discrete diffraction diagram.*

In this terminology, the set of amorphous materials has a fundamental subset called glasses. A glass is an amorphous solid which exhibits a glass transition. If a liquid (melt) is cooled very rapidly so that crystallization is bypassed, then a disordered structure can be frozen in. This disordered condensed phase is known as a glass. Such a glass-forming process involves supercooling of a liquid below its normal freezing point. The transformation from a melt to a glassy phase is a transition where a discontinuity can be observed in the second-order thermodynamic variables, such as the calorimetric heat capacity at constant pressure, $C_p = T(dS/dT)_p = -T(d^2G/dT^2)_p$, where G is the Gibbs free energy; $G(T, p, N) = E - TS + pV$.

3.2 Optical Spectroscopy

3.2.1 Raman Scattering

Raman scattering measurements provide us with useful information about the degree of disorder and about chemical bonding. Figure 3.1 shows the Raman spectrum of hydrogenated amorphous silicon (a-Si:H), where the Raman spectrum of crystalline silicon (c-Si) is also shown for comparison. A peak at

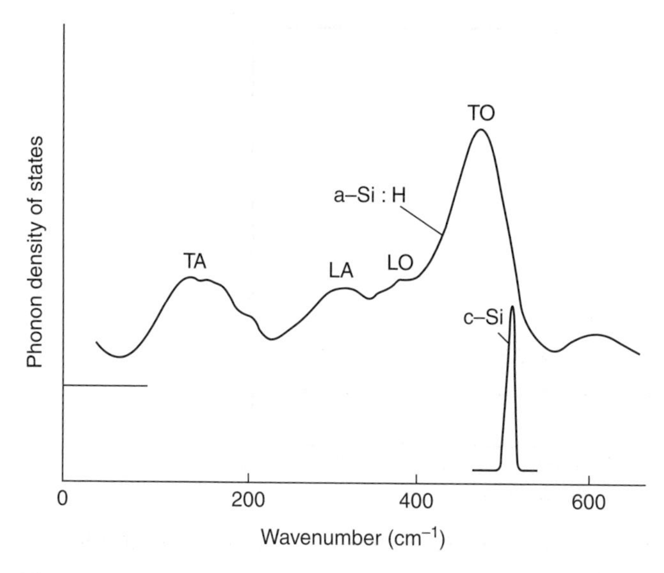

Figure 3.1 Raman spectra of a-Si:H and crystalline silicon. *Source:* R.A. Street 1991 [5].

480 cm^{-1} is associated with the transverse optical (TO) mode of the network vibration and corresponds to a peak at 520 cm^{-1} in c-Si. The shift of the TO band peak in a-Si:H to a lower frequency and its broadening compared with c-Si are related to the relaxation of the momentum selection rule caused by disorder in the network. A quantitative measure of the disorder can be obtained from the shift and broadening of the TO band compared with c-Si [4]. Raman spectroscopic measurements and Hartree–Fock *ab initio* calculations of Raman spectra have been carried out for amorphous selenium in order to identify the characteristic vibrational mode due to sigma bonds [5]. In these Raman spectra, the peaks around a wavenumber of 250 cm^{-1} correspond to vibrational modes of 0.234 nm covalent bonds in the amorphous material, as displayed in Figure 3.2.

3.2.2 Infrared Absorption

Infrared (IR) absorption measurements can be used to estimate the hydrogen content and the Si–H bonding scheme in a-Si:H. The Si–H bonding scheme, that is, the content of SiH, SiH_2, and SiH_3 groups, is obtained by observing IR absorption bands associated with vibration modes, that is, stretching, bending (rocking or wagging), and scissor bending modes, as shown in Figure 3.3 [6]. The vibration frequencies of each mode are illustrated in Figure 3.4. The IR absorption intensities depend on the characteristics of the samples, as shown in Figure 3.3. The device quality sample exhibits only IR vibrations due to SiH

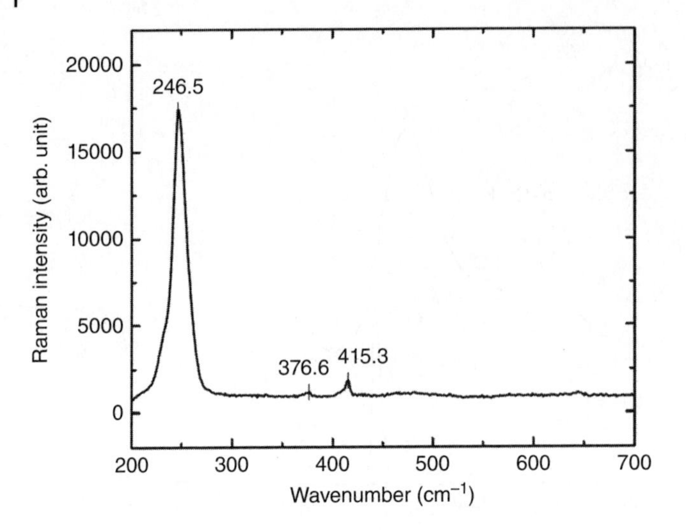

Figure 3.2 Raman spectrum of a-Se. *Source:* R. Lukacs *et al.* 2010 [5]. Reproduced with permission from AIP Publishing LLC.

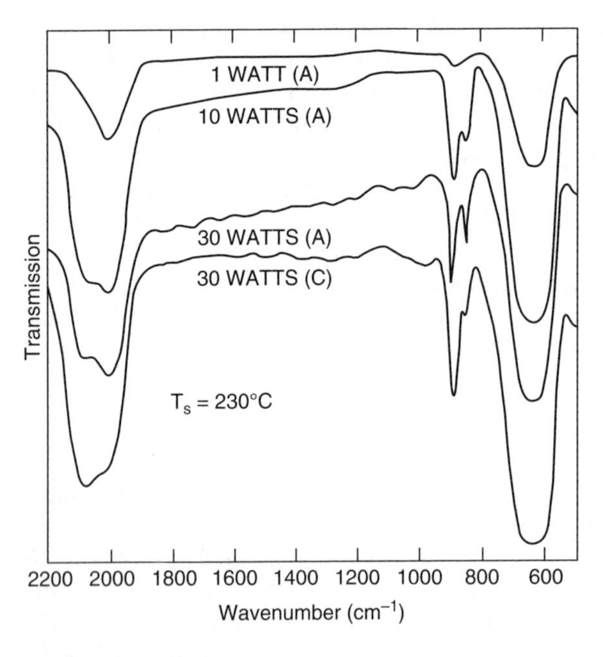

Figure 3.3 IR absorption bands of a-Si:H associated with vibration modes, that is, stretching, bending (rocking or wagging), and scissor bending. *Source:* G. Lucovsky *et al.* 1979 [6]. Reproduced with permission from the American Physical Society.

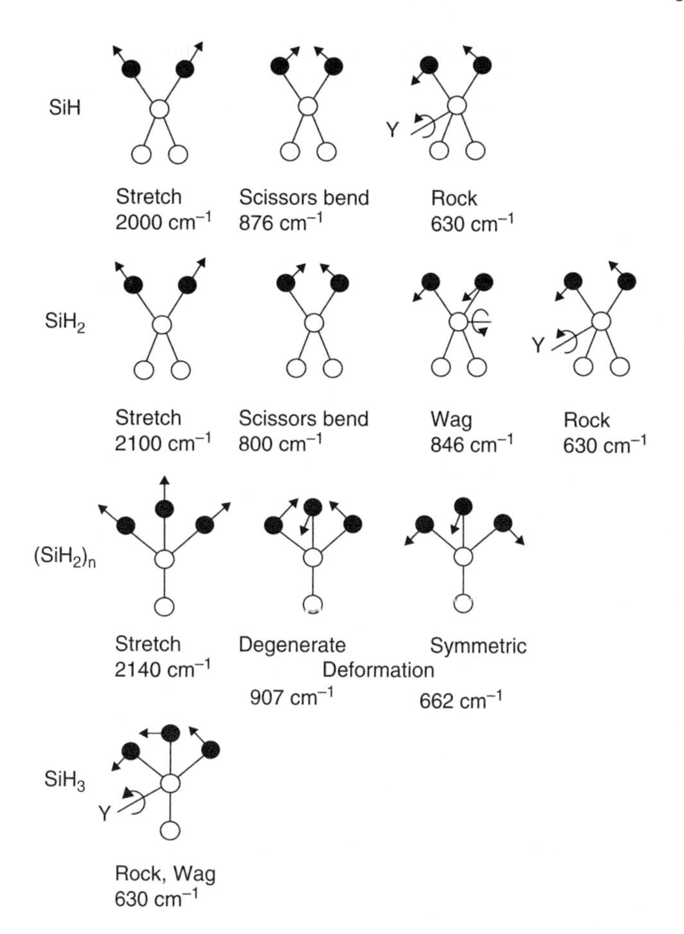

Figure 3.4 Vibration frequencies of each mode. *Source:* G. Lucovsky *et al.* 1979 [6]. Reproduced with permission from the American Physical Society.

groups. The Si–H bonding scheme plays an important role in the disorder in the network and also in the electronic properties of a-Si:H.

3.3 Neutron Diffraction

The atomic-scale structure of materials is the most essential information for deriving various physical properties of samples of those materials. The main result of any diffraction measurement is a projection from a three-dimensional atomic structure to a one-dimensional function. The probability of finding a particle at a distance *r* when another particle is located at the origin provides a possible characterization of the geometric structure at the microscopic level.

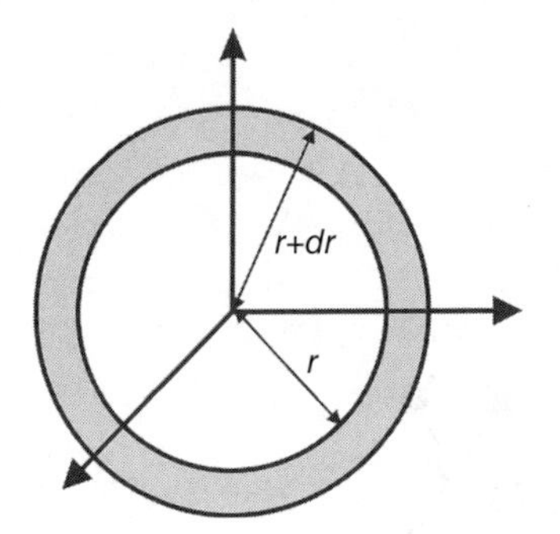

Figure 3.5 Schematic illustration of the radial distribution function: r and $r + dr$ are the distances between which the atomic distribution is calculated.

In the crystalline case, diffraction experiments produce diffractograms with characteristic Bragg reflections, evidenced by well-defined sharp peaks at given angular positions. Amorphous semiconductors do not possess long-range translational order in their atomic sites and therefore have no discrete diffraction pattern.

Two one-dimensional functions are usually applied to describe three-dimensional atomic distributions. When the atomic distribution in a given direction is being investigated, the pair correlation (distribution) function $g(r)$ is the most convenient function. $g(r)\,dV$ is proportional to the probability of finding an atom inside a volume dV. The radial distribution function (RDF), $J(r)$, represents the number of atoms between distanced of r and $r+dr$ away from a given particle (Figure 3.5). Amorphous materials are isotropic, and hence both functions provide us with the same information because of the relationship $J(r)\,dV = g(r)4\pi r^2\,dr$. These functions exhibit alternating maxima and minima following the coordination shells, and they must be zero below a given distance related to atomic repulsion. The correlation with the atom at the origin is lost at large r. This implies that $g(r)$ tends to unity or to the density, depending on the normalization, while the RDF has a parabolic envelope as r goes to infinity. The positions of the peaks in $J(r)$ provide the radii of the different coordination shells of atoms surrounding an average atom. The areas under the peaks can be used to derive the coordination numbers of the shells of atoms.

The structures of amorphous semiconductors are typically determined by neutron, electron, or X-ray diffraction experiments. Neutrons, having no charge, interact with the nuclei of the atoms, while X-rays and electrons are sensitive to the electron distribution in amorphous materials. A neutron beam is capable of penetrating well beyond the surface of a sample, to a depth of centimeters in a condensed phase. Thermal neutrons, having a de Broglie wavelength of 0.1–0.2 nm, are very convenient for atomic-scale structural investigations with usable resolution. In contrast to X-ray and electron diffraction, neutrons are scattered by nuclei and there is no relationship between atomic number and cross section; that is, different isotopes of the same element can have widely different cross sections. Neutron diffraction is a nondestructive technique and is a form of elastic scattering. Neutron cross sections can be divided into two parts: the coherent and incoherent cross sections. The elastic part of the coherent scattering gives information about the structure. In a standard neutron diffraction measurement, $g(r)$ and $J(r)$ can be derived from the measured scattered intensity by Fourier transforming the static structure factor $S(\boldsymbol{Q})$, where $\boldsymbol{Q} = \boldsymbol{k}_{\text{initial}} - \boldsymbol{k}_{\text{final}}$, and where $\boldsymbol{k}_{\text{initial}}$ and $\boldsymbol{k}_{\text{final}}$ are the initial and final wave vectors of the neutrons:

$$\rho\left(g(\boldsymbol{r})-1\right)=\frac{1}{(2\pi)^3}\int d\boldsymbol{Q}\left(S(\boldsymbol{Q})-1\right)e^{-i\boldsymbol{Q}\boldsymbol{r}}$$

It is assumed that the scattering is isotropic in amorphous materials, and hence this three-dimensional Fourier transformation can be reduced to one-dimensional sine transformations:

$$\rho\left(g(r)-1\right)=\frac{1}{2\pi^2 r}\int_0^\infty Q\left[S(Q)-1\right]\sin(Qr)\,dQ$$

where $Q = (4\pi/\lambda)\sin\theta$ and r are already scalar values; here λ is the de Broglie wavelength of the neutrons, and 2θ is the scattering angle. An excellent review [7] of neutron diffraction in noncrystalline materials has been published by Cuello.

3.3.1 Diffraction Measurements on Amorphous Silicon

The first electron diffraction measurements on amorphous silicon (a-Si) were carried out by Moss and Graczyk in 1969 [8] and by Barna *et al.* in 1977 [9]. Some years later, a pure evaporated a-Si sample (weighing 0.5 g) was prepared at the Central Research Institute for Physics, Budapest, and a neutron diffraction experiment was performed on this sample at the 7C2 spectrometer installed on the hot source at the Orphée reactor at Saclay, France [10]. Using an incident wavelength of $\lambda = 0.0706$ nm, a momentum transfer range of 5 to 160 nm^{-1} was covered. A second neutron diffraction measurement was performed on the same a-Si sample using the D4 twin-axis diffractometer at the high-flux reactor at the Institut Laue-Langevin, Grenoble, France [11]. The incident neutron wavelength of 0.04977 nm and the angular ranges of 1.5–65° and 46–131° covered by the two multidetectors provided a larger momentum transfer range of 3.3 to 230 nm^{-1}. The results of these experiments demonstrated that the covalently bonded a-Si films were not completely disordered. The bond lengths and coordination numbers were similar to those in the crystalline phase. Compared with a perfect crystal, the first- and second-neighbor peaks for a-Si were broadened but the positions were in the same locations, while the third peak completely disappeared in the measured RDF. This fact is one of the most important properties of column IV amorphous semiconductors. The measured mean square width of the first peak suggests that the bond length fluctuations are around 1–2%. The broadening of the second peak reflects a bond angle fluctuation of around 10%. The absence of the third peak confirms that there is no characteristic dihedral angle.

Neutron diffraction measurements were performed at the Special Environment Powder Diffractometer at the Argonne National Laboratory Intense Pulsed Neutron Source, USA, on an a-Si sample, as deposited and annealed at 600 °C [12]. This sample was prepared by RF sputtering under a low Ar pressure. The radial distribution functions indicated values of 108.4° and 108.6° for the tetrahedral angle in the as-deposited and the annealed film, respectively, which are smaller than the ideal angle of 109.47°. The estimated bond-angle widths were 9.7° and 11.3° for the as-deposited and annealed films, respectively. The coordination number in the annealed sample was 3.90, whereas for the other sample, a value of 3.55 was estimated, which is smaller than the expected value of 4.

In addition to neutron diffraction measurements, the structure factor $S(Q)$ of high-purity a-Si membranes prepared by self-ion implantation was measured over a large, extended Q range of 0.3–550 nm^{-1} by X-ray diffraction using a Huber six-circle diffractometer at the Cornell High Energy Synchrotron Source, Ithaca, USA, using the A2 wiggler beam line [13]. The sample was kept at 10 K to minimize thermal effects on the diffraction pattern. Mean values of the coordination number of 3.79 for the as-implanted sample and 3.88 for the annealed sample were observed. In these nanovoid-free samples the density deficit of a-Si relative to c-Si is due to a fundamental undercoordination; that is, dangling bonds can decrease the local density.

3.3.2 Diffraction Measurements on Hydrogenated Amorphous Silicon

A combined investigation using the complementary features of steady-state and pulsed neutron sources was carried out on a-Si:H [14]. Hydrogen has an incoherent cross section an order of magnitude larger than that of silicon. This causes serious difficulty, but substituting H with deuterium (D) eliminates this problem. The aim of this study was to provide accurate data concerning the short-range order in a-Si:H. Samples of a-Si:H and a-Si:D were prepared by sputtering and by a glow discharge method. Samples weighing 1–1.5 g were measured on the D4 twin-axis spectrometer in the steady state at ILL, Grenoble, and on the LAD spectrometer at the ISIS spallation neutron source, Rutherford Appleton Laboratory, Chilton, UK. Values of 0.148 and 0.321 nm were identified for the Si–H distances and 0.234 and 0.375 nm for the Si–Si first- and second-neighbor distances, as can be seen in Figure 3.6.

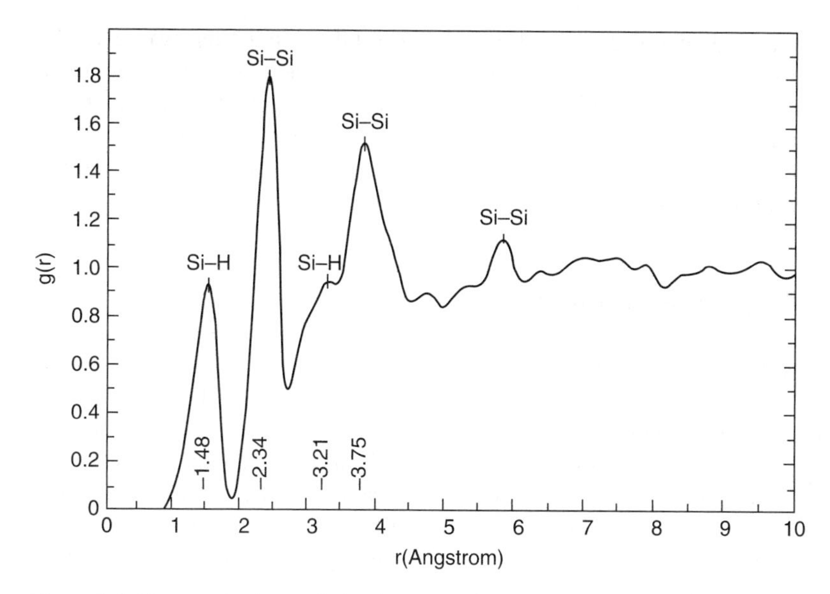

Figure 3.6 Measured pair correlation function of a-Si:H. *Source:* R. Bellissent *et al.* 1989 [14]. Reproduced with permission of Elsevier.

3.3.3 Diffraction Measurements on Amorphous Germanium

A large volume, weighing 9.6 g, of evaporated amorphous germanium, was contained in an 11.7 mm diameter thin-walled vanadium can for a neutron diffraction experiment. The experiment was carried out at ILL on the D4 twin-axis diffractometer [15]. The neutron wavelength was 0.05057 nm. The observed nearest-neighbor coordination number, $n_1 = 3.68$, was significantly less than the value of 4 expected for tetrahedral bonding. The second peak yielded a coordination number $n_2 = 12.11$, which is somewhat greater than that expected for tetrahedral bonding. The most important conclusion was that the structure of amorphous Ge is not adequately described by quasi-crystalline models based on the diamond, wurtzite, Si III, or Ge III polymorphs. (The last two are metastable crystalline phases with distorted tetrahedral bonding, produced by subjecting the normal forms to high pressure.) A much better fit can be achieved with random network models. The structure of amorphous Ge is accurately described by a tetrahedrally bonded random network.

3.3.4 Diffraction Measurements on Amorphous Selenium

Neutron diffraction measurements were performed on the SLAD instrument at Studsvik NFL, Nyköping, Sweden [16]. 3 g of crystalline Se powder was ball milled under an argon atmosphere for 6 h in a Spex mixer/mill. The milling procedure consisted of milling for 15 min followed by 45 min of rest to avoid heating, and then repeating the cycle. The total milling time was 6 h. The amorphous Se powder was contained in a thin-walled vanadium container. The experimental structure factor $S(Q)$ obtained was compared with three other values measured earlier in the interval of 8–120 nm^{-1} and pair correlation functions were presented, derived from unconstrained reverse Monte Carlo computer simulations.

3.4 Computer Simulations

The results that can be obtained from diffraction measurements are only a one-dimensional representation of the spatial atomic distribution in three dimensions. A knowledge of the three-dimensional atomic arrangement is an essential prerequisite for understanding the physical and chemical properties of materials. The main problem is the derivation of the three-dimensional structure from a one-dimensional function. This projection causes an information loss because an unlimited number of different possible disordered atomic structures can display the same one-dimensional pair correlation function. Unfortunately, no experimental technique has been discovered so far for the determination of microscopic atomic distributions. Atomic-resolution holography is a special method for direct imaging of the structure around a selected atom. Holograms with atomic resolution have been recorded using various kinds of de Broglie waves, for example electrons [17, 18] and neutrons [19, 20]. Recently, a method of magnetic neutron holography has been developed [21]. Despite these pioneering steps, the disordered atomic arrangement still remains a problem. Efforts to develop modeling techniques for analyzing structures with atomic resolution are therefore continually being made. In the last few decades, research has focused on the construction of realistic atomic configurations for amorphous materials using computer simulations. Computations cannot substitute for the lack of knowledge of the atomic-scale configuration but are a useful tool for constructing three-dimensional atomic structure models. Two main possible types of methods for structural modeling of disordered configurations exist: stochastic methods, namely Monte Carlo (MC)-type methods, and deterministic methods, namely molecular dynamics (MD) simulations.

3.4.1 Monte Carlo-Type Methods for Structure Derivation

The name "Monte Carlo" arises from the famous casino in Monaco, where probability plays an important role. Monte Carlo methods use random numbers to govern the atomic displacements during a simulation processes. Several different algorithms are called MC methods: the traditional MC, reverse MC, quantum MC, kinetic MC, and path integral MC methods, and so on. In this section, only the first two techniques are treated. An important condition is that only systems with a large number of degrees of freedom can be investigated by MC methods! The traditional MC method for generating three-dimensional configurations of particles in amorphous materials searches for a local minimum of the total energy on the energy hypersurface. In order to calculate the energy, we need to describe the influence of surrounding atoms on each atom. Two different interactions are usually applied in the calculations: classical empirical potentials, and others based on various quantum mechanical approaches such as the tight-binding model, density functional theory, or Hartree–Fock approximations.

The initial configuration can be any atomic arrangement except a perfect crystal, which provides the absolute minimum of the total energy on the energy hypersurface. A randomly chosen atom is displaced to a new position determined by means of a random number generator. A commonly used constraint in these simulations is the maximum displacement. The energy variation is the most important parameter in the MC method. If an MC step provides a downhill motion on the energy hypersurface, then the new configuration is accepted and this reordered structure becomes the initial atomic arrangement for the next MC step. In the uphill case, the new position is rejected conditionally. This decision is the essential part of MC simulations. A simple explanation for the selection rule is the following: if uphill motion is always rejected, than the procedure reaches the first shallow energy minimum and there is no chance of finding a deeper minimum, which probably belongs to a more realistic atomic configuration. A correct description of this MC philosophy can be found in any elementary textbook on statistical physics.

3.4.2 Atomic Interactions

Atomic interactions are needed for such simulations. The simplest versions of atomic interactions are the classical empirical potentials. The use of empirical potentials is computationally the cheapest method. The parameters describing the atomic interactions can be derived either from experimental observations or from advanced quantum mechanical calculations. For semiconductor structures, simple pair potentials such as Morse or Lennard-Jones potentials are not suitable because the directional nature of sigma bonds must be taken into account. Therefore, the potential must have at least two- and three-body terms. A short-range potential means that the total potential energy of a given particle is dominated by interactions with neighboring atoms that are closer than a cut-off distance. A well-known, very frequently used empirical potential is the Keating potential [22] for a three-atom local interaction, which has two macroscopically measurable parameters. It was originally designed to describe defect-free diamond-type crystals, such as silicon, germanium, and diamond-structure carbon, where the atoms have four nearest neighbors. The bond-stretching and bond-bending force constants α and β are determined by fitting to the macroscopic elastic properties of the crystal phase. The bond length in the equilibrium state is equal to r_0 (0.155, 0.235, and 0.245 nm for C, Si, and Ge, respectively) and the bond angle is the tetrahedral angle $\Theta = \arccos(-1/3) = 109.47°$. In the Keating potential, bonding geometries which deviate from these equilibrium values are assigned an energy penalty

$$U = \frac{3}{8r_0^2}\left[\frac{\alpha}{2}\sum_{ij}^{2N}\left(r_{ij}^2 - r_0^2\right)^2 + \beta\sum_{jik}^{6N}\left(r_{ij}\,r_{ik} + r_0^2/3\right)^2\right]$$

where r_{ij} is the vector pointing from atom i to j and $r_{ij}r_{ik}$ is the scalar product of two vectors. This formula is practically a Taylor expansion where the linear terms vanish and only the quadratic terms are kept. For silicon, $\alpha = 296.5\,\text{eV/nm}^2$ and $\beta = 0.285\alpha$ are obtained. The validity of a large deviation from the equilibrium values is questionable; this potential is suitable only for small displacements from the ideal positions. In particular, it should not be used for very disordered systems. Another disadvantage is that the adjacency matrix of the topological structure must be known for calculation of the total energy. This fact prevents the application of this potential to problems such as diffusion and other dynamical processes.

Several modified versions of the Keating potential have been developed. Fullerenes, nanotubes, and graphenes are formed from threefold-coordinated carbon atoms. Overney, Zhong, and Tománek reparameterized the potential [23]. Instead of the last term $r_0^2/3$, we must put $r_0^2/2$ because the bond angle is equal to $\Theta = 120°$ here. The bond-stretching and bond-bending force constants (α, β) were determined. Rücker and Methfessel obtained a generalization of the Keating model for group IV semiconductors and their alloys [24]. First, the energy of mixing relative to the pure materials is separated into two terms, namely, U_{chemical} and U_{strain}. The chemical term arises because the strength of an AB bond is different from the average of the AA and BB bond strengths. Secondly, higher-order anharmonic terms are included in order to describe strongly distorted systems. A modified Keating potential can be found in the literature even for group III–V ($Al_xGa_{1-x}As$) alloys [25]. The anharmonic model potential contains Coulomb interaction terms in the formulas because of charge transfer between different types of atoms.

The Stillinger–Weber empirical potential [26], which was developed for the liquid and crystalline phases of silicon, is based on two and three-body interactions:

$$U = A\varepsilon\left\{\sum_{ij} v_{ij}^{(2)}\left(r_{ij}\right) + \frac{\lambda}{A}\sum_{jik} v_{jik}^{(3)}\left(r_{ij}, r_{ik}\right)\right\}$$

This formula contains two functions and several fitted parameters, which were fitted to the crystalline and liquid phases of silicon. A modified Stillinger–Weber potential was fitted directly to the amorphous phase of silicon by Vink *et al.* [27]. It turned out that the strength of the three-body interaction had to be boosted by approximately 50% to describe this phase correctly. Recently, a new parameterization of this potential has been presented for defects in and plasticity of silicon-based materials [28]. An improved version of the potential was suggested by Justo *et al.* [29]. The environment-dependent interatomic potential in this case contains similar two- and three-body terms, which depend on the local environment of atom i through its effective coordination number.

Another class of empirical potentials for silicon and multicomponent covalent systems was developed by Tersoff [30–33]. Sometimes these potentials are called Abell–Tersoff potentials, recalling the pioneering work on empirical pseudopotential theory by Abell [34]. The pair potential is

$$U = \frac{1}{2}\sum_{i,j\neq i} f(r_{ij})\left[A\exp(-\lambda_1 r_{ij}) - B_{ij}\exp(-\lambda_2 r_{ij})\right]$$

where r_{ij} is the distance between atoms i and j, and A, B, λ_1, and λ_2 are all positive, and $f(r_{ij})$ is a cutoff function to restrict the range of the potential. The first term is repulsive, while the second is interpreted as representing bonding as in a Morse-type potential. B_{ij} implicitly includes the bond order and depends upon the local environment. Deviations from the simple pair potential are ascribed to the dependence of B_{ij} upon the local environment. The bonding strength B_{ij} should be a monotonically decreasing function of the number of competing bonds, the strength of the bonds, and the cosines of the angles with the competing bonds. The mathematical terms in the potential contain 64 fitting parameters.

The other important class of amorphous semiconductors is chalcogenides. The model element for these is selenium. An empirical three-body potential has been developed by Oligschleger *et al.* [35] with the intention of providing a description of the selenium–selenium interaction that is both realistic and simple. The most common selenium crystals are built up from either selenium rings or infinite helical chains; in both structures every selenium atom has two nearest neighbors, and this unique property has to be reproduced by the interatomic potential. Oligschleger *et al.* took both small selenium clusters and crystalline phases into account during the parameter fitting of the potential. Its analytical form is

$$U = \sum_{i<j} V_2(r_{ij}) + \sum_{i<j<k} h(r_{ij}, r_{kj}, \theta_{ijk}) + \text{cyclic permutations}$$

In the two functions here, altogether 23 parameters are fitted to density functional calculations. The potential has been carefully tested and compared with experimental results.

3.4.2.1 Quantum Mechanical Energy Calculations

Tight-binding models offer a semiempirical quantum chemical approach to describing the interaction between atoms. Such models are useful for generating and describing the atomic structure of amorphous semiconductors. We focus only on the orthogonal tight-binding models that have been developed for silicon and selenium. The total energy of the system is written as

$$E_{\text{tot}} = E_{\text{b}} + E_{\text{rep}}$$

where E_{b} is the sum of the electronic eigenvalues ϵ_i over all occupied electronic states, given by

$$E_{\text{b}} = 2\sum_{i}^{n_{\text{occ}}} \epsilon_i$$

and E_{rep} is the short-range repulsive energy. The electronic eigenvalues are obtained by diagonalizing an empirical tight-binding Hamiltonian. The on-site elements are the atomic orbital energies of the corresponding atoms, whereas the off-diagonal elements of the tight-binding Hamiltonian are described by a set of orthogonal sp^3 two-center hopping parameters $Vss\sigma$, $Vsp\sigma$, $Vpp\sigma$, and $Vpp\pi$ scaled with the interatomic separation r as a function $s(r)$. The remainder of E_{tot} is modeled by a short-range repulsive term E_{rep} given by

$$E_{\text{rep}} = \sum_{\text{i}} f\left(\sum_{\text{j}} \Phi(r_{ij})\right)$$

where $\Phi(r_{ij})$ is a pairwise potential between atoms i and j, and f is expressed as a fourth-order polynomial. In order to model an amorphous structure, the tight-binding model has to be transferable. This can be achieved by adopting a suitable functional form, in this case the one suggested by Goodwin, Skinner, and Pettifor [36]. Thus the scaling functions $s(r)$ and $\Phi(r)$ are given by

$$s(r) = (r_0/r)^n \exp\left\{n\left[-\left(\frac{r}{r_{\text{c}}}\right)^{n_{\text{c}}} + (r_0/r_{\text{c}})^{n_{\text{c}}}\right]\right\}$$

and

$$\Phi(r) = \Phi_0 (d_0/r)^m \exp\left\{m\left[-\left(\frac{r}{d_{\text{c}}}\right)^{m_{\text{c}}} + (d_0/d_{\text{c}})^{m_{\text{c}}}\right]\right\}$$

where r_0 denotes the nearest-neighbor atomic separation in the diamond structure. Parameters for silicon can be found in a paper by Kwon *et al.* [37]. A similar formalism has been used for selenium by Molina *et al.* [38]. A more sophisticated tight-binding model for silicon has been developed by Wang, Pan, and Ho [39].

A more accurate approach is to derive the atomic interaction directly from the electronic ground state, where the total energy functional is calculated using density functional theory (DFT). The difficulty of DFT is in obtaining an accurate exchange and correlation interaction in a many-body system. Nowadays this is a rapidly developing and very popular method in solid state physics. The computational cost is much higher than for the tight-binding method and is comparable to that of the Hartree–Fock *ab initio* method, which is not the best for structure investigation.

3.4.3 a-Si Models Constructed by Monte Carlo Simulation

After some model construction by hand, the first traditional MC simulation for pure a-Si and a-Ge was performed by Wooten, Winer, and Weaire (WWW) [40]. The main purpose was to construct a defect-free network. In this model, every atom had four nearest neighbors. The initial configuration was simply a cubic diamond structure, and some bond transpositions were carried out (Figure 3.7). This process destroyed crystal symmetries, and the result included fivefold and sevenfold rings. After this process, the structure was relaxed using a Keating empirical potential and the Metropolis MC algorithm. The continuous random

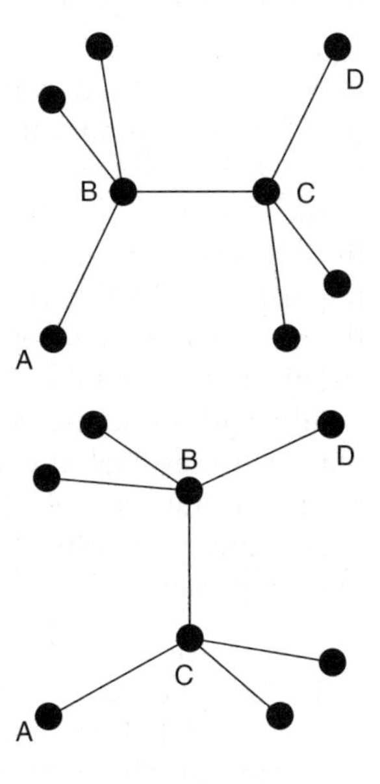

Figure 3.7 Bond transposition process. The top panel shows the initial diamond crystal configuration and the bottom demonstrates the structure after bond transposition.

network obtained, with 216 atoms in, what was called a supercell, had periodic boundary conditions in three dimensions. The topology included fivefold, sixfold, sevenfold, and eightfold rings. In the diamond structure, only sixfold and eightfold rings can be found. Based on the close analogy between silicon and germanium, and even saturated carbon compounds, this model can be rescaled by the nearest-neighbor distances and be applied to a-Si, a-Ge, and tetrahedrally bonded amorphous carbon configurations. Since that time, several computer-generated models have been constructed using various classical empirical potentials [27, 41–43] or by applying quantum mechanical methods [44–51]. However, the WWW model is still considered to be the best defect-free three-dimensional atomic-scale representation of covalently bonded a-Si (and a-Ge). The WWW model and modified WWW models have been applied in electronic structure calculations and the results suggest that covalently bonded a-Si structures are not completely disordered. The bonds, the coordination numbers, and the local arrangements are similar to those in the crystalline phase.

3.4.4 Reverse Monte Carlo Methods

Reverse Monte Carlo (RMC) simulation was developed by McGreevy and Pusztai (1988) for investigation of the structure of disordered condensed phases and liquids [52]. This method of solution of an inverse problem is convenient for modeling amorphous structures. RMC methods are based on the results of diffraction measurements and are free from any description of the local interaction between atoms. Theoretically, both the structure factor and the radial distribution function (or pair correlation function) can be used as initial data for an RMC simulation but the use of $S(Q)$ has an important advantage. The main problem with the application of the radial distribution function, the fact that Q is measured only in a limited interval but the Fourier transformation is done in a infinite interval, is bypassed. To speed up the simulation and/or to reach more realistic configurations, constraints are usually applied. To start an RMC simulation, a given number of particles must be confined in a box. The density, which can be derived from this number, is the first important constraint. Another constraint that has been used several times is the coordination number. It can easily be prescribed that particles must have strictly a given number of particles as first neighbors. A lower limit on the bond length (a hard-core model) can also be imposed in order to eliminate physically meaningless configurations. Upper and lower limits on bond angles are also useful constraints for covalently bonded materials.

The use of constraints modifies the original RMC algorithm in the following way:

1) The algorithm starts with an initial particle configuration. Its radial distribution function or structure factor is calculated. The difference between the simulated and experimental RDFs or structure factors (χ_0^2) is also calculated. The comparison with experiment is quantified using a function

of the form $\chi^2 = \Sigma(y_{exp} - y_{sim})^2/\sigma^2$, where y_{exp} and y_{sim} are the measured and calculated quantities, and σ is a measure of the accuracy of the measurement. The sum includes all points in a function such as the RDF or the structure factor.

2) A new trial configuration is generated by random motion of a particle.
3) The algorithm checks whether the new configuration satisfies the constraint(s) applied. If not, it starts again from step 2. This is an additional step with respect to the standard RMC simulation.
4) Provided that the constraints are satisfied, χ_n^2 is calculated for the new trial configuration.
5) If $\chi_n^2 < \chi_{(n-1)}^2$, the new configuration becomes the starting configuration, that is, the move is accepted, otherwise it is accepted with a probability that follows a normal distribution.
6) The process is repeated from step 2 until χ_n^2 converges to its "equilibrium" value.

The latest developments in RMC software are aimed particularly at (i) covalent glasses [53, 54] and (ii) molecular liquids [55]. For the former, constraints on bond angle distributions have been introduced and the coordination constraints have been elaborated on [53]: in this way, for example, it is now possible to require that a given type of central atom should have a given number of neighbors of given types, within given distance ranges, so that a given distribution of bond (i.e., neighbor–center–neighbor) angles is realized. In this version, called RMC++, the modeling of extended X-ray absorption fine structure (EXAFS) data is also greatly facilitated. Such a combination has proven to be extremely useful in the investigation of multicomponent glasses [56]. For molecular liquids, the application of classical interatomic potential functions (including multibody terms) has been made possible by the code RMC_POT [55]. This code, first of all, allows one to handle flexible molecules of arbitrary complexity [57], but also, with three-body potential functions, covalent networks may be studied in the future to an advanced level.

The most recent free version of RMC++ can be downloaded from the following web page: http://www.szfki.hu/~nphys/rmc++/opening.html

3.4.4.1 Constraints

Information about local order in amorphous structures may be provided by analyzing embedded fragments in large molecules. The Cambridge Structural Database (CSD) is the world's largest database of experimentally determined crystal structures and includes the results of X-ray and neutron diffraction studies. The CSD provides a good chance to obtain measured lower and upper limits on bond lengths, bond angles, second-neighbor distances, and so on if the database contains a large number of relevant diffraction records. A systematic search of structural data can be carried out for X–X and X–X–X fragments in this database, where X means the element forming the disordered system of

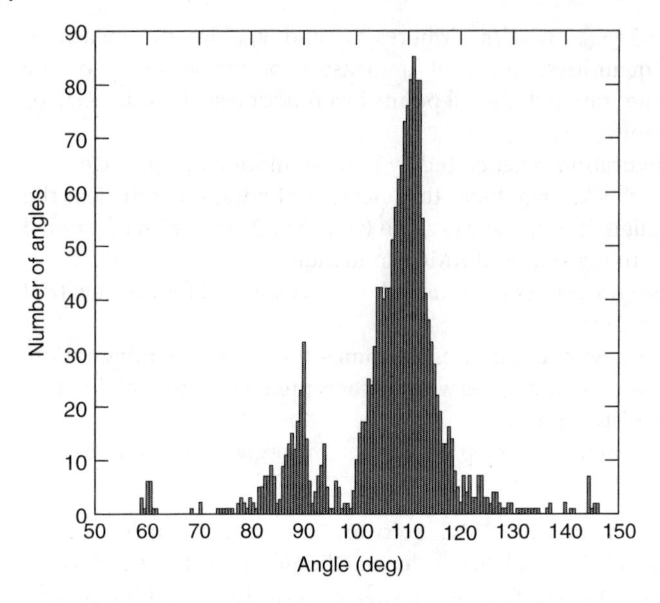

Figure 3.8 Histogram of bond angle distribution of Si–Si–Si fragments collated from the Cambridge Structural Database. *Source:* Kugler and Varallyay 2001 [58]. Reproduced with permission of Taylor & Francis Ltd.

interest. We have collected the experimentally determined structural data for molecules containing Si–Si–Si [58] and Se–Se–Se [59] fragments. Figure 3.8 displays the results for the bond angle distribution in a-Si. The results of the search are surprising. The majority of the points fall in the expected region, that is, around 0.235 nm and 109.47°. The minimum bond length is about 0.22 nm and the maximum about 0.27 nm. The average bond length and angle are 0.237 nm and 106.3°, respectively. There are some extrema. Two well-defined, unexpected regions can also be found. There are some significant angles in an interval of 75–96°, in an unexpected region. These angles belong to a nearly planar square arrangement. Several squares were found in the database. Bond angles around 60° can also be found. The second conclusion from the bond angle analysis is that nearly equilateral Si_3 triangles are present among the fragments. This corresponds to the other unexpected region, and most theoretical models for a-Si do not contain such components of the structure as three- and four-membered rings.

3.4.5 a-Si Model Constructed by RMC Simulation

The first RMC simulations including constraints were performed for pure a-Si by Kugler *et al.* [11]. A neutron diffraction measurement on a pure evaporated amorphous Si sample was performed at the high-flux reactor at the Institut

Laue-Langevin, Grenoble (see Section 3.3.1). The experimentally measured structure factor was used as input data for the RMC calculations. Three structural models were constructed by the RMC method with and without constraints. In all cases simulated, $N = 1728$ particles were confined in a cubic box of side length $L = 3.2\,\text{nm}$. This setup (constraint) gave the experimental microscopic number density $\rho = 0.0505\,\text{Å}^{-3}$. A hard-core diameter (the lowest limit on the bond length) of 0.220 nm was also applied in order to efficiently avoid physically meaningless configurations. About 1 million accepted steps were completed in the three different runs.

Model 3 satisfied a complicated constraint. Here, it was required that all the atoms had four neighbors. In addition to this coordination constraint, if an attempted move resulted in new bond angles with an average that did not fit a normal-like distribution centered on the tetrahedral angle, the move was immediately rejected. In systems such as a-Si, it is straightforward to require that the most of the angles should be roughly tetrahedral. Angles far from the tetrahedral value are less probable but permitted. The function $g(r)$ for Model 3 is displayed in Figure 3.9. Figure 3.10 shows $g(r)$ obtained by using the Fourier transform of the measured $S(Q)$. The characteristics of these functions are rather similar. The distributions of the cosines of the bond angles ($\cos\theta$) show significant differences between the microscopic structures obtained with three different constraints (Figure 3.11). This makes it clear that a diffraction experiment by itself provides insufficient information for determining unambiguously the microscopic local structure. Identical $S(Q)$ and/or $g(r)$ functions can be consistent with substantially different bond angle distributions. The distribution of cosines for Model 1, which does not contain any constraints, has a large peak at around 60° and, also, a large proportion of angles near 180° were found. However, an intense peak around 109.5° is provided by the model. These results are completely different from those for the other two models. Model 2 reflects probably the most fundamental requirement, that the atoms in any solid form of Si should be fourfold coordinated. This simple constraint greatly reduces the freedom for moving atoms around. The main problem with this picture is the probable overrepresentation of angles higher than 130°. Model 3 was introduced in order to further narrow the distribution of cosines, but no substantial narrowing could be achieved compared with Model 2.

Wooten *et al.*'s model of a-Si [40] contains bond angles in an interval between about 90° and 150°. The classical empirical Keating potential was applied in the construction of the model, in which the interaction has an energy term that is quadratic in the difference between the cosine of the bond angle and the cosine of the ideal bond angle. This term prevents large deviations from the canonical value of the bond angle. Note that RMC simulations do not take account of the energy. The existence of angles smaller than 90° or 60° is in harmony with the CSD too.

A comparison between the experimental structure factors of evaporated and ion-implanted amorphous silicon has been done using modeling by reverse

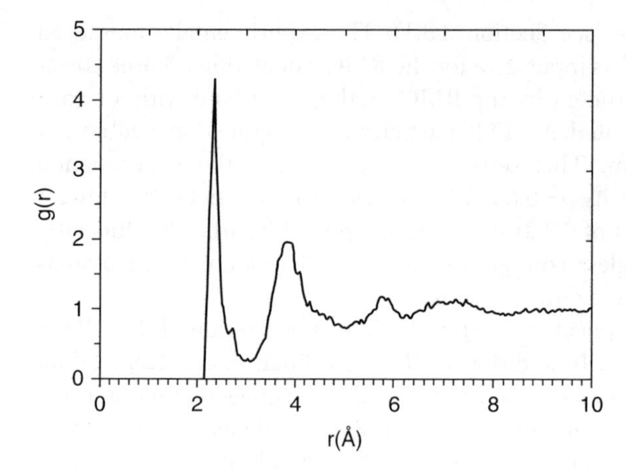

Figure 3.9 Pair correlation function for Model 3 of a-Si obtained by RMC simulation. *Source:* S. Kugler *et al.* 1993 [11]. Reproduced with permission from the American Physical Society.

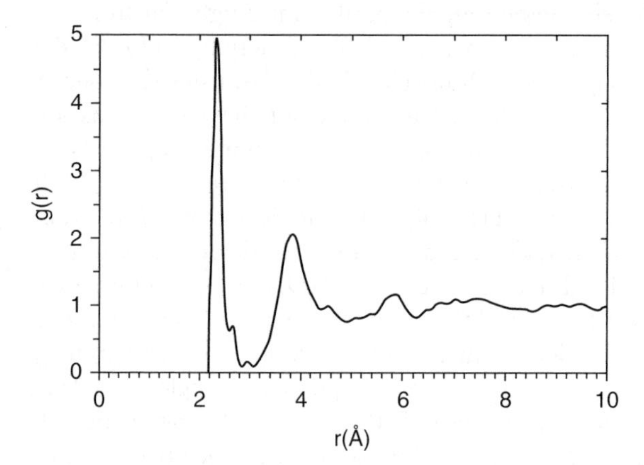

Figure 3.10 Pair correlation function of a-Si obtained by Fourier transformation of measured *S*(*Q*). *Source:* S. Kugler *et al.* 1993 [11]. Reproduced with permission from the American Physical Society.

Monte Carlo simulation [60]. A detailed comparison in terms of the pair correlation function and the distribution of the cosines of the bond angles was reported for the two materials. It was found that for an acceptable RMC reproduction of the measured structure factors, the models of the evaporated material had to contain more "small" bond angles than was necessary in the models of the implanted material.

3.4.6 a-Se Model Constructed by RMC Simulation

Reverse Monte Carlo simulation was carried out on a model material for chalcogenide glasses by Jóvári *et al.* [16]. Neutron diffraction measurements were performed on the SLAD instrument at Studsvik NFL on a ball-milled amorphous selenium sample. The structure factor obtained was interpreted by means of an RMC simulation. Selenium has several crystalline forms, made up of Se_8 rings or chains. Jóvári *et al.* concluded from the diffraction data that

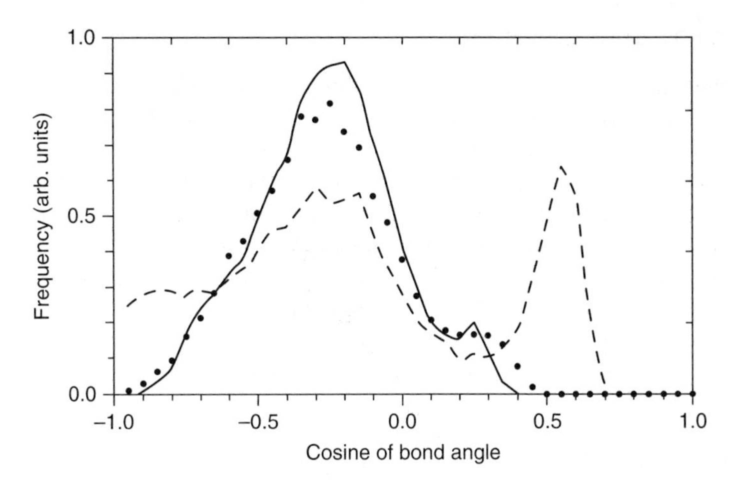

Figure 3.11 Distribution of the cosines of bond angles due to three different a-Si models obtained by RMC simulations. The dashed line represents the unconstrained model (Model 1). *Source:* S. Kugler *et al.* 1993 [11]. Reproduced with permission from the American Physical Society.

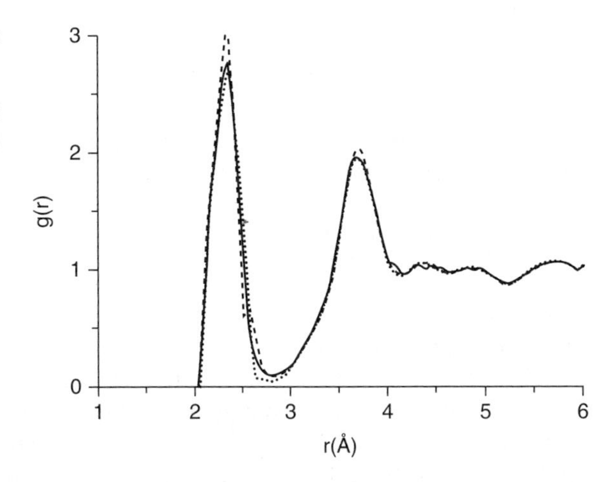

Figure 3.12 Pair correlation functions of a-Se obtained from three different RMC simulations. *Source:* P. Jóvári *et al.* 2003 [16]. Reproduced with permission from the American Physical Society.

it was not possible to determine whether the amorphous selenium consisted of long chains, Se_8 rings, or a mixture of the two. Both types of model were possible to construct whenever the structure factors agreed quantitatively with the measured data. The sample measured was made from crystalline Se consisting of long chains, and so it was suggested that this sample contained primarily chainlike structures. Pair correlation functions of a-Se obtained from three different RMC simulations are displayed in Figure 3.12.

3.4.7 Molecular Dynamics Simulation

The use of a deterministic MD method for describing noncrystalline systems is an alternative method for construction of 3D structure models. The primary application of MD is to follow the preparation process in situ and to reach a condensed phase. In this process, all that is assumed is the validity of classical mechanics. The Newtonian equations of motion can be solved for a small time interval dt if the interaction potential and the position and velocity of N particles are given at time t. The simplest and best-known method to integrate the MD equations of motion is the Verlet algorithm. The atomic motion can be followed using an MD algorithm. The basic idea is to make third-order Taylor expansions for $\boldsymbol{r}(t+dt)$ and $\boldsymbol{r}(t-dt)$, where $\boldsymbol{r}(t)$ is the atomic position at time t:

$$\boldsymbol{r}(t+dt)=\boldsymbol{r}(t)+\boldsymbol{v}(t)dt+\frac{1}{2}\boldsymbol{a}(t)dt^2+\frac{1}{6}\boldsymbol{b}(t)dt^3+O(dt^4)$$

$$\boldsymbol{r}(t-dt)=\boldsymbol{r}(t)-\boldsymbol{v}(t)dt+\frac{1}{2}\boldsymbol{a}(t)dt^2-\frac{1}{6}\boldsymbol{b}(t)dt^3+O(dt^4)$$

Adding the two expressions gives

$$\boldsymbol{r}(t+dt)=2\boldsymbol{r}(t)-\boldsymbol{r}(t-dt)+\boldsymbol{a}(t)dt^2+O(dt^4)$$

The acceleration $\boldsymbol{a}(t)$ at time t can be calculated from the atomic interactions. This is the basic equation of the Verlet algorithm and, by subtracting, one has

$$\boldsymbol{v}(t)=\frac{\boldsymbol{r}(t+dt)-\boldsymbol{r}(t-dt)}{2\,dt}+O(dt^3).$$

On the basis of these equations, therefore, knowing the positions of each particle at time t and, earlier, at time $(t-dt)$, and also knowing the potential, $V(r)$, we can derive the new configuration at time $(t+dt)$. The need for knowledge of $\boldsymbol{r}(t-dt)$ causes a small difficulty in the first MD step but this is easy to handle, knowing the initial velocity. Application of the Hellmann–Feynman theorem is needed in the calculation of the forces in quantum mechanical cases. The local error in a given position is proportional to $O(dt^4)$. The cumulative error in the positions over a given time period is

$$\text{error}(T)=O(dt^2)$$

A modified version, the velocity Verlet algorithm, is used more frequently. This new method uses a similar approach but explicitly incorporates the velocity, solving the problem of the first time step in the basic Verlet algorithm. The error in the velocity Verlet algorithm is of similar order to that in the basic Verlet algorithm. The same amount of computer memory is needed for this modified version.

In the leapfrog integration algorithm, the velocity is calculated in two steps. The new positions are calculated as in the velocity Verlet algorithm. Then, the velocities at half of the time step are calculated from the old velocities and accelerations:

$$\boldsymbol{v}\left(t+\frac{1}{2}dt\right)=\boldsymbol{v}(t)+\frac{1}{2}\boldsymbol{a}(t)dt$$

The values of $(t+dt)$ are then computed and, finally, the velocity can be obtained in the following way:

$$\boldsymbol{v}(t+dt)=\boldsymbol{v}\left(t+\frac{1}{2}\right)+\frac{1}{2}\boldsymbol{a}(t+dt)\,dt$$

The leapfrog algorithm is numerically more stable than a version in which the velocities are calculated in one step.

For an amorphous sample, the preparation temperature plays a crucial role and so it must be controlled. Both of the main preparation techniques involve amorphization at a relatively low temperature. Each of the events that occurs is highly nonequilibrium in nature. A heat bath removes energy from another set of atoms. So, at every MD simulation step, the temperature must be monitored. During the process of film growth, particles bombard the growing film. Deposition on a substrate requires the substrate temperature to be kept constant. The instantaneous kinetic energies of the substrate atoms and the temperature of the substrate, however, may increase during an MD simulation owing to the particle impacts, depending on the energy of the incoming particles. In the rapid-cooling technique, the characteristic rate of quenching in experiments is around 10^6 K/s or lower. A variety of thermostat methods to control the temperature in simulations can be found in the literature, but the simplest velocity rescaling methods are the most popular.

An MD computer code (ATOMDEP) has been developed to simulate the preparation procedure of amorphous semiconductors, which are usually grown by a vapor deposition technique on a substrate [62]. It should be noted that, experimentally, no rapid-quenching preparation method has yet been reported for column IV amorphous materials. To integrate the MD equations of motion, the velocity Verlet algorithm was included in the code, with a time step equal to about a femtosecond. The substrate temperature was fixed by velocity rescaling. During the growth procedure, an atom started to move forward toward the target surface. In order to avoid surface effects, periodic boundary conditions in two dimensions were applied. The initial x, y coordinates and the deposition energies were randomly distributed. The bottom layer of the substrate was fixed at its ideal lattice sites in order to stabilize the substrate mechanically.

The typical deposition rate during sample preparation in the laboratory is around 10^{14} atoms s^{-1} cm^{-2}. The surface of the substrate in the computer experiments had an area of around 10^{-14} cm^2; that is, one incoming atom should reach the substrate in one second. This deposition rate causes difficulty in the simulation, and it is impossible to handle it easily by MD with today's computer facilities within a tight-binding atomic interaction model. Despite this disadvantage, the high theoretical deposition rate applied was low enough to allow the relaxation of the previous atom before the next deposition. The computer code developed was successfully applied to the preparation of a-Si and a-Se, which will be discussed below.

3.4.8 a-Si Model Construction by Molecular Dynamics Simulation

Amorphous silicon networks were constructed by atomic deposition (and also by rapid quenching) [51]. In the simulations, the transferable tight-binding Hamiltonian [37] was applied to describe the interaction between silicon atoms. The group that developed this Hamiltonian had already developed an excellent tight-binding potential for carbon systems. All the parameters and functions of the interatomic potential for silicon were fitted to the results of local density functional calculations. This tight-binding model reproduces the energies of several different cluster structures in crystalline silicon, the elastic constants of crystalline silicon, and the formation energies of vacancies and interstitials in crystalline silicon. According to the authors of [37], the only disadvantage of this tight-binding model is that the bond lengths inside small clusters are a slightly longer than those derived from other theoretical calculations and from experimental results.

A surprising result was found in the ring statistics. The networks prepared using the models have a significant number of squares. Furthermore, triangles are also present in the atomic arrangements, as can be seen in Figure 3.13. Most of the theoretical models for a-Si do not contain such structural components. This seems to be an important result, although a neutron diffraction measurement carried out on a pure evaporated a-Si sample and evaluated using the reverse MC method had already reached a similar conclusion. Later on, the simulation was repeated using another tight-binding model [63] and again similar conclusions were obtained.

3.4.9 a-Si:H Model Construction by Molecular Dynamics Simulation

There are many models for pure a-Si but only a few models can be found for the electronically more important a-Si:H in the literature. Hydrogen causes difficulties. In the MD process, all that is assumed is the validity of classical mechanics. An indicator of the validity of classical mechanics is the de Broglie wavelength $\lambda = h/\sqrt{2mE}$. For a particle moving with a kinetic energy of 1 eV, the de Broglie wavelength of the particle is 0.3 nm for hydrogen and 0.005 nm

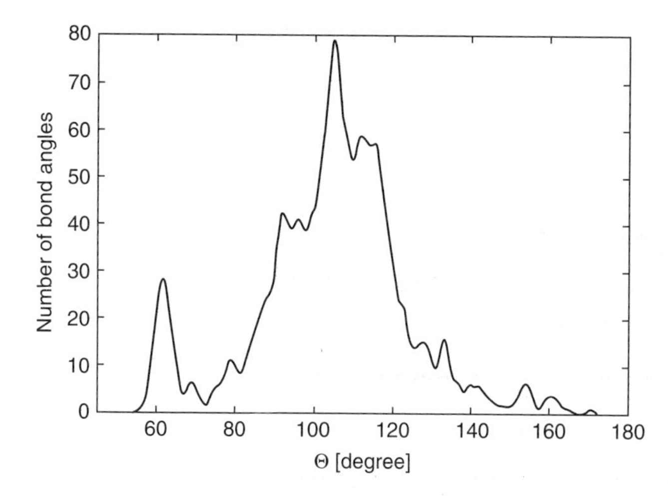

Figure 3.13 Bond angle distribution in an a-Si model obtained by MD simulations. There is a significant peak at an angle of 60°. A large number of bond angles can be found around 90° too. *Source:* K. Kohary 2004 [51]. Reproduced with permission of Taylor & Francis.

for Si. The characteristic distance is the interatomic spacing, which is in the range 0.1–0.3 nm. The motion of a 1 eV hydrogen atom in the solid state can therefore only described by including quantum effects in the dynamics. Another difficulty is the interaction of H with bulk silicon. A relatively large (242 atoms) a-Si:H model was constructed by Tuttle and Adams [64] using a Harris functional MD method. The hydrogen concentration was 11%. The model obtained contains microclusters of 2–4 H in addition to H passivating the surface of large cavities.

The structure of (hydrogenated) amorphous silicon seems to be well understood, but that is not exactly true. One of the main problems in several different models is the coordination number. The models contain a large number of overcoordinated and undercoordinated Si atoms which cannot be identified as dangling or floating bonds. There is no acceptable explanation of the Staebler–Wronski effect in a-Si:H, even though this effect was discovered in 1977! An additional problem is the local structure of impurities [65]. The structure is therefore still an open question.

3.4.10 a-Se Model Construction by Molecular Dynamics Simulation

We have used the tight-binding model developed by Molina and coworkers [38] for the description of the interaction between selenium atoms. Self-consistency was taken into account via the usual on-site Hubbard term and was found to reduce any large charge transfer [66]. The time step was equal to 2 fs. Several different glassy selenium networks were prepared in a rectangular box

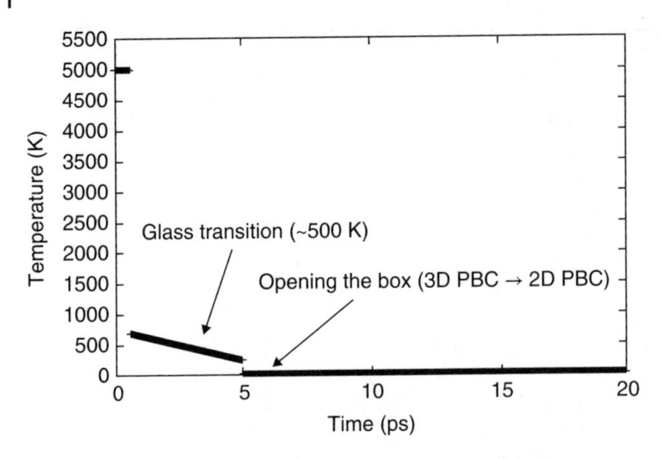

Figure 3.14 The "cook and quench" sample preparation procedure.

with periodic boundary conditions, and the samples contained 162 atoms. The "cook and quench" sample preparation procedure was the following, shown in Figure 3.14. First, the temperature of the system was kept at 5000 K. During the subsequent steps, we decreased the temperature linearly from 700 to 250 K, driving the sample through the glass transition and reaching the condensed phase. Then the final temperature of 20 K was reached and the sample was relaxed for 500 MD steps (1 ps). The periodic boundary conditions were lifted along the z direction at this point. This procedure provided a slab geometry with periodic boundary conditions in only two dimensions. The system was then relaxed at 20 K. The short quenching times in the simulation compared with those in experiments might lead to many liquid-state defects being retained in the amorphous structure. Therefore, the Hubbard parameter U was taken to be 5 eV for the first 4000 MD steps during quenching in order to avoid a large number of coordination defects, especially onefold- and threefold-coordinated atoms. Then U was changed to its accepted value of 0.875 eV for selenium [67]. The radial distribution function for one of our selenium glassy networks at 20 K is shown in Figure 3.15. The first main peak, at 0.24 nm, belongs to covalently bonded atoms. The crystalline nearest-neighbor distance is 0.237 nm. The second peak, at 0.36 nm, corresponds to the intrachain second-nearest-neighbor distance as expected. The prepeak at 0.33 nm reflects the smallest interchain atomic distances in amorphous selenium. In simulations at higher temperatures ($T = 300$ K), these two peaks merge, as observed in Figure 3.15.

Another class of modeling has been applied to two different basic preparation methods for amorphous selenium, liquid quenching and evaporation, which are available experimentally in the laboratory. The MD computer code ATOMDEP was used for modeling the a-Se structure. A classical empirical

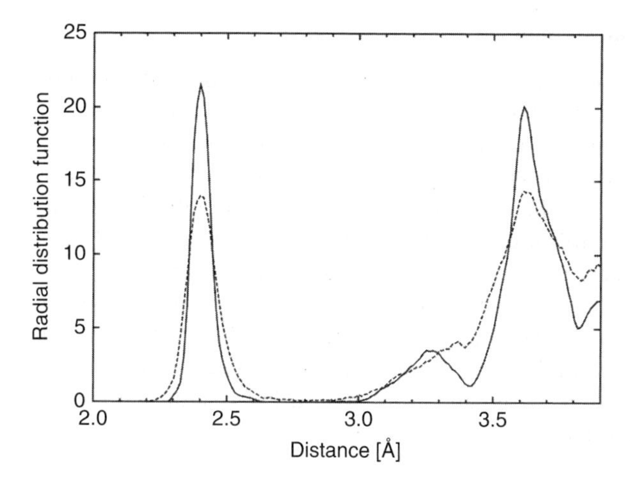

Figure 3.15 Radial distribution functions of selenium glassy networks at $T = 20\,K$ (solid line) and at $T = 300\,K$ (dashed line). The first peak, at 0.24 nm, belongs to covalently bonded atoms. The second peak, at 0.36 nm, corresponds to the intrachain second-nearest-neighbor distance. The prepeak around 0.33 nm represents distances between two atoms in different chains (interchain bonds). These two peaks merge at higher temperature. *Source:* J. Hegedüs *et al.* 2005 [67]. Reproduced with permission from the American Physical Society.

three-body potential [35] was included in the computer code. A crystalline lattice cell containing 324 selenium atoms was employed to mimic the substrate. There were 108 fixed atoms at the bottom of the substrate. The remaining atoms could move with full dynamics. The simulation cell was open along the positive z direction and periodic boundary conditions were applied in the x and y directions. Three different structures were constructed using average bombarding energies of 0.1, 1, and 10 eV. At the end of the deposition process, the atomic networks contained about 1000 selenium atoms. Rapid cooling is also frequently used to prepare glassy structures. The system is usually cooled down to room temperature at a rate of 10^{11}–10^{16} K/s in computer simulations, but the cooling rate is some orders of magnitude smaller in the experimental techniques. In order to obtain information about rapid cooling, we prepared a computer model in the following way. The temperature of one of the deposited films (average bombarding energy 1 eV) was increased to 900 K to provide an initial state (a liquid phase). After this melting, the trajectories of the selenium atoms were followed by full dynamics for 100 ps. The substrate temperature was kept at 100 K and led to cooling of the film above the substrate. This technique can be considered as a computer simulation of real splat cooling, where small droplets of a melt are brought into contact with a chill-block. Materials prepared by rapid quenching are more homogeneous than their deposited counterparts. An observable difference was obtained between the two different preparation techniques.

3.4.11 Car and Parrinello Method

Although simulations based on empirical potentials and on tight-binding models have achieved remarkable success, the extension of simulation methods to more accurate quantum mechanical methods would be important. Unfortunately, DFT within the most popular local-density approximation is very computer-time-consuming; it is difficult to perform a separate self-consistent electronic minimization at every MD step. An approach has been developed by Car and Parrinello [68]. This method overcomes this difficulty and has been successfully applied in MD simulations of covalently bonded amorphous semiconductors, yielding valuable structural information about small systems. Structural simulations have been carried out on the basis of the Car–Parrinello method for a-Si [69] and a-Si:H [70] using systems with much fewer than a hundred atoms.

References

1 Chittik, R.C., Alexander, J.H., and Sterling, H.J. (1969) The preparation and properties of amorphous silicon. *J. Electrochem. Soc.*, **116**, 77–81.
2 Spear, W.E. and Le Comber, P.G. (1975) Substitutional doping of amorphous silicon. *Solid State Commun.*, **17**, 1193–1196.
3 Shechtman, D., Blech, I. Gratias, D., and Cahn, J.W. (1984) Metallic phase with long-range orientational order and no translational symmetry. *Phys. Rev. Lett.*, **53**, 1951–1953.
4 Street, R.A. (1991) *Hydrogenated Amorphous Silicon*, Cambridge University Press.
5 Lukacs, R., Veres, M., Shimakawa, K., and Kugler, S. (2010). On photoinduced volume change in amorphous selenium: Quantum chemical calculation and Raman spectroscopy. *J. Appl. Phys.*, **107**, 073517-1-5.
6 Lucovsky, G., Nemanich, R.J., and Knight, J.C. (1979) Structural interpretation of the vibrational spectra of a-Si: H alloys. *Phys. Rev. B*, **19**, 2064–2073.
7 Cuello, G.J. (2008) Structure factor determination of amorphous materials by neutron diffraction. *J. Phys.: Condens. Matter*, **20**, 244109-1-9.
8 Moss, S.C. and Graczyk, J.F. (1969) Evidence of voids within the as-deposited structure of glassy silicon. *Phys. Rev. Lett.*, **23**, 1167–1171.
9 Barna, A., Barna, P.B., Radnoczi, G., Toth, L., and Thomas, P. (1977) A comparative study of the structure of evaporated and glow discharge silicon. *Phys. Stat. Sol. A*, **41**, 81–84.
10 Kugler, S., Molnar, G., Peto, G., Zsoldos, E., Rosta, L., Menelle, A., and Bellissent, R. (1989) Neutron-diffraction study of the structure of evaporated pure amorphous silicon. *Phys. Rev. B*, **40**, 8030–8032.

11 Kugler, S., Pusztai, L., Rosta, L., Chieux, P., and Bellissent, R. (1993) Structure of evaporated pure amorphous silicon: Neutron-diffraction and reverse Monte Carlo investigation. *Phys. Rev. B*, **48**, 7685–7688.
12 Fortner, J. and Lannin, J.S. (1989) Radial distribution functions of amorphous silicon. *Phys. Rev. B*, **39**, 5527–5530.
13 Laaziri, K., Kycia, S., Roorola, S., Chicoine, M., Robertson, J.L., Wang, L. and Moss, S.C. (1999) High resolution radial distribution function of pure amorphous silicon. *Phys. Rev. Lett.*, **82**, 3460–3463.
14 Bellissent, R., Menelle, A., Howells, W.S., Wright, A.C., Brunier, T.M., Sinclair, R.N., and Jansen, F. (1989) The structure of amorphous Si:H using steady state and pulsed neutron sources. *Physica B*, **156–157**, 217–219.
15 Etherington, J.H., Wright, A.C., Wenzel, J.T., Dore, J.C., Clarke, J.H., and Sinclair, R.N. (1982) A neutron diffraction study of the structure of evaporated amorphous germanium. *J. Non-Cryst. Solids*, **48**, 265–289.
16 Jóvári, P., Delaplane, R.G., and Pusztai, L. (2003) Structural models of amorphous selenium. *Phys. Rev. B*, **67**, 172201-1–4.
17 Harp, G.R., Saldin, D.K., and Tonner, B.P. (1990) Atomic-resolution electron holography in solids with localized sources. *Phys. Rev. Lett.*, **65**, 1012–1015.
18 Tegze, M. and Faigel, G. (1996) X-ray holography with atomic resolution. *Nature*, **380**, 49–51.
19 Hayashi, K., Ohoyama, K., Orimo, S., Nakamori, Y., Takahashi, H., and Shibata, K. (2008) Neutron holography measurement using multi array detector. *Japan. J. Appl. Phys.*, **47**, 2291–2293.
20 Cser, L., Török, Gy., Krexner, G., Sharkov, I., and Faragó, B. (2002) Holographic imaging of atoms using thermal neutrons. *Phys. Rev. Lett.*, **89**, 175504-1–4.
21 Szakal, A. Marko, M., and Cser, L. (2015) Local magnetic structure determination using polarized neutron holography. *J. Appl. Phys.*, **117**, 17E132-1–5.
22 Keating, P.N. (1966) Effect of invariance requirements on the elastic strain energy of crystals with application to the diamond structure. *Phys. Rev.*, **145**, 637–645.
23 Overney, G., Zhong, W., and Tománek, D. (1993) Structural rigidity and low frequency vibrational modes of long carbon tubules. *Z. Phys. D*, **27**, 93–96.
24 Rücker, H. and Methfessel, M. (1995) Anharmonic Keating model for group-IV semiconductors with application to the lattice dynamics in alloys of Si, Ge, and C. *Phys. Rev. B*, **52**, 11059–11072.
25 Sim, E., Beckers, J., Leeuw, S., Thorpe, M., and Ratner, M.A. (2005) Parameterization of an anharmonic Kirkwood–Keating potential for $Al_xGa_{(1-x)}As$ alloy. *J. Chem. Phys.*, **122**, 174702-1-6.
26 Stillinger, F.H. and Weber, T.A. (1985) Computer simulation of local order in condensed phases of silicon. *Phys. Rev. B.*, **31**, 5262–5271.

27 Vink, R.L.C., Barkema, G.T., van der Weg, W.F., and Mousseau, N. (2001) Fitting the Stillinger–Weber potential to amorphous silicon. *J. Non-Cryst. Solids*, **282**, 248–255.

28 Pizzagalli, L., Godet, J., Guénolé, J., Brochard, S., Holmstrom, E., Nordlund, K., and Albaret, T. (2013) A new parametrization of the Stillinger–Weber potential for an improved description of defects and plasticity of silicon. *J. Phys.: Condens. Matter*, **25**, 055801-1-12.

29 Justo, J.F., Bazant, M.Z., Kaxiras, E., Bulatov, V.V., and Yip, S. (1998) Interatomic potential for silicon defects and disordered phases. *Phys. Rev. B*, **58**, 2539–2550.

30 Tersoff, J. (1986) New empirical model for the structural properties of silicon. *Phys. Rev. Lett.*, **56**, 632–635.

31 Tersoff, J. (1988) New empirical approach for the structure and energy of covalent systems. *Phys. Rev. B*, **37**, 6991–7000.

32 Tersoff, J. (1988) Empirical interatomic potential for silicon with improved elastic properties. *Phys. Rev. B*, **38**, 9902–9905.

33 Tersoff, J. (1989) Modeling solid-state chemistry: Interatomic potentials for multicomponent systems. *Phys. Rev. B*, **39**, 5566–5568. (Erratum: *Phys. Rev. B*, **41**, 3248 (1990)).

34 Abell, G.C. (1985) Empirical chemical pseudopotential theory of molecular and metallic bonding. *Phys. Rev. B*, **31**, 6184–6196.

35 Oligschleger, C., Jones, R.O., Reimann, S.M., and Schober, H.R. (1996) Model interatomic potential for simulations in selenium. *Phys. Rev. B*, **35**, 6165–6173.

36 Goodwin, L., Skinner, A.J., and Pettifor, D.G. (1989) Generating transferable tight-binding parameters: Application to silicon. *Europhys. Lett.*, **9**, 701–706.

37 Kwon, I., Biswas, R., Wang, C.Z., Ho, K.M., and Soukoulis, C.M. (1994) Transferable tight-binding models for silicon. *Phys. Rev. B*, **49**, 7242–7250.

38 Molina, D., Lomba, E., and Kahl, G. (1999) Tight-binding model of selenium disordered phases. *Phys. Rev. B*, **60**, 6372–6382.

39 Wang, C.Z., Pan, B.C., and Ho, K.M. (1999) An environment-dependent tight-binding potential for Si. *J. Phys.: Condens. Matter*, **11**, 2043–2049.

40 Wooten, F., Winer, K., and Weaire, D. (1985) Computer generation of structural models of amorphous Si and Ge. *Phys. Rev. Lett.*, **54**, 1392–1395.

41 Kelires, P.C. and Tersoff, J. (1988) Glassy quasithermal distribution of local geometries and defects in quenched amorphous silicon. *Phys. Rev. Lett.*, **61**, 562–565.

42 Ishimaru, M., Munetoh, S., and Motooka, T. (1997) Generation of amorphous silicon structures by rapid quenching: A molecular-dynamics study. *Phys. Rev. B*, **56**, 15133–15138.

43 Ishimaru, M. (2001) Molecular-dynamics study on atomistic structures of amorphous silicon. *J. Phys.: Condens. Matter*, **13**, 4181–4189.

44 Stich, I., Car, R., and Parrinello, M. (1991) Amorphous silicon studied by *ab initio* molecular dynamics: Preparation, structure, and properties. *Phys. Rev. B*, **44**, 11092–11104.

45 Drabold, D.A., Fedders, P.A., Klemm, S., and Sankey, O.F. (1991) Finite-temperature properties of amorphous silicon. *Phys. Rev. Lett.*, **67**, 2179–2182.

46 Toth, G. and Naray-Szabo, G. (1994) Novel semiempirical method for quantum Monte Carlo simulation: Application to amorphous silicon. *J. Chem. Phys.*, **100**, 3742–3746.

47 Hensel, H., Klein, P., Urbassek, H.M., and Frauenheim, T. (1996) Comparison of classical and tight-binding molecular dynamics for silicon growth. *Phys. Rev. B*, **53**, 16497–16503.

48 Yang, R. and Singh, J. (1998) Study of the stability of hydrogenated amorphous silicon using tight-binding molecular dynamics. *J. Non-Cryst. Solids*, **240**, 29–34.

49 Cooper, N.C., Goringe, C.M., and McKenzie, D.R. (2000) Density functional theory modelling of amorphous silicon. *Comput. Mater. Sci.*, **17**, 1–6.

50 Valladares, A.A., Alvarez, F., Liu, Z., Sticht, J., and Harris, J. (2001) *ab initio* studies of the atomic and electronic structure of pure and hydrogenated a-Si. *Eur. Phys. J. B*, **22**, 443–453.

51 Kohary, K. and Kugler, S. (2004) Growth of amorphous silicon: Low energy molecular dynamics simulation of atomic bombardment. *Mol. Simul.*, **30**(1), 17–22.

52 McGreevy, R.L. and Pusztai, L. (1988) Reverse Monte Carlo simulation: A new technique for the determination of disordered structures. *Mol. Simul.*, **1**, 359–367.

53 Evrard, G. and Pusztai, L. (2005) Reverse Monte Carlo modelling of the structure of disordered materials with RMC++: A new implementation of the algorithm in C++. *J. Phys.: Condens. Matter*, **17**, S1–S13.

54 Gereben, O., Jóvári, P., Temleitner, L., and Pusztai, L. (2007) A new version of the RMC++ reverse Monte Carlo programme, aimed at investigating the structure of covalent glasses. *J. Optoelectron. Adv. Mater.*, **9**, 3021–3027.

55 Gereben, O. and Pusztai, L. (2012) RMC_POT, a computer code for Reverse Monte Carlo modeling the structure of disordered systems containing molecules of arbitrary complexity. *J. Comput. Chem.*, **33**(29), 2285–2291.

56 Pethes, I., Kaban, I., Wang, R.-P., Luther-Davies, B., and Jóvári, P. (2015) Short range order in Ge–As–Se glasses. *J. Alloys Compounds*, **623**, 454–459.

57 Gereben, O. and Pusztai, L. (2013) Conformational analyses of bis(methylthio) methane and diethyl sulfide molecules in the liquid phase: Reverse Monte Carlo studies using classical interatomic potential functions. *J. Phys.: Condens. Matter*, **25**, 454201.

58 Kugler, S. and Varallyay, Z. (2001) Possible unusual atomic arrangements in the structure of amorphous silicon. *Philos. Mag. Lett.*, **81**, 569–574.

59 Hegedüs, J. and Kugler, S. (2005). Growth of amorphous selenium thin films: Classical versus quantum mechanical molecular dynamics simulation. *J. Phys.: Condens. Matter*, **17**, 6459–6468.
60 Pusztai, L. and Kugler S. (2005). Comparison of the structures of evaporated and ion-implanted amorphous silicon samples. *J. Phys.: Condens. Matter*, **17**, 2617–2624.
61 Swope, W.C., Andersen, H.C., Berens, P.H., and Wilson, K.R. (1982) A computer simulation method for the calculation of equilibrium constants for the formation of physical clusters of molecules: Application to small water clusters. *J. Chem. Phys.*, **76**, 637–649.
62 Kohary, K. and Kugler, S. (2001). Growth of amorphous carbon. Low energy molecular dynamics simulation of atomic bombardment. *Phys. Rev. B*, **63**, 193404-1–4.
63 Lenovsky, T., Kress, J.D., Kwon, I., Voter, A.F., Edwards, B., Richards, D.F., Yang, S., and Adams, J.B. (1997) Highly optimized tight-binding model of silicon. *Phys. Rev. B*, **55**, 1528–1544.
64 Tuttle, B. and Adams, J.B. (1996) Structure of *a*-Si:H from Harris-functional molecular dynamics. *Phys. Rev. B*, **53**, 16265–16271.
65 Cai, B. and Drabold, D.A. (2011) Theoretical studies of structure and doping of hydrogenated amorphous silicon. *Mater. Res. Soc. Symp. Proc.*, **mrss11-1321**, a10-01.
66 Lomba, E., Molina, D., and Alvarez, M. (2000) Hubbard corrections in a tight-binding Hamiltonian for Se: Effects on the band structure, local order, and dynamics. *Phys. Rev. B*, **61**, 9314–9321.
67 Hegedüs, J., Kohary, K., Pettifor, D.G., Shimakawa, K., and Kugler, S. (2005) Photoinduced volume changes in amorphous selenium. *Phys. Rev. Lett.*, **95**, 206803-1–4.
68 Car, R. and Parrinello, M. (1985) Unified approach for molecular dynamics and density-functional theory. *Phys. Rev. Lett.*, **55**, 2471–2474.
69 Stich, I., Car, R., and Parrinello, M. (1991) Amorphous silicon studied by *ab initio* molecular dynamics: Preparation, structure, and properties. *Phys. Rev. B*, **44**, 11092–11104.
70 Buda, F., Chiarotti, G.L., Car, R., and Parrinello, M. (1991) Structure of hydrogenated amorphous silicon from *ab initio* molecular dynamics. *Phys. Rev. B*, **44**, 5908–5911.

4

Electronic Structure of Amorphous Semiconductors

A simple definition of semiconductors is that these materials have an energy gap of around 1–2 eV in the electronic density of states and their resistivity falls between that of insulators and that of good conductors. A more accurate definition is associated with the temperature dependence of the resistivity,

$$\rho(T) = \rho_0 \exp(\varepsilon_0 / k_B T) \tag{4.1}$$

where ρ_0 and ε_0 are constants, k_B is the Boltzmann factor, and T is the temperature. The resistivity ρ decreases with increasing T, which is opposite to the behavior of conventional metals, where ρ is proportional to T.

Pure amorphous semiconductor materials are located in the even-numbered groups of the periodic table, namely group IV (carbon [in the amorphous phase], silicon, and germanium) and group VI (selenium and tellurium). In group VI, the stable allotropes of sulfur are excellent electrical insulators. Amorphous semiconducting alloys contain at least one of the pure elements that are amorphous, and are combined with elements from the same or another column. Amorphous semiconductors have a covalently bonded network.

4.1 Bonding Structures

The $(8 - N)$ rule suggests that the local coordination number in covalently bonded semiconductors is equal to eight minus the column number in the periodic table. This rule can be understood within a valence bond framework by assuming that single saturated covalent bonds are formed with the nearest neighbors, establishing a closed, stable octet shell of electrons.

Amorphous Semiconductors: Structural, Optical, and Electronic Properties, First Edition.
Kazuo Morigaki, Sándor Kugler, and Koichi Shimakawa.

4.1.1 Bonding Structures in Column IV Elements

In order to understand the electronic structure and bonds in these materials, the framework of the linear combination of atomic orbitals (LCAO) approach can be used. Two-center localized molecular orbitals (MO) are constructed using LCAO to form sigma bonds. We now illustrate this methodology for silicon atoms. An isolated silicon atom has an electron configuration of $1s^2 2s^2 2p^6 3s^2 3p^2$. Therefore, one might expect that the deeper states (the 1s, 2s, 2p, and 3s states) would not be involved in the bonding structure, and twofold chemical bonds would be formed using only the 3p states. The situation is, however, more complex and the 3s states also contribute to bond formation; this mechanism is known as hybridization.

In covalently bonded semiconductors, the valence electrons are localized and form chemical bonds. Therefore the valence electron wave functions are similar to the bonding orbitals found in molecules. For a tetrahedrally coordinated silicon atom with four hydrogen atoms, for example the small stable molecule silane (SiH_4), the silicon atom needs to have four orbitals with the correct symmetry to bond to the four hydrogen atoms. Therefore, four orbitals are required in the case of tetrahedral silicon. The solution is linear combinations of the 3s and 3p wave functions, called hybridized orbitals. The 3s orbitals are mixed with the three 3p (p_x, p_y, and p_z) orbitals to form four sp^3 hybrids:

$$\begin{aligned}
sp_1^3 &= \frac{1}{2}s + \frac{1}{2}p_x + \frac{1}{2}p_y + \frac{1}{2}p_z \\
sp_2^3 &= \frac{1}{2}s - \frac{1}{2}p_x - \frac{1}{2}p_y + \frac{1}{2}p_z \\
sp_3^3 &= \frac{1}{2}s - \frac{1}{2}p_x + \frac{1}{2}p_y - \frac{1}{2}p_z \\
sp_4^3 &= \frac{1}{2}s + \frac{1}{2}p_x - \frac{1}{2}p_y - \frac{1}{2}p_z
\end{aligned} \tag{4.2}$$

These four hybrids are orthonormalized states with orbitals that are 25%s and 75% p in the crystalline case. In SiH_4, the four sp^3 hybridized orbitals overlap with hydrogen 1s orbitals, yielding four σ bonds (that is, four single covalent bonds containing two electrons each). An important point to note is that the four equivalent covalent bonds have the same length and strength. When atoms from column IV of the periodic table combine to form a condensed phase, the interaction splits the valence states into a bonding electron level and a higher-energy antibonding level. Between these two levels a gap can be found

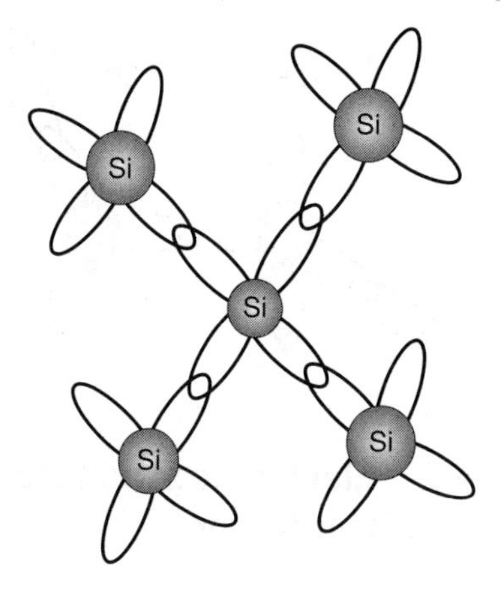

Figure 4.1 sp^3 orbitals overlap with similar orbitals of four adjacent atoms. Four sigma bonds containing two electrons each are formed at each atom in the condensed phase.

where no electron states exist. Inside a crystal, sp^3 orbitals overlap with similar orbitals of four adjacent atoms (Figure 4.1). This theory fits fourfold-coordinated column IV elements (such as diamondlike carbon, silicon, and germanium) in noncrystalline arrangements, too. However, the prefactors of ½ have to be modified in order to account for deviations from the ideal local arrangement of four nearest-neighbor atoms.

4.1.2 Bonding Structures in Column VI Elements

Pure noncrystalline selenium is a model material for chalcogenide glasses. Se atoms contain six electrons ($4s^2 4p^4$) in the highest energy level. Only four independent orbitals can be constructed from s, p_x, p_y, and p_z; that is, theoretically, two out of these four should be doubly occupied. The simple molecule hydrogen selenide has an H–Se–H atomic structure with a bond angle of 91°. The bonding process can be explained in the following way: two near-perpendicular sigma bonds are formed by putting one electron in the p_x and another in the p_y atomic orbital, which overlap with the 1s orbital of each hydrogen atom. The other two unshared electrons occupy the p_z atomic orbital, forming a nonbonding lone pair state, opposite to the case of sp^3 hybridization. The s atomic orbitals are untouched! In the condensed phase, selenium atoms form a chain-like structure (Figure 4.2). It can be concluded that only two electrons on the p_x and p_y orbitals form sigma bonds, overlapping with similar orbitals on adjacent atoms. The bond angles are slightly larger than 90°. This means that the s atomic orbital is not exactly untouched: a small contribution of the s orbital is needed to form near-perpendicular sigma bonds and a lone pair. The energy of these nonbonding lone pair states lies between the bonding and antibonding electron energy levels. Note that the coordination number $Z = 2 < 2.4$, so the structure is "underconstrained" or soft.

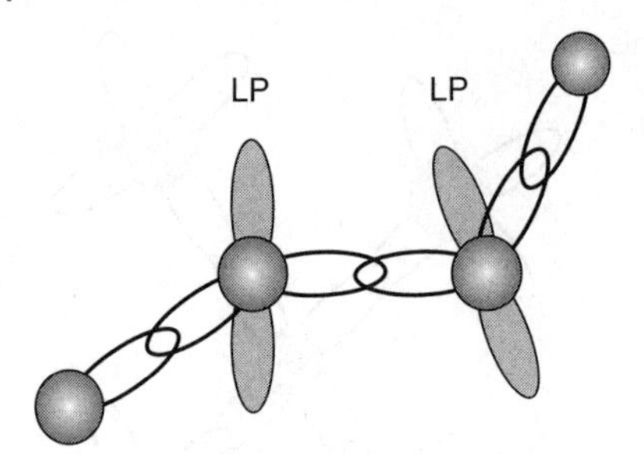

Figure 4.2 p_x and p_y orbitals overlap with similar orbitals of adjacent atoms and form sigma bonds. The remaining two unshared electrons occupy the p_z atomic orbital, forming a nonbonding lone pair (LP) state.

4.2 Electronic Structure of Amorphous Semiconductors

In a real crystal, the atoms are arranged periodically in three dimensions to a finite extent only, and the crystal must contain some imperfections such as vacancies, interstitial atoms, dislocations, and impurities at temperatures above zero. When these defects can be neglected and periodic boundary conditions in three dimensions are applied, we obtain a model of the atomic configuration which provides us with simple methods for calculating various properties of condensed matter, including the electronic density of states. The presence of perfect translational symmetry simplifies the mathematical treatment of these crystalline materials. The electron states can be described by Bloch wave functions extending throughout the crystal,

$$\varphi(\boldsymbol{k},\boldsymbol{r}) = u(\boldsymbol{k},\boldsymbol{r})\exp(i\boldsymbol{kr}) \tag{4.3}$$

where $u(\boldsymbol{k}, \boldsymbol{r})$ has the periodicity of the crystal lattice, that is, $u(\boldsymbol{k}, \boldsymbol{r}) = u(\boldsymbol{k}, \boldsymbol{r} + \boldsymbol{R})$; $\boldsymbol{R}$ is a lattice translation vector, and the exponential term represents a plane wave. The wave vectors $\boldsymbol{k}$ of the electrons are related to the translation symmetry. In realistic cases, defects can usually be handled as perturbations. Unfortunately, however, this tool is useless for noncrystalline materials. The lack of periodicity does not allow us to define the wave vector $\boldsymbol{k}$, and therefore a classification of the electronic band structure $E(\boldsymbol{k})$ also cannot be made. Furthermore, the traditional use of the definition of the effective mass is prevented [1]. However, although the atoms are not arranged periodically in amorphous materials, they do not have random locations as in a gas. In amorphous semiconductors, the atoms retain short-range order similar to that in the crystalline phase.

Instead of the function $E(\boldsymbol{k})$, the number of electronic states at an energy E per unit energy, $N(E)$, is one of the most important relationships that characterize the electronic properties of amorphous semiconductors. The occupancy of $N(E)$ can be derived from the Fermi distribution function at any given temperature T; this function is nearly a step function at room temperature. As the temperature rises above absolute zero, there is more energy to spend on atomic vibrations and on lifting some electrons into energy states in the conduction band. Electrons excited into the conduction band leave behind holes in the valence band. Both the conduction band electrons and the valence band holes contribute to the electrical conductivity.

4.3 Fermi Energy of Amorphous Semiconductors

The Fermi energy ε_F is defined as the chemical potential $\mu(T)$ at a temperature of 0 K. An important property of semiconductors is the existence of a band gap separating the valence and conduction bands. Note that $k_B T \approx 0.025$ eV at room temperature. This implies that thermal excitation of electrons is possible only from the top part of the valence band and that the bottom part does not play any important role in this excitation. To derive the Fermi energy of a pure amorphous semiconductor, we can assume that the occupation of the density of states in the conduction band at temperature T is the following:

$$n_c(T) = \int_{\varepsilon_c}^{\infty} \rho_c(\varepsilon) \frac{1}{e^{(\varepsilon - \mu(T))/k_B T} + 1} d\varepsilon \tag{4.4}$$

where $\rho_c(\varepsilon)$ is the density of states due to the conduction band and ε_c is the bottom point of the conduction band; $(\varepsilon_c - \varepsilon_v)$ defines the gap in the electronic density of states, where ε_v is the highest energy in the valence band. It must be noted, however, that the band gap of a disordered system is not properly defined. Tail states in the gap cause this problem.

At the top of the valence band, an analogous expression describes the occupation:

$$n_v(T) = \int_{-\infty}^{\varepsilon_v} \rho_v(\varepsilon) \frac{1}{e^{(\varepsilon - \mu(T))/k_B T} + 1} d\varepsilon \tag{4.5}$$

where $\rho_v(\varepsilon)$ is the density of states in the valence band or highest occupied band. From this latter equation, we can calculate the hole distribution:

$$p_v(T) = \int_{-\infty}^{\varepsilon_v} \rho_v(\varepsilon) \left(1 - \frac{1}{e^{(\varepsilon - \mu(T))/k_B T} + 1}\right) d\varepsilon = \int_{-\infty}^{\varepsilon_v} \rho_v(\varepsilon) \frac{1}{e^{-(\varepsilon - \mu(T))/k_B T} + 1} d\varepsilon \tag{4.6}$$

The chemical potential is located around the middle of the gap. This means that $\varepsilon_c - \mu(T) \gg k_B T$ and $\mu(T) - \varepsilon_v \gg k_B T$ when $\varepsilon_c - \varepsilon_v \gg k_B T$ is satisfied (that is, in the semiconductor case). Using these relations, we can make an approximation:

$$\frac{1}{e^{(\varepsilon-\mu(T))/k_B T}+1} \cong e^{-(\varepsilon-\mu(T))/k_B T} \tag{4.7}$$

and

$$\frac{1}{e^{(\mu(T)-\varepsilon)/k_B T}+1} \cong e^{-(\mu(T)-\varepsilon)/k_B T} \tag{4.8}$$

Applying these approximations, we get

$$n_c(T) = \int_{\varepsilon_c}^{\infty} \rho_c(\varepsilon) \frac{1}{e^{(\varepsilon-\mu(T))/k_B T}+1} d\varepsilon \cong \int_{\varepsilon_c}^{\infty} \rho_c(\varepsilon) e^{-(\varepsilon-\mu(T))/k_B T} d\varepsilon \tag{4.9}$$

Instead of $(\varepsilon - \mu(T))$, we can put $(\varepsilon - \varepsilon_c) + (\varepsilon_c - \mu(T))$ in the exponent, and rewrite this formula as

$$n_c(T) \cong e^{-(\varepsilon_c-\mu(T))/k_B T} \int_{\varepsilon_c}^{\infty} \rho_c(\varepsilon) e^{-(\varepsilon-\varepsilon_c)/k_B T} d\varepsilon \tag{4.10}$$

Analogously, after some algebra, we obtain

$$p_v(T) \cong e^{-(\mu(T)-\varepsilon_v)/k_B T} \int_{-\infty}^{\varepsilon_v} \rho_v(\varepsilon) e^{-(\varepsilon_v-\varepsilon)/k_B T} d\varepsilon \tag{4.11}$$

In such a system the number of excited electrons must be equal to the number of holes, that is, $n_c(T) = p_v(T)$. We obtain

$$e^{-(\varepsilon_c-\mu(T))/k_B T} \int_{\varepsilon_c}^{\infty} \rho_c(\varepsilon) e^{-(\varepsilon-\varepsilon_c)/k_B T} d\varepsilon = e^{-(\mu(T)-\varepsilon_v)/k_B T} \int_{-\infty}^{\varepsilon_v} \rho_v(\varepsilon) e^{-(\varepsilon_v-\varepsilon)/k_B T} d\varepsilon \tag{4.12}$$

Rearranging this equation, we obtain the following relationship:

$$\mu(T) = \frac{\varepsilon_c + \varepsilon_v}{2} + k_B T \ln \frac{\int_{-\infty}^{\varepsilon_v} \rho_v(\varepsilon) e^{-(\varepsilon_v-\varepsilon)/k_B T} d\varepsilon}{\int_{\varepsilon_c}^{\infty} \rho_c(\varepsilon) e^{-(\varepsilon-\varepsilon_c)/k_B T} d\varepsilon} \tag{4.13}$$

In pure amorphous (and also crystalline) semiconductors, the Fermi energy is at the middle of the gap because

$$\varepsilon_{\mathrm{F}} = \lim_{T\to 0} \mu(T) = \frac{\varepsilon_{\mathrm{c}} + \varepsilon_{\mathrm{v}}}{2} \tag{4.14}$$

At low temperature the variation of $\mu(T)$ as a function of T is very small because the prefactor $k_{\mathrm{B}}T$ in the second term is very small and the remaining part is not extremely large.

4.4 Differences between Amorphous and Crystalline Semiconductors

In the case of crystalline semiconductors, there are two basic types of material, indirect- and direct-band-gap semiconductors. In an indirect-band-gap material, such as silicon, to move into the valence band, an electron must undergo a change in both energy and momentum. The change in energy releases a photon, while the change in momentum produces a phonon, which is a mechanical vibration which heats the crystal lattice. In a direct-band-gap material such as GaAs, only a change in energy is required. For this reason, GaAs is very efficient at producing light without heating the environment, although it does so in the infrared spectrum. Amorphous semiconductors cannot be classified in this way, because of the lack of a $\boldsymbol{k}$ vector.

4.5 Charge Distribution in Pure Amorphous Semiconductors

The atoms in monatomic crystalline semiconductors have no charge at their equilibrium positions because of the symmetric structural arrangement, but in the amorphous case they carry charges owing to geometric distortions. This atomic charge accumulation results in a chemical shift. In the 1980s this phenomenon was an intensively investigated topic in amorphous semiconductor physics [2–11]. The atomic net charges are not observable directly, but experimental determination of fluctuations in them is possible.

A simple model based on the deviations from the ideal bond angle was developed to derive the net atomic charges in fourfold-coordinated atomic networks such as amorphous silicon (a-Si) [9] and diamondlike amorphous carbon (da-C) [11]. Consider an elementary triad of atoms, denoted by K, L, and M, forming two bonds (KM and LM) with an angle KML between them denoted by θ (Figure 4.3). The net charge on the atoms of the triad depends

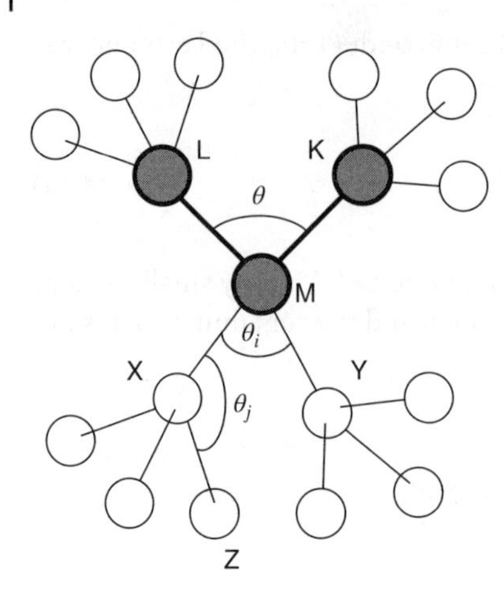

Figure 4.3 Local arrangement of fourfold-coordinated network in a monatomic amorphous semiconductor.

linearly on the deviation of θ from the ideal tetrahedral value, that is, $d\theta = \theta - 109.47°$:

$$q_{\mathrm{M}} = 2A\,d\theta \text{ and } q_{\mathrm{K}} = q_{\mathrm{L}} = -A\,d\theta \tag{4.15}$$

where A is a fitting parameter. The total net charge on atom M, $q_{\mathrm{M}}^{\mathrm{total}}$, is a sum of contributions originating from all triads containing atom M. Since in the distorted tetrahedral model each atom is at the center of six triads and at an end of 12 triads, as displayed in Figure 4.3, the following relation can be obtained:

$$q_{\mathrm{M}}^{\mathrm{total}} = A\left(2\sum_{i=1}^{6} d\theta_i - \sum_{j=1}^{12} d\theta_j\right) \tag{4.16}$$

Values of $A = -0.69$ millielectrons/deg for a-Si [9] and −0.51 millielectrons/deg for da-C (Figure 4.4) have been obtained from a semiempirical Hartree–Fock cluster calculation [11]. By applying this formula for charge accumulation and using the geometric model of tetrahedrally bonded amorphous semiconductors proposed by Wooten, Winer, and Weaire [12], we obtained $dq_{\mathrm{Si}} = 0.021$ electrons for the rms charge fluctuation in a-Si and $dq_{\mathrm{C}} = 0.015$ electrons for da-C. A revised analysis of a core-level spectroscopic measurement [2] estimated that the charge fluctuation in a-Si must be lower than 0.030 electrons.

In hydrogenated amorphous silicon, hydrogen generates a larger charge transfer in the H–Si bonds: negative charge accumulation occurs on the hydrogen atom and the Si atom becomes positively charged. This implies that Si–Si bonds adjacent to an Si–H bond elongate and become weaker. Sometimes such

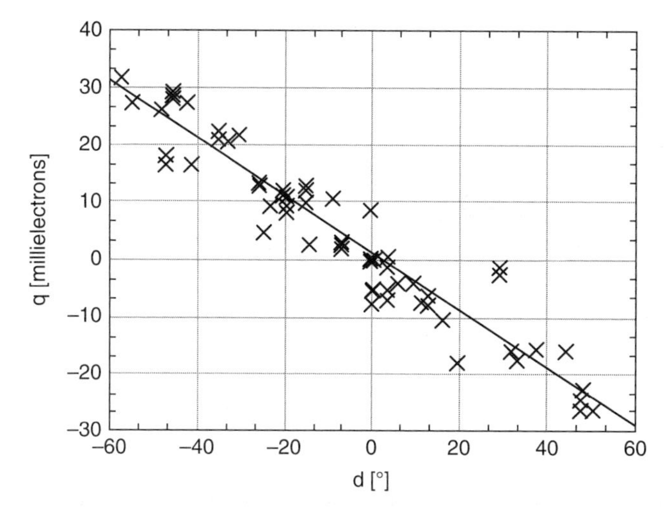

Figure 4.4 Comparison of net atomic charges in diamondlike amorphous carbon obtained from a semiempirical Hartree–Fock calculation (crosses) and from the model presented in the text. *d* denotes the quantity in parentheses in Equation (4.16). *Source:* Kugler and Naray-Szabo 1991 [11]. Reproduced with permission of The Japan Society of Applied Physics.

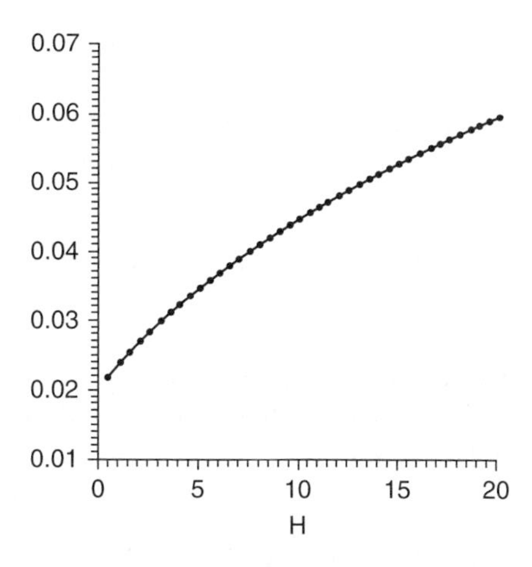

Figure 4.5 Charge fluctuation in a-Si, measured in electrons, versus hydrogen content in atom percent. *Source:* Kugler and Naray-Szabo 1991 [10]. Reproduced with permission of Elsevier.

Si–Si bonds are called "weak bonds" or "back bonds." A rough estimation of the charge fluctuation as a function of hydrogen content has been done [10] (Figure 4.5).

A similar theoretical model of atomic charge accumulation was reported for distorted, long, disordered selenium chains [13]. The net charge on the

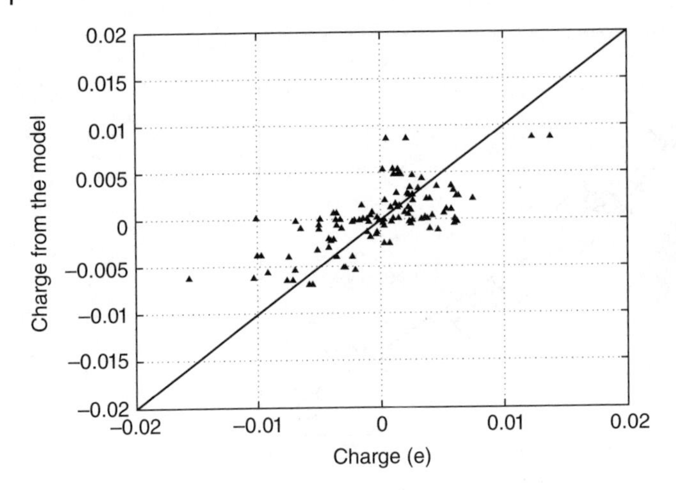

Figure 4.6 Comparison of net atomic charges calculated by Hartree–Fock *ab initio* method (line) and from the model presented in the text (symbols).

twofold-coordinated atoms in a triad depends on the deviation of θ from the average value, that is, $d\theta = \theta - 101°$:

$$q_M = 2\left(A\,d\theta + B\,d\theta^2\right) \quad \text{and} \quad q_K = q_L = -\left(A\,d\theta + B\,d\theta^2\right) \tag{4.17}$$

The total net charge on atom M, q_M^{total}, is a sum of contributions originating from the three triads containing atom M:

$$q_M^{total} = 2\left(A\,d\theta_M + B\,d\theta_M^2\right) - A\left(d\theta_{M-1} + d\theta_{M+1}\right) - B\left(d\theta_{M-1}^2 + d\theta_{M+1}^2\right) \tag{4.18}$$

Based on a Hartree–Fock *ab initio* calculation, $A = -0.45$ millielectrons/deg and $B = -0.0089$ millielectrons/deg^2 were obtained as the best fit, as can be seen in Figure 4.6.

4.6 Density of States in Pure Amorphous Semiconductors

Another important general difference between amorphous and crystalline semiconductors arises from Anderson's theory of localization [17]. A simple description of this theory is that an increase in the disordered potential causes electron localization; that is, the wave function becomes confined in a small

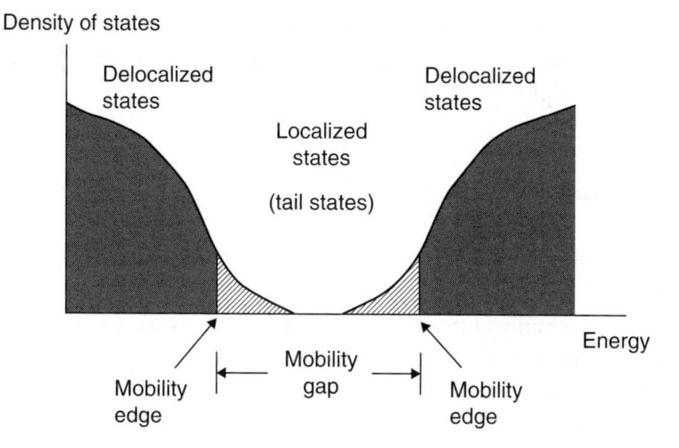

Figure 4.7 Electronic density of states. The mobility edge divides the occupied and unoccupied states into localized and delocalized parts.

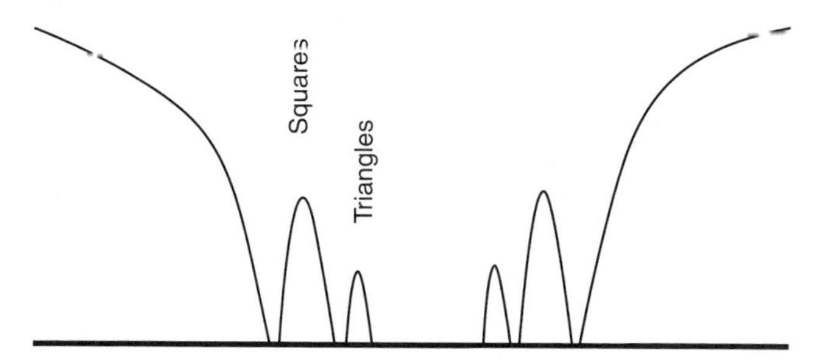

Figure 4.8 Electronic density of states of an amorphous semiconductor, containing two significant peaks associated with triangles and squares.

volume and the electronic density of states shows the structure shown in Figure 4.7. This characteristic difference between the electronic structures of crystalline and noncrystalline solids plays an essential role in their transport and optical properties, and so on. Further investigation provides a more detailed description of this localization. The diamond crystal structure consists of sixfold atomic rings, whereas the corresponding amorphous diamond-like structure is made of fivefold, sixfold, and sevenfold rings; threefold and fourfold rings are atypical. Triangles and squares break the local order; that is, they can be considered as a type of defect. Such defects have been found in a-Si [14]. As a consequence, two characteristic peaks can be found in the tails of the band structure [15]. The first, larger peak corresponds to square atomic arrangements, while the peak at larger energy is due to triangles (Figure 4.8). In earlier investigations of electron transport, hopping conductivity, optical properties, and so on, the tails have usually been considered as exponential or

Gaussian decaying functions. Triangles and squares have never been considered in any band structure calculations, although they probably play an important role in several phenomena.

4.7 Dangling Bonds

Broken or unsaturated bonds which break the local topological order are observed in tetrahedrally bonded amorphous semiconductors (Figure 4.9). A simple dangling bond normally contains one electron and is electrically neutral. Under certain circumstances (chemical or electronic doping), however, the electron occupancy changes from D^0 (neutral) to D^+ (positive) or D^- (negative), where D denotes a dangling bond, as shown in Figure 4.10. As D^- has an extra electron, its energy level is raised by an amount U_c (the coulombic repulsive energy at a dangling bond). However, D^+ should have the same energy level as D_0. D^- does not have spin. To reduce the number of dangling bonds, amorphous silicon (and germanium) is passivated by hydrogen, reducing the

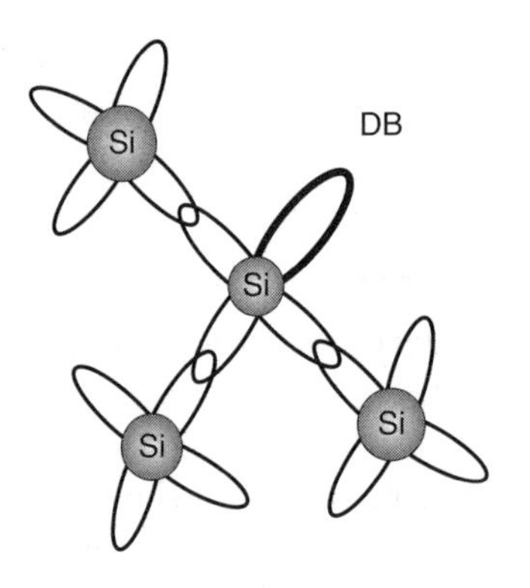

Figure 4.9 Here, an sp^3 orbital remains unpaired and has an unsatisfied valence. This dangling bond (DB) is a localized state.

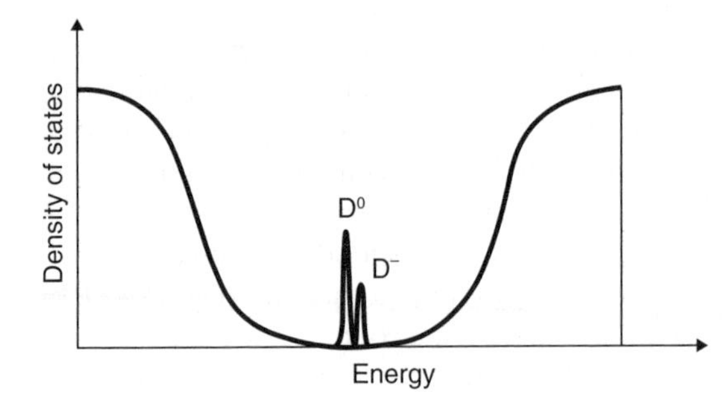

Figure 4.10 Model of defects in a-Si. D^0, D^+ (not shown), and D^- indicate neutral, positively charged, and negatively charged dangling bonds, respectively. D^+ and D^- have no spin.

dangling-bond density by several orders of magnitude. These intrinsic under-coordination defects form electronic states inside the gap, and have a significant electron spin resonance (ESR) signal with $g = 2.0055$ in a-Si, which is the most important experimental evidence for dangling bonds. The spin density is estimated to be around $10^{19}\,\mathrm{cm}^{-3}$ in amorphous silicon but much less, around $10^{15}\,\mathrm{cm}^{-3}$, in hydrogenated amorphous silicon; the latter value is highly dependent on the preparation technique and contamination of the hydrogen. In crystalline Si such defects do not exist, because removing one atom produces a vacancy with four dangling bonds. This arrangement has too high an energy and bond reconstruction occurs. Pantelides [16] has suggested other coordination defects, namely overcoordinated atoms (floating bonds). Whether floating bonds exist or not is still not clear, and answering this question is not easy. If we consider the fourth-nearest-neighbor fourfold-coordinated atom of a threefold-coordinated atom where the direction of dangling is toward this atom, a weak interaction probably exists between the two atoms. This geometrical arrangement provides a fivefold-coordinated atom. This weak interaction, let us say bond, does not contain two localized electrons. It is not a normal sigma bond, but it is a bond in some sense.

A similar dangling bond is theoretically also a possible defect in chalcogenide glasses. At the ends of chains, unpaired spins are expected. However, experiments show different results compared with a-Si. Amorphous selenium does not show any ESR signal in dark conditions; that is, there are no neutral dangling bonds with an unpaired spin. Anderson proposed a negative-U concept [17], and this idea was applied to chalcogenide glasses. Normally, a repulsive Coulomb interaction should occur between two electrons. In the Anderson theory, the phonon–electron interaction gives rise to an attraction, effectively. The Hamiltonian describing the phenomenon has electron, phonon, and electron–phonon parts. If the coupling constant in the electron–phonon part is large enough, the effective correlation between electrons is negative (there is a negative-U center). Instead of C^0, two charged states, C^+ and C^-, are formed, which implies that there are no ESR-active spins:

$$2C^0 \rightarrow C^+ + C^-$$

This means that the charged states C^+ and C^- are more stable than the neutral state C^0. Kastner, Adler, and Fritzsche [18] showed that these are a threefold-coordinated state C_3^+ and a onefold-coordinated state C_1^- on the basis of chemical bonding; these states are called valence alternation pairs (VAPs). Furthermore, another defect, also a threefold-coordinated Se atom (a junction), can be formed in the network.

A result contradictory to the negative-U concept has been reported by Lukacs, Hegedus, and Kugler [19]. The charge distribution in a-Se was calculated by two different methods: a density functional method and a

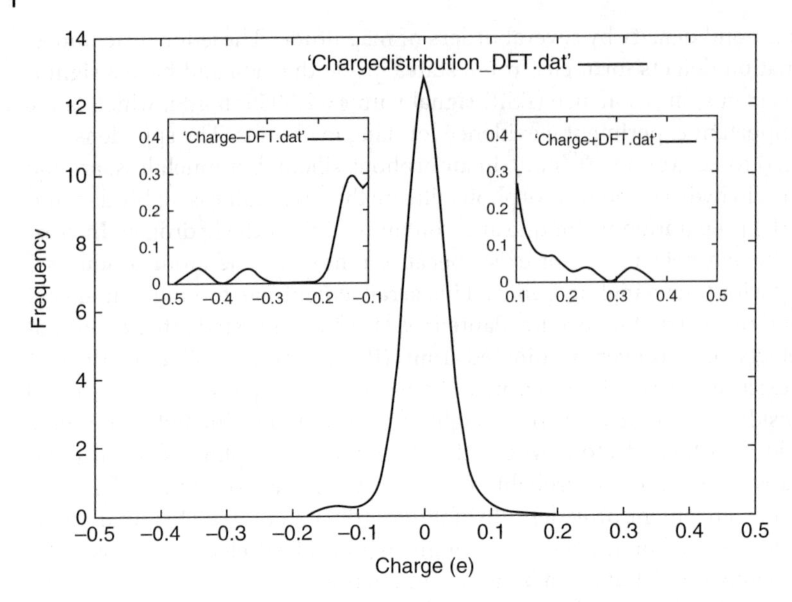

Figure 4.11 Charge distributions obtained from density functional GGA calculations on 162-atom clusters. Basically, the absolute value of the charge accumulation on the twofold-coordinated atoms is less than 0.1 electron units. Additionally, larger charge accumulations in both directions are observed (see the two insets). *Source:* Lukacs *et al.* 2009 [19]. Reproduced with permission from Springer Science and Business Media.

tight-binding model. Figure 4.11 displays the charge distributions obtained from density functional GGA calculations on 162-atom clusters. Basically, the absolute values of the charge accumulation on the twofold-coordinated atoms are less than 0.1 electron units. However, in addition, larger charge accumulations in both the positive and the negative direction are observed (see the two insets in Figure 4.11). The larger positive charge accumulations are associated with threefold-coordinated atoms, while the negatively charged atoms are onefold coordinated. A similar effect was observed using a tight-binding model calculation. These calculations do not contain any phonon–electron interaction, but the results suggest that the threefold-coordinated atoms have lost an electron and these electrons have been transferred to the ends of chains. This looks similar to the picture of VAPs, but these defects are not negative-U defects caused by phonon–electron interaction.

The pure chalcogenide glasses have a slightly different electronic density of states from column IV amorphous semiconductors. As shown in Figure 4.12, an extra peak appears in the middle of the gap; this is attributed to so-called nonbonding lone pair states (or a lone pair band) between the bonding and antibonding states (or bands). In these cases, the Fermi energy $E_F = \mu(T = 0)$ lies between the top of the lone pair band and the conduction band.

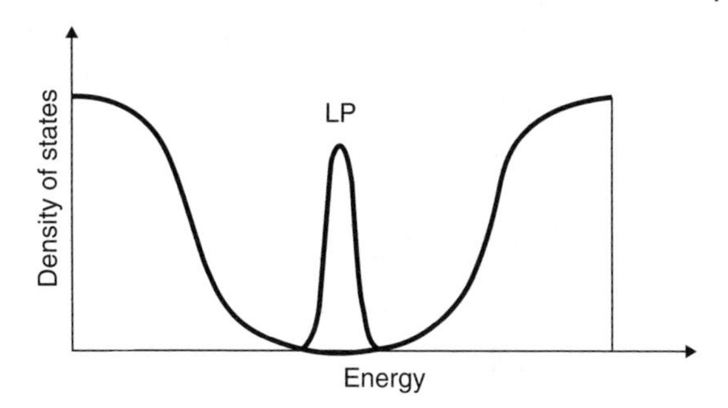

Figure 4.12 Nonbonding lone pair (LP) states occur in the middle of the gap in chalcogenide glasses.

4.8 Doping

The situation is somewhat different for the Fermi energy in doped semiconductors. In amorphous semiconductor networks, normally coordinated impurity atoms that have no unsatisfied bonds are electronically neutral. This means that Mott's famous $(8 - N)$ rule [20] is valid, and a further conclusion is that amorphous semiconductors can *not* be doped! The first reported successful experiment on doping was carried out on amorphous silicon by Chittick, Alexander, and Sterling at Standard Telephone Laboratories, Harlow, UK [21], although this was not exploited. Chittick *et al.* included the following sentence in the abstract of their paper: "The effects of heat-treatment, ageing, and doping on the properties of amorphous silicon are reported." Some years later Spear and Le Comber [22] showed that plasma glow-discharge deposition of silicon, using a mixture of silane and either phosphine or diborane, enabled electrically active pentavalent P (Figure 4.13) and trivalent B (Figure 4.14) impurities to be incorporated into the resulting films, making them n-type and p-type, respectively. These authors could prepare unsatisfied bonds in the network which were electronically active. After this second piece of experimental

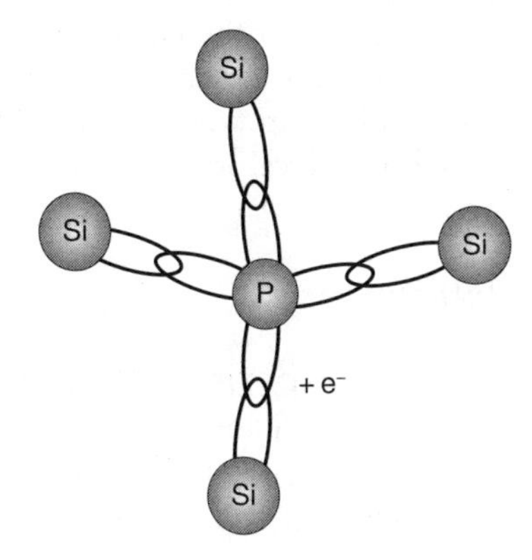

Figure 4.13 An electron excess around a doping P atom in a network.

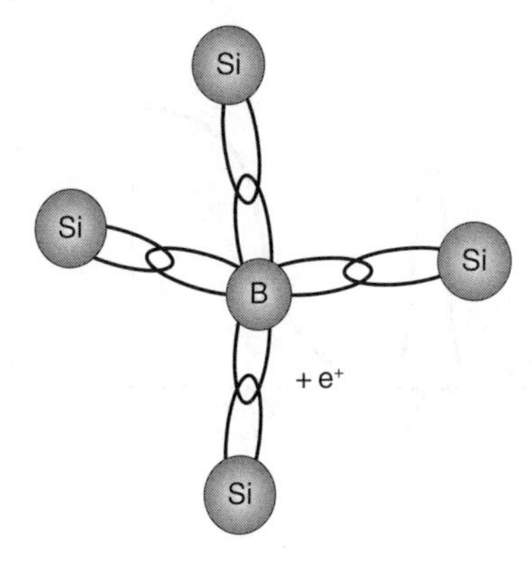

Figure 4.14 An electron deficiency around a doping B atom in a network.

evidence, the "amorphous society" accepted the breaking of Mott's (8 – *N*) rule, and intensive research started on doping effects and their applications.

A hydrogen-atom-type model seems to be a good approach for estimating the energy levels introduced by doping. For example, the core charge of a phosphorus atom is higher than the core charge of the atoms connected to it. The extra fifth electron, which does not take part in any sigma bonds, is weakly bound to the phosphorus atom by a screened Coulomb field. To calculate of this interaction, one must take account of the fact that here a free electron moves in a dielectric medium instead of a vacuum. Furthermore, in heavily doped semiconductors, two doping atoms can be located near to each other, that is, in first-, second-, and third-neighbor positions and so on, and there is an interaction between them [23]. A usable approach to calculation of the resulting energy levels might be to apply a hydrogen-molecule-type of model. The conclusion from these considerations is that extra doping energy levels (donor levels, ε_d) occupied by additional electrons (n-type doping) lie below the conduction band near its bottom. The energy levels of holes due to p-type doping (acceptor levels, ε_a) occur just above the valence band. Doping atoms change the Fermi energy drastically. In the n-type doping case, instead of ε_v one must put ε_d into the equation $\varepsilon_F = \lim_{T \to 0} \mu(T) = (\varepsilon_c + \varepsilon_v)/2$; that is, the Fermi energy lies between the donor level and the conduction band. A similar situation occurs for p-type doping, where the Fermi energy can be found between the acceptor level and the valence band.

References

1 Singh, J. (2002) Effective mass of charge carriers in amorphous semiconductors and its applications. *J. Non-Cryst. Solids*, **299–302**, 444–448.
2 Ley, L., Reichardt, J., and Johnson, R.L. (1982) Static charge fluctuations in amorphous silicon. *Phys. Rev. Lett.*, **49**, 1664–1667.

3 Klug, D.D. and Whalley, E. (1982) Effective charges of amorphous silicon, germanium, arsenic, and ice. *Phys. Rev. B*, **25**, 5543–5546.
4 Kramer, B., King, H., and Mackinnon, A. (1983) Charge fluctuations in hydrogenated amorphous silicon. *J. Non-Cryst. Solids*, **59–60**, 73–76.
5 Brey, L., Tejedor, C., and Verges, J.A. (1984) Comment on "Static charge fluctuations in amorphous silicon". *Phys. Rev. Lett.*, **52**, 1840.
6 Winer, K. and Cardona, M. (1986) Theory of infrared absorption in amorphous silicon. *Solid State Commun.*, **60**, 207–211.
7 Kugler, S. and Naray-Szabo, G. (1987) Charge distribution in amorphous silicon clusters: Quantum chemical study combined with ring statistics. *J. Non-Cryst. Solids*, **97–98**, 503–506.
8 Bose, S.K., Winer, K., and Andersen, O.K. (1988) Electronic properties of a realistic model of amorphous silicon. *Phys. Rev. B*, **37**, 6262–6277.
9 Kugler, S., Surjan, P.R., and Naray-Szabo, G. (1988) Theoretical estimation of static charge fluctuation in amorphous silicon. *Phys. Rev. B*, **37**, 9069–9071.
10 Kugler, S. and Naray-Szabo, G. (1991) Weak bonds and atomic charge distribution in hydrogenated amorphous silicon. *J. Non-Cryst. Solids*, **137–138**, 295–298.
11 Kugler, S. and Naray-Szabo, G. (1991). Atomic charge distribution in diamondlike amorphous carbon. *Japan. J. Appl. Phys.*, **30**, L1149–L1151.
12 Wooten, F., Winer, K., and Weaire, D. (1985) Computer generation of structural models of amorphous Si and Ge. *Phys. Rev. Lett.*, **54**, 1392–1395.
13 Lukacs, R. and Kugler, S. (2010) A simple model for the estimation of charge accumulation in amorphous selenium. *Chem. Phys. Lett.*, **494**, 287–288.
14 Kugler, S., Kohary, K., Kadas, K., and Pusztai, L. (2003) Unusual atomic arrangements in amorphous silicon. *Solid State Commun.*, **127**, 305–309.
15 Kugler, S. (2012) Advances in understanding the defects contributing to the tail states in pure amorphous silicon. *J. Non-Cryst. Solids*, **358**, 2060–2062.
16 Pantelides, S.T. (1986) Defects in amorphous silicon: A new perspective. *Phys. Rev. Lett.*, **57**, 2979–2982.
17 Anderson, P.W. (1975) Model for the electronic structure of amorphous semiconductors. *Phys. Rev. Lett.*, **34**, 953–955.
18 Kastner, M., Adler, D., and Fritzsche, H. (1976) Valence-alternation model for localized gap states in lone-pair semiconductors. *Phys. Rev. Lett.*, **37**, 1504–1507.
19 Lukacs, R., Hegedus, J., and Kugler, S. (2009) Microscopic and macroscopic models of photo-induced volume changes in amorphous selenium. *J. Mater. Sci.: Mater. Electron.*, **20**, S33–S37.
20 Mott, N.F. (1969) Conduction in non-crystalline materials. *Philos. Mag.*, **19**, 835–852.

21 Chittik, R.C., Alexander, J.H., and Sterling, H.J. (1969) The preparation and properties of amorphous silicon. *J. Electrochem. Soc.*, **116**, 77–81.
22 Spear, W.E. and Le Comber, P.G. (1975) Substitutional doping of amorphous silicon. *Solid State Commun.*, **17**, 1193–1196.
23 Kádas, K., Ferenczy, Gy.G., and Kugler, S. (1998) Theory of dopant pairs in four-fold coordinated amorphous semiconductors. *J. Non-Cryst. Solids*, **227–230**, 367–371.

5

Electronic and Optical Properties of Amorphous Silicon

5.1 Introduction

In this chapter, we deal with the electronic and optical properties of amorphous silicon, particularly hydrogenated amorphous silicon (a-Si:H), which is a typical amorphous semiconducting material from the viewpoint of physics and applications. As mentioned in Chapter 4, the band tail states of amorphous semiconductors are typical localized states arising from the disordered amorphous network, and affect the electronic and optical properties. In actual amorphous semiconductors, however, localized states are created by defects in the network, that is, so-called structural defects. Such localized states affect the properties as well. In Section 5.2, band tails and structural defects in a-Si:H are overviewed. In Section 5.3, recombination processes are treated, and are related particularly to photoluminescence and photoconduction. The electrical and optical properties of a-Si:H are treated in Sections 5.4 and 5.5, respectively. Section 5.6, on electron magnetic resonance and spin-dependent properties, is included in this chapter because experimental methods such as electron magnetic resonance and related magnetic resonance methods provide us with important information concerning the microscopic nature of band tails and structural defects. In particular, recent investigations using pulsed electron magnetic resonance techniques are introduced and discussed in more detail. Furthermore, spin-dependent properties of a-Si:H can be explored using combinations of electron magnetic resonance and various types of phenomena. In Section 5.7, light-induced phenomena in a-Si:H are dealt with, because they are associated with the disordered nature of a-Si:H. Since these generally occur as a result of light-induced defect creation, the mechanism of light-induced defect creation is also described, as well as experimental evidence for those mechanisms and the nature of light-induced defects.

Amorphous Semiconductors: Structural, Optical, and Electronic Properties, First Edition.
Kazuo Morigaki, Sándor Kugler, and Koichi Shimakawa.

5.2 Band Tails and Structural Defects

5.2.1 Introduction

As discussed in Chapter 4, band tails are formed between two mobility edges. Their states arise from weak Si–Si bonds whose binding energy is weaker than the normal binding energy. Furthermore, structural defects are formed in real a-Si:H samples owing to misbonding between two silicon atoms. An ideal amorphous silicon network would be bonded together by silicon atoms to form a continuous random network. Defects generated by misbonding are called structural defects. In this section, we consider the nature of band tails and structural defects, that is, their microscopic properties.

5.2.2 Band Tails

In this section, we consider the electronic states of tail electrons and tail holes on the basis of magnetic resonance measurements, particularly light-induced electron spin resonance (LESR), optically detected magnetic resonance (ODMR), and optically detected electron–nuclear double resonance (ODENDOR) measurements.

5.2.2.1 LESR Measurements

Under illumination, two electron spin resonance (ESR) signals with $g = 2.004$ and 2.012 have been observed in a-Si:H [1]. These signals are called LESR signals and have been identified as being due to tail electrons and tail holes, respectively, on the basis of the g-shift δg, that is, the deviation of the g-value from the free-electron g-value of 2.0023 [1]. Generally, $\delta g < 0$ and $\delta g > 0$ suggest that the magnetic centers responsible for these signals may be electrons and holes, respectively. In the following, we first consider tail electrons. Their LESR signal shows a long decay time [2], as shown in Figure 5.1. This result suggests that the magnetic center responsible for $g = 2.004$ should be a deep center. This is contrary to the expectation for shallow tail electrons. An ESR signal of tail electrons has been observed at $g = 1.987$ in time-resolved ODMR measurements, using a pulsed microwave source and a pulsed laser [3, 4]. On the other hand, we have observed a tail hole resonance at $g = 2.0065$, arising from a hole trapped at a specific weak Si–Si bond, in ODMR measurements [3, 4]. This specific weak Si–Si bond is illustrated in Figure 5.2; it is a weak Si–Si bond adjacent to an Si–H bond. This identification has been confirmed by ENDOR measurements, by monitoring the ODMR signal due to the self-trapped holes and scanning the radio frequency to excite nuclear transitions [5–8]. The observed ENDOR spectrum is shown in Figure 5.3, where the arrows indicate the natural nuclear frequencies of the ^{29}Si (2.8 MHz) and ^{1}H (13.9 MHz) nuclei. The ENDOR signal at 13.9 MHz can be decomposed into two components, a narrow component (normal ENDOR) and a broad component (matrix

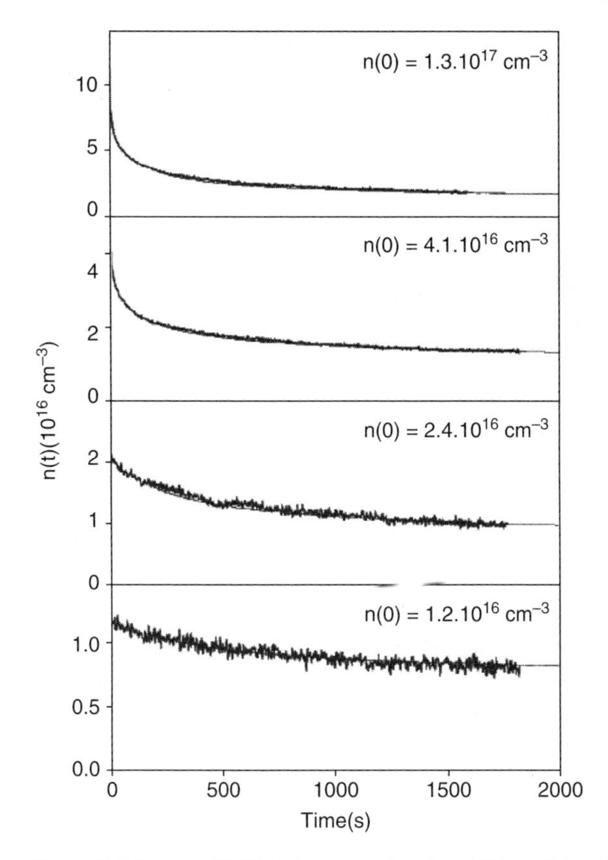

Figure 5.1 Decay of LESR (spin-carrier density $n(t)$) in a-Si:H at 40 K after saturation at excitation intensities, from top to bottom, of 19.2×10^{-1}, 1.6×10^{-3}, and 3.6×10^{-4} mW/cm^2, respectively. The solid lines are theoretical fits to the data. *Source:* Yan *et al.* 2000 [2]. Reproduced with permission from the American Physical Society.

ENDOR). The results of the ODENDOR measurements can be summarized as follows:

1) The matrix ENDOR due to the ^{1}H nucleus observed at 13.9 MHz allows us to estimate the distance between the spin of a self-trapped hole and a nearby hydrogen nucleus as about 4 Å.
2) The local ENDOR due to the ^{29}Si nucleus allows us to estimate the wave function of the self-trapped hole in terms of the tight-binding approximation from the isotropic and anisotropic hyperfine interactions with the ^{29}Si nucleus as follows:

$$\psi = \sum_j \eta_j \psi_j \tag{5.1}$$

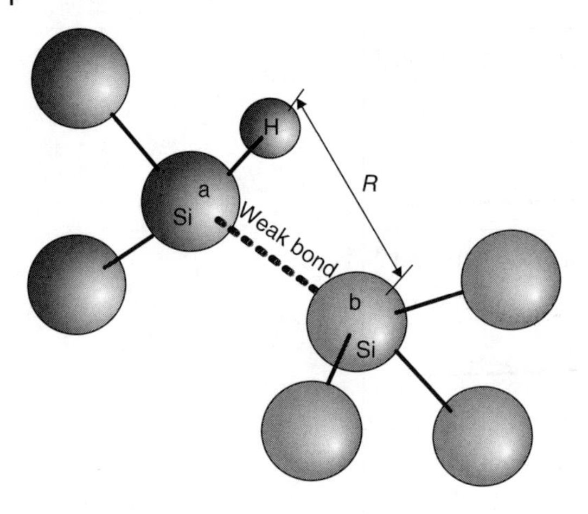

Figure 5.2 Atomic configuration around a specific weak Si–Si bond consisting of a weak Si–Si bond and an Si–H bond.

$$\psi_j = \alpha_j \left(\psi_{3s}\right)_j + \beta_j \left(\psi_{3p}\right)_j \tag{5.2}$$

where the wave function at the jth site is composed of atomic 3 s and 3p orbitals ψ_{3s} and ψ_{3p} at the jth site, whose percentage 3 s and 3p characters are designated by α_j^2 and β_j^2, respectively. The isotropic and anisotropic hyperfine interaction constants arising from the unpaired electron at the jth site, A_{iso} and A_{aniso}, are given by

$$A_{iso,j} = (8\pi/3) g_e g_n \mu_B \mu_n \alpha_j^2 \eta_j^2 \left|\Psi_{3s}(0)_j\right|^2 \tag{5.3}$$

$$A_{aniso,\,j} = (2/5) g_e g_n \mu_B \mu_n \beta_j^2 \eta_j^2 \left\langle r_{3p}^{-3} \right\rangle_j \tag{5.4}$$

where g_e, g_n, μ_B, and μ_n are the electronic and nuclear g-values, the Bohr magneton, and the nuclear magneton, respectively, and $\langle r_{3p}{}^{-3} \rangle$ is the average value of $\langle r^{-3} \rangle$ for the 3p state on the jth site. The wave function may be estimated from an analysis of the hyperfine structure. A detailed analysis is given in [8]. From the fitting of the hyperfine pattern to the observed ENDOR spectrum, as shown in Figure 5.3, we obtain $\eta_0{}^2 = 0.13$, $\alpha_0{}^2 = 0.03$, and $\beta_0{}^2 = 0.97$, where the 0th site is the central silicon atom, that is, the b site in Figure 5.2. This shows the hole nature of the magnetic centers responsible for the enhanced ODMR signal at $g = 2.0065$ and is consistent with a self-trapped hole. The result $\eta_0{}^2 = 0.13$ shows that the wave function is rather extended compared with the dangling bond electron, whose energy level is located in the middle of the band gap. The depth of the self-trapped hole center has been estimated from photoinduced absorption measurements [9, 10] as 0.25 eV from the mobility edge of the valence band.

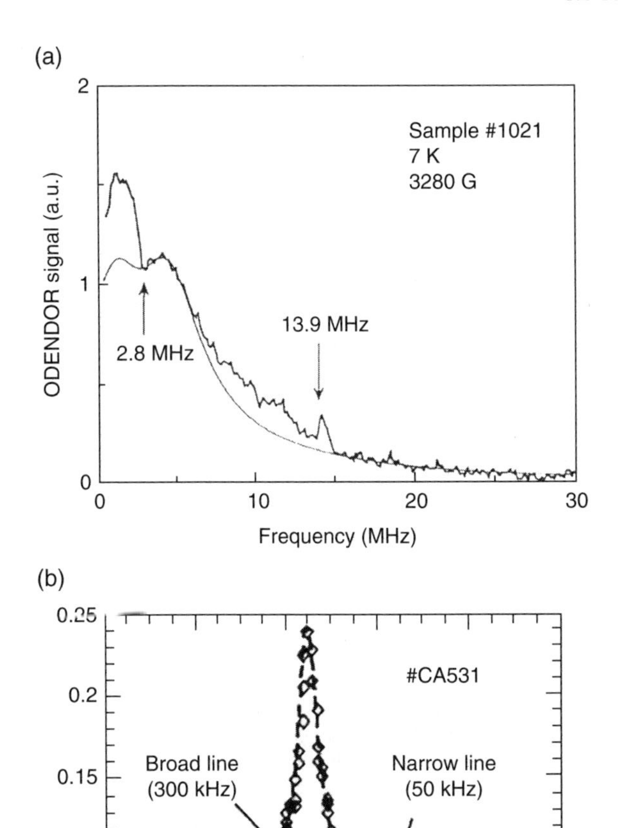

Figure 5.3 (a) ODENDOR spectrum of an a-Si:H sample prepared at 250 °C. Its hydrogen content was about 15 at.%. The solid curve is a calculated spectrum (see text). The positions of the natural nuclear frequencies of ^{29}Si and ^{1}H are shown at 2.8 and 13.9 MHz, respectively. (b) ODENDOR spectrum of an a-Si:H sample prepared at 300 °C. Its hydrogen content was about 3%. The dotted and dashed curves correspond to the broad line and the narrow line, respectively. The full width at half maximum (FWHM) of the lines is 300 kHz and 50 kHz, respectively. *Source:* Kondo and Morigaki 1991 [6]. Reproduced with permission of Elsevier.

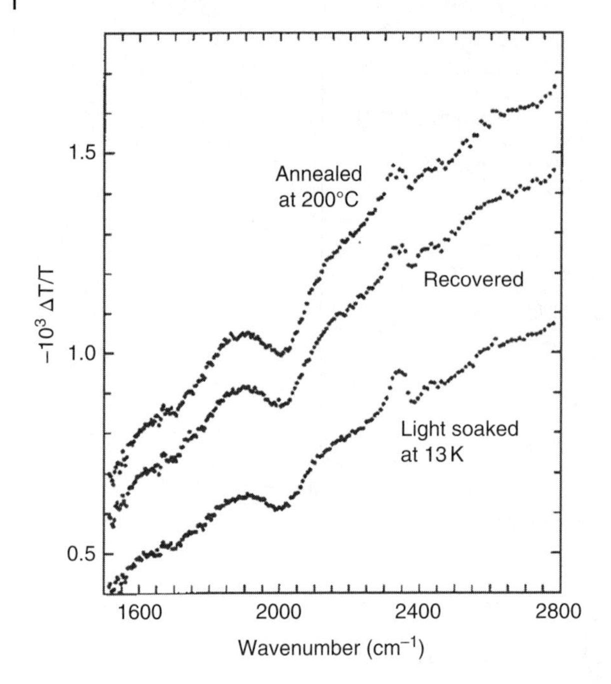

Figure 5.4 Photomodulation spectra of a-Si:H in the annealed and light-soaked states (at 13 K for 20 h by Kr^+ laser light at 1.9 eV and 2 W cm^{-2}). The annealed state was measured after keeping the sample at room temperature for 2 days after the light soaking at 13 K. *Source:* Oheda 1999 [12]. Reproduced with permission from the American Physical Society.

The self-trapping of holes at specific weak Si–Si bonds was also inferred from photomodulated infrared (IR) measurements by Vardeny and Olszakier [11] and by Oheda [12]. Vardeny and Olszakier observed that the IR peak intensity at 2000 cm^{-1} decreases under optical excitation. This result can be understood by assuming that the dipole moment for the IR mode (the Si–H bond-stretching mode) becomes small because of self-trapping of a hole at a weak Si–Si bond adjacent to the Si–H bond responsible for the IR mode. Oheda found that the photomodulated IR peak intensity around 2000 cm^{-1} was affected by prolonged illumination at 297 and 13 K, that is, the amount of the decrease in the IR peak intensity decreases after prolonged illumination with Kr^+ laser light of energy 1.92 eV, as shown in Figure 5.4. This is also consistent with the model of self-trapping of holes at a weak Si–Si bond adjacent to an Si–H bond. A detailed account is given in [13].

5.2.3 Structural Defects

It is well known that structural defects are present even in device quality a-Si:H samples, because those samples do not have an ideal continuous

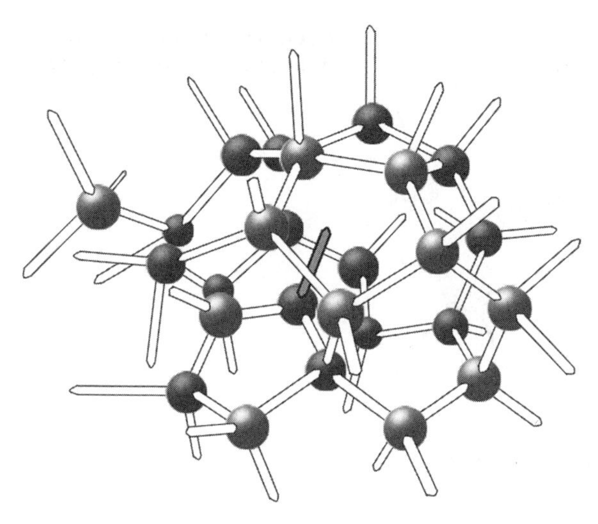

Figure 5.5 Schematic view of a normal dangling bond in a-Si:H.

network, and so some of the constituent atoms are "deficient"; that is, Si dangling bonds (Figure 5.5) (so-called normal dangling bonds) are formed. The structural and electronic properties of Si dangling bonds in a-Si:H have been extensively investigated, using various types of measurements. First, we consider their electronic states, as investigated using magnetic resonance techniques. From ENDOR measurements [14–17], the wave function of dangling bond electrons can be determined in terms of the tight-binding approximation [16], giving

$$\eta_0^2 = 0.54, \quad \alpha_0^2 = 0.10, \quad \text{and} \quad \beta_0^2 = 0.90$$

From a comparison with the self-trapped hole mentioned above, the wave function of dangling bond electrons is more localized than that of self-trapped holes.

Furthermore, it has been confirmed by ENDOR measurements [14–17] that hydrogen-related dangling bonds with hydrogen in a nearby site are present (Figure 5.6). This is also suggested by deconvolution of ESR spectra into two components [18, 19], that is, normal dangling bonds and hydrogen-related dangling bonds, as shown in Figure 5.7. The presence of hydrogen-related dangling bonds was first suggested by a model of light-induced defect creation in a-Si:H [20]. See Section 5.6 for further details.

A defect similar to hydrogen-related dangling bonds, called the vacancy–hydrogen complex (Figure 5.8), has been observed at low temperatures ($\leq$100 K) by Nielsen *et al.* [21] in hydrogen-implanted crystalline silicon.

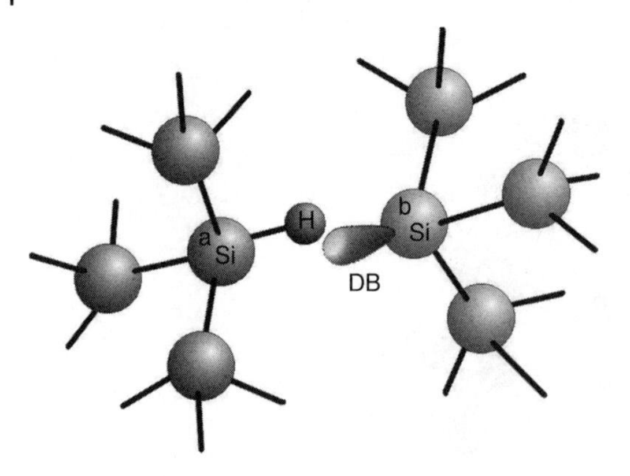

Figure 5.6 Schematic view of a hydrogen-related dangling bond.

5.3 Recombination Processes

5.3.1 Introduction

After creation of free electrons and free holes by optical excitation, they quickly move down to the bottom of their respective bands to thermalize, and then they become trapped in tail states and gap states and recombine, either with each other or through defects. In this section, we treat radiative recombination, in which light (photons) is emitted, and nonradiative recombination, in which phonons are emitted.

5.3.2 Radiative Recombination

5.3.2.1 Geminate Electron–Hole Pair Recombination

If a electron and a hole are created simultaneously at a certain site by optical excitation and are coupled by a Coulomb attractive interaction, then such a pair is called a geminate electron–hole pair [22]. Geminate electron–hole pair recombination occurs under weak optical excitation and at low temperatures, as will be discussed in Section 5.5.4.

5.3.2.2 Nongeminate Electron–Hole Pair Recombination

When the optical excitation becomes strong, that is, the free-carrier generation rate increases sufficiently, the density of electrons and holes becomes high, and then an electron can be coupled with a hole closer to the electron. These electron–hole pairs are called nongeminate electron–hole pairs or distant electron–hole pairs. Radiative recombination of nongeminate electron–hole pairs occurs in the following way. The distance between the electron and hole is denoted by R. The radiative recombination rate, P_R, is given by

$$P_R = P_0 \exp\left(-\frac{2R}{r_e}\right) \tag{5.5}$$

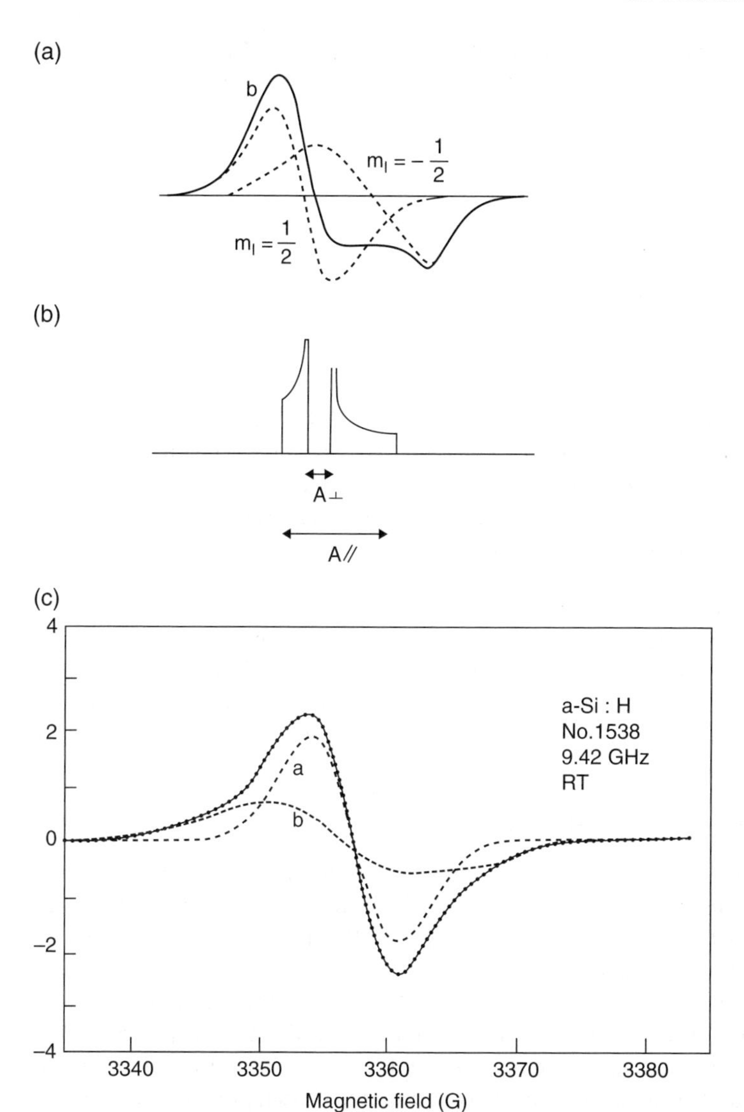

Figure 5.7 (a) Hyperfine structure (HFS) of a hydrogen-related dangling bond due to the ^{1}H nucleus. Two components, $m_I = 1/2$ and $-1/2$, are shown. (b) Schematic diagram of HFS with axial symmetry due to the ^{1}H nucleus. Source: Hikita *et al.* 1997 [18]. (c) Deconvolution of the observed ESR spectrum of a-Si:H into two components due to normal dangling bonds a and hydrogen-related dangling bonds b. *Source:* Morigaki *et al.* 1998 [19]. Reproduced with permission of Elsevier.

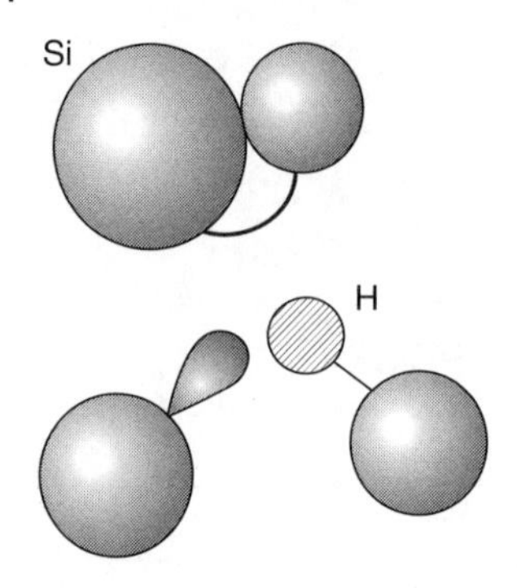

Figure 5.8 Atomic configuration of the vacancy–hydrogen complex in crystalline silicon.

where r_e is the radius of a trapped electron. Here, it is assumed that the radius of a trapped hole is smaller than r_e. The radiative lifetime, τ, is given by P_R^{-1}. For a-Si:H, τ_0 ($\equiv P_0^{-1}$) is equal to 10^{-8} s. The value of τ normally ranges between 10^{-6} and 10^{-2} s in a-Si:H. The model of radiative recombination expressed by Equation (5.5) is often called the radiative tunneling model [23].

5.3.2.3 Exciton Recombination

Excitons are generally observed in semiconductors and ionic crystals. In a-Si:H, singlet exciton recombination with $\tau \risingdotseq 10^{-8}$ s has been observed in photoluminescence (PL) spectra under pulsed optical excitation, while triplet excitons have been observed in ODMR measurements [24–26], as will be mentioned in Section 5.6.

5.3.3 Nonradiative Recombination

When an electron and a hole recombine with each other at a deep defect, for example a Si dangling bond in a-Si:H, the electron–phonon interaction causes the recombination to be of nonradiative nature, that is, an electron recombines nonradiatively with a hole to emit phonons.

First, we treat the case of strong electron–phonon interaction. In this case, the configuration coordinate model is used to consider the deep center that acts as the nonradiative recombination center. As illustrated in Figure 5.9, a photogenerated electron is captured by a trapping center, jumps over a barrier of height E_A, and then recombines with a hole at a deep nonradiative center.

Following Englman and Jortner [27], the nonradiative recombination rate, P_{NR}, is given by

$$P_{NR} = W_0 \left(\frac{\hbar\Omega_0}{k_B T^*} \right)^{1/2} \exp\left(-\frac{2R}{r_0} - \frac{E_A}{k_B T^*} \right) \tag{5.6}$$

$$T^* = \frac{\hbar\Omega_0}{2k_B} \coth\left(\frac{\hbar\Omega_0}{2k_B T} \right) \tag{5.7}$$

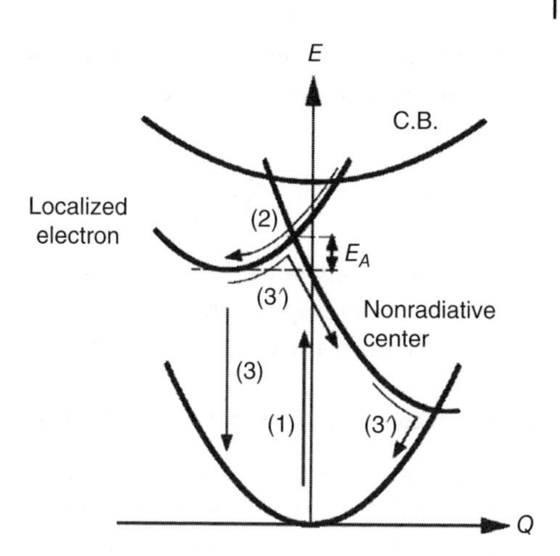

Figure 5.9 Schematic diagram of the configuration coordinate model in the case of strong electron–phonon coupling: (1) optical excitation, (2) thermalization of the electron, (3) radiative recombination, (3′) nonradiative recombination. C.B., conduction band.

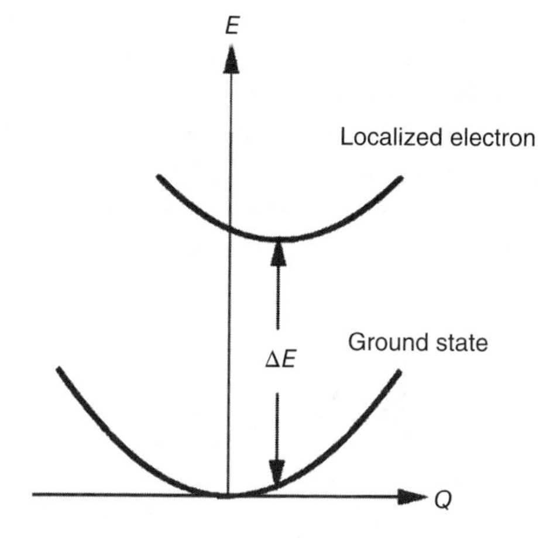

Figure 5.10 Schematic diagram of the configuration coordinate model in the case of weak electron–phonon coupling.

where W_0 is the order of $10^{13}\,\mathrm{s}^{-1}$ and $\hbar\,\Omega_0$ is the optical phonon energy, and T^* is approximated by

$$T^* = \hbar\Omega_0/2k_\mathrm{B} \quad \text{for} \quad T \ll \left(\hbar\Omega_0/2k_\mathrm{B}\right) \tag{5.8}$$

and

$$T^* = T \quad \text{for} \quad T \gg \left(\hbar\Omega_0/2k_\mathrm{B}\right) \tag{5.9}$$

Next, we treat the case of weak electron–phonon interaction. In this case, multiphonon emission occurs between two curves, as shown in Figure 5.10. The number of emitted phonons is approximately given by $\Delta E/\hbar\Omega_\mathrm{M}$, where ΔE

is the difference in energy between the two curves, as shown in Figure 5.10, and Ω_M is the highest vibration frequency. P_{NR} is expressed by

$$P_{NR} = W_0 \exp\left(-\frac{\gamma' \Delta E}{\hbar \Omega_M}\right) \tag{5.10}$$

where γ' is a constant of the order of 1. As shown in Equation (5.10), P_{NR} is independent of temperature. This case applies to the self-trapping of holes at a weak Si–Si bond adjacent to an Si–H bond.

5.3.4 Recombination Processes and Recombination Centers in a-Si:H

The diagram shown in Figure 5.11 represents the recombination processes in a-Si:H. Radiative recombination occurs mainly between tail electrons and tail holes at temperatures higher than 100 K and mainly between tail electrons and self-trapped holes (A centers) at temperatures lower than 100 K. Such recombination contributes to the principal component of the PL spectrum. This conclusion has been obtained from ODMR measurements [28–30], as will be mentioned in Section 5.3.5. The low-energy PL mentioned in Section 5.5.3 arises from radiative recombination of an electron at a T_3^+–N_2^- pair defect with a self-trapped hole at lower temperatures and with a tail electron at higher temperatures; here, T_3^+ and N_2^- designate a positively charged threefold silicon center and negatively charged twofold nitrogen center,

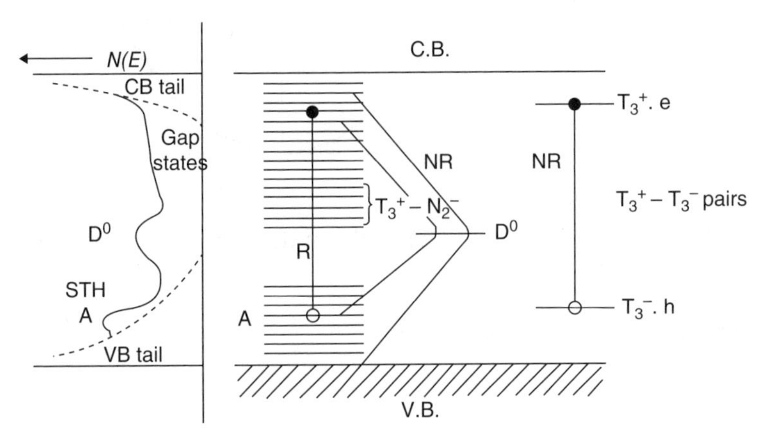

Figure 5.11 Schematic diagram of the tail and gap states and the recombination processes in a-Si:H. R and NR designate radiative and nonradiative recombination, respectively. A diagram of the density-of-states spectrum is illustrated on the left side. A (STH), self-trapped holes at specific Si–Si weak bonds; D, neutral dangling bonds; T^{3+}, T^{3-}, positively and negatively charged threefold-coordinated Si atoms, respectively; N^{2-}, negatively charged twofold-coordinated nitrogen atom; C.B., conduction band; V.B., valence band.

respectively. Nitrogen is incorporated into the silicon network as an impurity from unintentional contamination. This model of low-energy PL was suggested by ODMR measurements. Neutral Si dangling bonds, electrons captured by T_3^+, and holes captured by T_3^- act as nonradiative centers, as has been suggested by ODMR measurements. These states are illustrated in Figure 5.11.

5.3.5 Spin-Dependent Recombination

First, we consider distant-pair recombination of radiative nature. The distant pair consists of two separate carriers, that is, an electron and a hole, each having spin 1/2. Recombination between an electron and a hole that have the same spin direction with respect to a magnetic field is forbidden, assuming that these two spins are in a static magnetic field that is strong enough compared with their mutual dipolar interaction. Their energy levels in a static magnetic field, called Zeeman levels, are shown in Figure 5.12, as well as transition processes between them and the ground state, that is, the state before the optical excitation that generates electrons and holes. When the Zeeman levels are unthermalized, that is, the spin–lattice relaxation time T_1 is longer than the lifetime of an electron–hole pair, the populations of the Zeeman levels deviate from the Boltzmann distribution, and then the populations of the antiparallel spin states become smaller than those of the parallel spin states in the steady state under continuous optical excitation [31, 32]. When an ESR transition occurs for either electrons or holes, the population of the antiparallel spin state becomes larger. As a result, when PL arises from radiative recombination between electrons and holes, the PL intensity I increases at resonance.

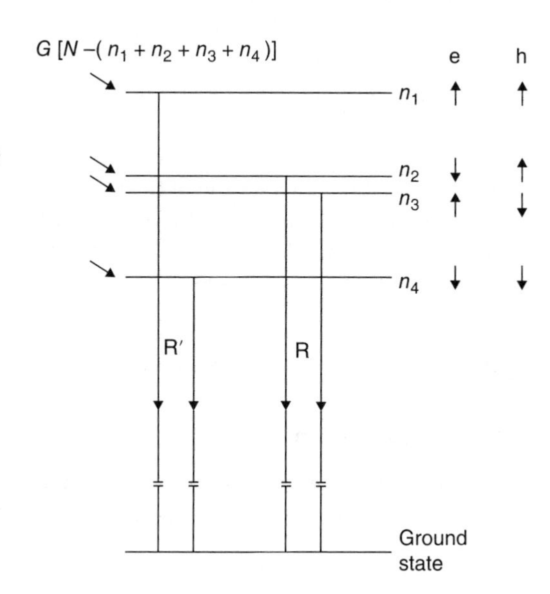

Figure 5.12 Zeeman level diagram of an electron–hole pair in the presence of a static magnetic field. G, N, and n_i designate the generation rate, the total number of pairs of electron and hole traps, and the population of the ith level, respectively.

Thus, the ODMR signal associated with resonance is called an enhancing signal. On the other hand, in the case of radiative recombination of geminate pairs, the ODMR signal appears as a quenching of the PL intensity, because the geminate pairs consisting of antiparallel spins decrease in number owing to resonance of either electrons or holes, so that the PL due to geminate radiative pairs decreases in intensity.

Secondly, we consider nonradiative recombination of electrons and holes through dangling bonds in a-Si:H. The neutral dangling bond center acts as a deep nonradiative center in a-Si:H because its energy level is almost in the middle of the band gap, as shown in Figure 5.11. When a tail electron is captured by a dangling bond, then the dangling bond becomes negatively charged and, as a result, the electron recombines with a tail hole, a self-trapped hole, or a free hole. This process is enhanced by resonance of dangling bonds in a static magnetic field, because the spin of the dangling bond is inverted at resonance, and then capture of an electron by the dangling bond is enhanced. This means that the nonradiative recombination process through dangling bonds is emphasized compared with the radiative recombination process of tail electrons and self-trapped holes (A centers) or tail holes (see Figure 5.11 again). As a result, the PL intensity decreases at resonance of the dangling bonds. Hence, the ODMR signal of dangling bonds is observed as a quenching signal.

As seen above, the ODMR measurement technique is very useful for identifying whether recombination centers act as radiative or nonradiative centers. Detailed accounts of ODMR measurements are given in [29–33].

By monitoring the photocurrent, ESR can be detected with high sensitivity. This method is also called electrically detected magnetic resonance (EDMR). This technique uses spin-dependent photoconductivity [34] in the following way. For instance, let us think about a system consisting of a free carrier and a trap. For the trap, we take a neutral dangling bond. When ESR of dangling bonds occurs, their spin is inverted, and then free carriers are captured by dangling bonds. Thus, the lifetime of the free carriers is shortened so that the photocurrent decreases in magnitude. The ESR of dangling bonds is detected through spin-dependent photoconductivity. Detailed accounts of EDMR measurements are given in [35–37].

5.4 Electrical Properties

5.4.1 DC Conduction

DC conduction in a-Si consists of band conduction and hopping conduction. The electrical conductivity is generally described as follows:

$$\sigma = ne\mu \tag{5.11}$$

where n and μ are the density and mobility, respectively, of the carriers. In amorphous semiconductors, electrical conduction near the mobility edge can be treated in terms of a random walk model, because the mean free path is comparable to the interatomic distance in this case. Calculation of the electrical conductivity has been performed on the basis of the random phase model [38, 39]. Here, we assume that the memory of the phase of the electron's wave function is lost from site to site. Furthermore, a wave function based on the tight-binding approximation is used in the following way. The DC electrical conductivity associated with an energy E is described in terms of the Kubo–Greenwood formula as follows:

$$\sigma_E(0) = \left(2\pi e^2 \hbar V / 3m^2\right) \left\langle |k'|p|k|^2 \right\rangle_{\mathrm{AV}} \{N(E)\}^2 \tag{5.12}$$

where V and $N(E)$ are the volume of the sample and the density of states at E, respectively. $\langle k'|p|k\rangle$ is the matrix element of the momentum p between two states k' and k, and the configurational average of its square is obtained in terms of the random phase model as follows:

$$\left\langle |k'|p|k|^2 \right\rangle_{\mathrm{AV}} = \left(a^2/V\right) \left\langle \sum_{n'} \left\langle |n'|p|n|^2 \right\rangle \right\rangle_{\mathrm{AV}} \tag{5.13}$$

where a and n are the atomic separation and the number of the site, respectively. The wave function is expressed by a superposition of a Wannier function and a plane wave, that is, $\exp(i\boldsymbol{k}\boldsymbol{R}_n)$, where $\boldsymbol{R}_n$ is the position vector at the nth site. Since $\langle n'|p|n\rangle$ is of the order of $\hbar/a$, we obtain

$$\left\langle \sum_{n'} \left\langle |n'|p|n|^2 \right\rangle \right\rangle_{\mathrm{AV}} = \lambda^2 \hbar^2 / a^2 \tag{5.14}$$

where λ is a number of order 1. Using Equations (5.4.2) –(5.4.4), we obtain

$$\sigma_E(0) = \left(2\pi e^2 \hbar^3 a \lambda^2 / 3m^2\right) \{N(E)\}^2 \tag{5.15}$$

5.4.1.1 Band Conduction

Band conduction takes place by way of minority carriers, either electrons or holes thermally excited into their respective bands. This type of conduction is expressed by the following equation for the conductivity:

$$\sigma = \sigma_0 \exp\left(-E_{\mathrm{a}} / k_{\mathrm{B}} T\right) \tag{5.16}$$

where σ_0 and E_{a} are a prefactor and an activation energy, respectively. For electrons, E_{a} is given by

$$E_{\mathrm{a}} = E_{\mathrm{C}} - E_{\mathrm{F}} \tag{5.17}$$

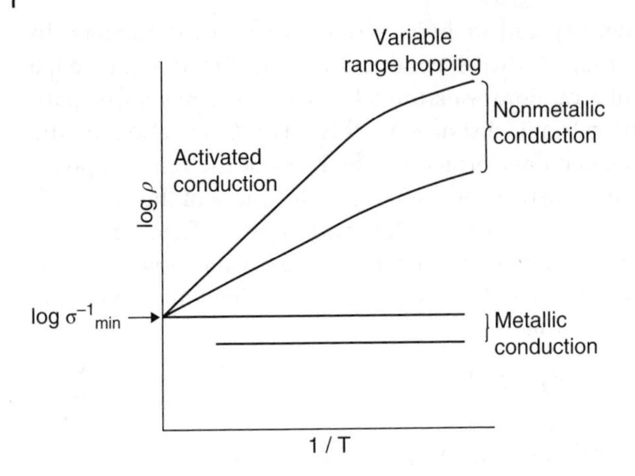

Figure 5.13 Schematic illustration of the temperature dependence of nonmetallic conduction (activated conduction and variable-range hopping) and metallic conduction.

where E_C and E_F are the conduction edge energy and the Fermi energy, respectively. When E_F approaches E_C from lower energies, E_a becomes smaller and, for $E_F \geqq E_C$, E_a becomes zero, that is metallic conduction occurs, as shown in Figure 5.13. Such a transition from nonmetallic to metallic conduction is called an Anderson transition. As an example, the temperature dependence of σ for amorphous $Si_{1-x}Au_x$ films is shown in Figure 5.14 [40]. In these alloy films, the Anderson transition occurs at $x = 0.13$ at low temperature, as will be mentioned again below.

5.4.1.2 Hopping Conduction

In hopping conduction, there are two types of hopping, namely, nearest-neighbor hopping and variable-range hopping.

5.4.1.2.1 Nearest-Neighbor Hopping

In compensated semiconductors, for example, n-type semiconductors in which some of the acceptors are ionized, as shown in Figure 5.15, the energy level of a donor electron fluctuates spatially because of the Coulomb potential due to ionized acceptors, and then such an electron may move to the nearest neighboring empty donor site by absorbing and/or emitting phonons. This type of hopping is called nearest-neighbor hopping (see, for example, [41]). The electrical conductivity is expressed by

$$\sigma_h = \sigma_{h0} \exp(-\varepsilon_3 / k_B T) \tag{5.18}$$

where the activation energy ε_3 corresponds almost to the magnitude of the potential fluctuation of the donor levels for hopping electrons.

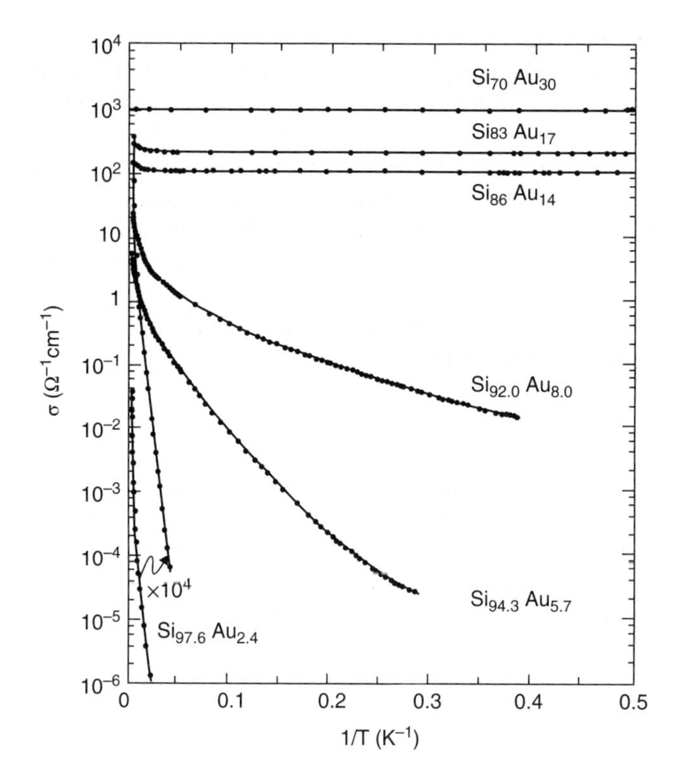

Figure 5.14 Conductivity plotted as a function of the inverse temperature for amorphous $Si_{100-x}Au_x$ systems with various Au concentrations. *Source:* Kishimoto and Morigaki 1979 [40]. Reproduced with permission from the Physical Society of Japan.

Figure 5.15 Schematic illustration of neutral and ionized donor levels and ionized acceptor levels in compensated semiconductors.

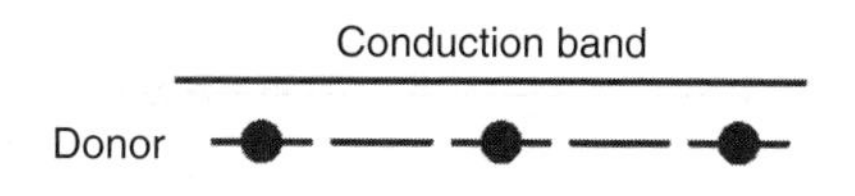

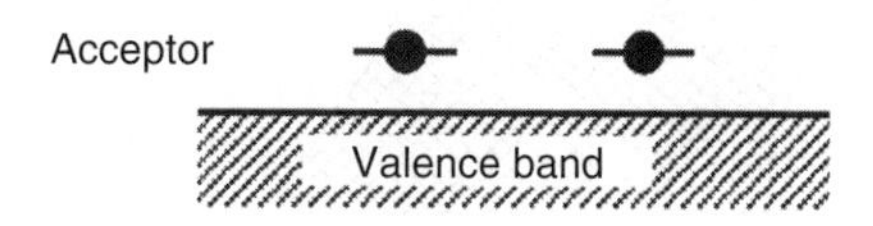

5.4.1.2.2 Variable-Range Hopping

Amorphous semiconductors exhibit a quasi-continuous density of states below the mobility edge, that is, in the tail states, as shown in Section 5.2. In this case, even though the overlap of wave functions is small between neighbors distant from each other, the energy difference is small, so that tail electrons may jump to their neighbors with the assistance of phonons at low temperatures, as shown in Figure 5.16. In general, when the temperature becomes sufficiently low, this type of hopping conduction, called variable-range hopping by Mott [42], occurs in place of nearest-neighbor hopping.

We consider variable-range hopping in the following way [42]. The jumping frequency p among a few sites is given by

$$p = \nu_{\mathrm{ph}} \exp(-2\alpha R - W/k_{\mathrm{B}}T) \tag{5.19}$$

where ν_{ph}, α, R, and W are the phonon frequency, the decay rate of the wave function, the distance between two sites, and the energy difference between those two sites, respectively. Using the following relationships, namely, a relationship between the diffusion coefficient D and p, and the Einstein relation between the mobility μ and D,

$$D \cong pR^2/6 \tag{5.20}$$

$$\mu = eD/k_{\mathrm{B}}T \tag{5.21}$$

the electrical conductivity, σ, is given by

$$\sigma = ne\mu \cong \left(e^2/6k_{\mathrm{B}}T\right)pR^2N\left(E_{\mathrm{F}}\right)k_{\mathrm{B}}T = (1/6)e^2pR^2N\left(E_{\mathrm{F}}\right) \tag{5.22}$$

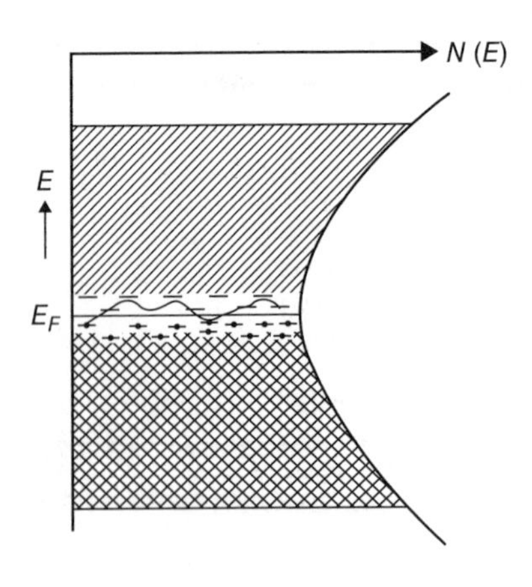

Figure 5.16 Schematic illustration of variable-range hopping of an electron located around the Fermi level. The singly shaded and doubly shaded regions represent occupied and unoccupied localized states, respectively, in the band gap region.

The energy difference, W, is estimated as follows. When an electron jumps from a certain site to another site at a distance R, the number of states situated between E and $E+\Delta E$ within a sphere of radius R is given by

$$\frac{4\pi}{3}R^3 N(E)\Delta E \tag{5.23}$$

This is the number of states in a width of ΔE, so that the energy difference W is given by

$$W = \frac{\Delta E}{(4\pi/3)R^3 N(E)\Delta E} = \frac{3}{4\pi R^2 N(E)} \tag{5.24}$$

The hopping electron lies at around $E = E_F$, and thus the following equation is obtained by substituting Equation (5.24) into Equation (5.19):

$$p \cong v_{ph} \exp\left(-2\alpha R - \frac{3}{4\pi R^3 N(E_F) k_B T}\right) \tag{5.25}$$

The most probable distance R is obtained from $\partial p/\partial R = 0$ as follows:

$$R = \left\{\frac{3}{2\alpha(4\pi/3)N(E_F)k_B}\right\}^{1/4} T^{-1/4} \tag{5.26}$$

Substituting Equation (5.26) into Equation (5.25), we obtain

$$p = v_{ph} \exp\left(-B/T^{1/4}\right) \tag{5.27}$$

$$B = 2.063\left\{\alpha^3/k_B N(E_F)\right\}^{1/4} \tag{5.28}$$

Substituting Equations (5.26) and (5.27) into Equation (5.22), σ is obtained as follows:

$$\sigma = \sigma_0 \exp\left(-B/T^{1/4}\right) \equiv \sigma_0 \exp\left\{-(T_0/T)^{1/4}\right\} \tag{5.29}$$

where

$$T_0 = \frac{18.1\alpha^3}{k_B N(E_F)} \tag{5.30}$$

$$\sigma_0 = \frac{1}{6} e^2 v_{\mathrm{ph}} \left(\frac{9}{8\pi k_{\mathrm{B}}} \right)^{1/4} \{N(E_{\mathrm{F}})\}^{3/4} T^{-1/4} \tag{5.31}$$

As seen in Equation (5.29), we have $\ln\sigma \propto T^{-1/4}$, so this is called a $T^{-1/4}$ law. The variable-range hopping conductivity has been calculated more rigorously by several authors, for example Ambegaokar, Halpern, and Langer [43] and Kirkpatrick [44]. The former authors give

$$T_0 = \frac{16\alpha^3}{k_{\mathrm{B}} N(E_{\mathrm{F}})} \tag{5.32}$$

and the latter author, taking into account an electron–phonon interaction applicable to Si and Ge, gives

$$\sigma_0 = 0.022 \left(\frac{E_1{}^2 \alpha}{\pi d s^5 \hbar^4} \right) \left(\frac{2e^3 \alpha}{3\kappa} \right) \left(\frac{T_0}{T} \right) \tag{5.33}$$

$$T_0 = \frac{60\alpha^3}{\pi k_{\mathrm{B}} N(E_{\mathrm{F}})} \tag{5.34}$$

where E_1, d, ρ, and κ are the deformational potential constant, density, sound velocity, and dielectric constant, respectively.

As an example, the temperature variation of the electrical conductivity of amorphous $Si_{1-x}Au_x$ alloy films has already been shown in Figure 5.14 as a function of T^{-1}. This alloy system exhibits a metal–nonmetal transition at $x = 0.13$. This transition can be seen in Figure 5.17, in which the value of σ extrapolated to $T = 0$, $\sigma(0)$, is plotted as a function of x for this alloy system. The details of this transition were reported by Nishida *et al.* [45, 46]. Band conduction occurs at high temperatures and variable-range hopping conduction at low temperatures. As shown in Figure 5.18, the $T^{-1/4}$ law can be seen in the temperature variation of the electrical conductivity.

5.4.2 AC Conduction

The AC conductivity is directly connected to the complex dielectric constant $\varepsilon^*(\omega)$ ($= \varepsilon_1 - i\varepsilon_2$) and is defined as $\sigma^*(\omega) = i\omega\varepsilon_0\varepsilon^*(\omega)$, where ε_0 is the electric constant and ω is the angular frequency of the external electric field. Thus the *real* part σ_1 of the AC conductivity is related to the *imaginary* part ε_2 of the dielectric constant, which is called the dielectric loss. Note here that the physical quantities are defined for an electric field with a time variation of $\exp(+i\omega t)$. As σ_1 expresses the energy loss in an AC electric field, it is often called the AC loss.

Figure 5.17 Zero-temperature-limit conductivity $\sigma(0)$ plotted as a function of Au content x; σ_{min} designates the minimum metallic conductivity. *Source:* Nishida *et al*. 1982 [45]. Reproduced with permission of Elsevier.

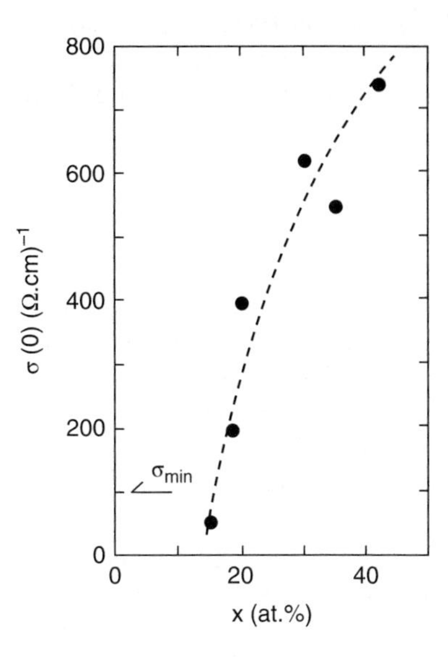

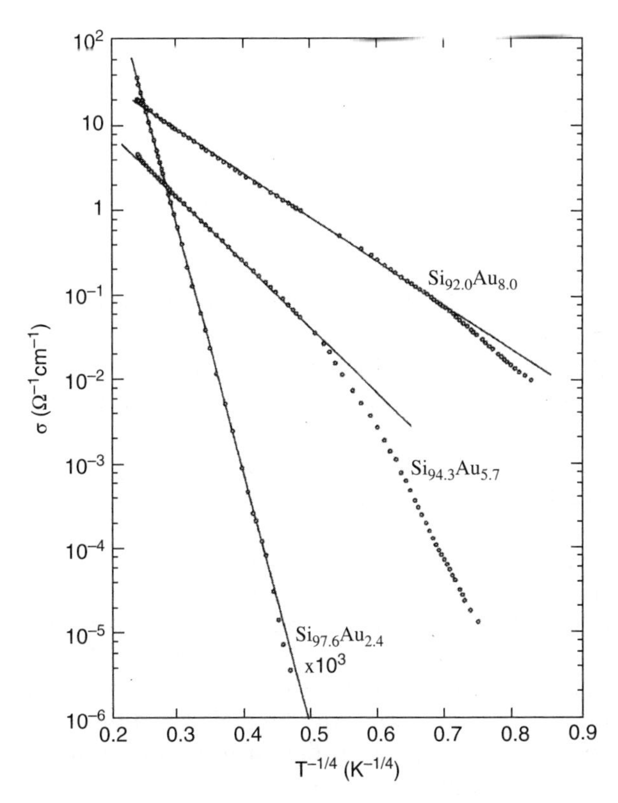

Figure 5.18 Conductivity plotted as a function of $T^{-1/4}$ for amorphous $Si_{100-x}Au_x$ systems with various Au concentrations. *Source:* Kishimoto and Morigaki 1979 [40]. Reproduced with permission from the Physical Society of Japan.

The AC loss in the audio (or radio) frequency range can be described empirically by

$$\sigma_1\left(\omega, T\right) = A\left(T\right)\omega^{s(T)} \tag{5.35}$$

where $A(T)$ and $s(T)$ (<1.0) are temperature-dependent parameters [47, 48]. This simple behavior, also called dispersive AC loss, was first discovered in doped crystalline Si [49] and then found to apply also to disordered matter [50].

The AC loss observed in disordered materials can be classified into the following three categories: (i) Debye-type dielectric relaxation arising from atomic or molecular dipoles, (ii) hopping or tunneling of electronic (or ionic) charge, and (iii) a Maxwell–Wagner-type macroscopic charge inhomogeneity. Before showing the experimental data, we must first discuss the principal models for AC loss mechanisms. Let us begin with the mechanisms in category (i). In this case, the response should have a Debye form, that is, $1/(1+i\omega\tau)$, where τ is the dielectric relaxation time of the atoms or molecules. The AC loss is given in general form as

$$\sigma_1\left(\omega\right) = N_\mathrm{p}\int \alpha\left(\tau\right)\frac{\omega^2\tau}{1+\omega^2\tau^2}P\left(\tau\right)d\tau \tag{5.36}$$

where N_p is the number of dipoles, $\alpha(\tau)$ is the polarizability, and $P(\tau)$ is the distribution function of the relaxation time. It is known that $\sigma_1(\omega)$ is nearly proportional to ω when $P(\tau)$ is proportional to $1/\tau$ [47, 48]. As shown by Pollak and Geballe [49], Equation (5.36) can be also applied to the electronic hopping case, with some restrictions. For the hopping case, N_p should be the number of carriers here, and the hopping carriers are assumed to be confined into certain pairs of sites (the pair approximation). The principal assumption in the pair approximation is that the confined charge motion within a pair of states is equivalent to atomic or molecular dipolar motion. The pair approximation is thus expected to be applicable at high frequency. Note that $\sigma_1(0) = 0$ for $\omega = 0$ in Equation (5.36).

It is known therefore that Equation (5.36) is not useful in the DC limit ($\omega = 0$) for hopping carriers: at low frequencies, hopping carriers are not expected to be confined to particular sites and multiple hopping may occur, suggesting that the pair approximation is not a good assumption for hopping carriers at lower frequencies. Equation (5.36) can therefore be applied only in the case where the DC and AC mechanisms are not the same, for example when there is a mixture of band transport (DC) and hopping or dipolar relaxation at higher frequencies (AC). In the earlier stages of the study of AC conductivity in disordered matter (see, for example, [47, 48]), the AC and DC losses were thought to have different origins; that is, free electrons in the conduction band contributed to the DC transport and hopping electrons followed a Debye response with a distribution of relaxation times (Equation (5.36)), which would induce AC loss [47, 48, 51].

Note again that the above hopping model is based on the pair approximation, in which $\sigma_1(\omega)$ cannot give a DC conductivity: $\sigma_1(0)$ as $\omega \to 0$. The model based on the pair approximation cannot account for the experimental data if both DC and AC transport occur by the same mechanism [47, 48, 52]. We then go to category (ii). An appropriate approach for overcoming this drawback of the pair approximation seems to be the *continuous-time random-walk* (CTRW) approximation, originally developed by Scher and Lax [53]. A simple form of the AC conductivity based on the CTRW approximation was presented by Dyre [52] as

$$\sigma^*(\omega) = \sigma(0)\frac{i\omega\tau}{\ln(1+i\omega\tau)} \tag{5.37}$$

where τ is the *maximum* hopping (or tunneling) relaxation time in the sequence of the random walk. It was shown that the AC conductivity is directly related to the DC conductivity $\sigma(0)$. Equation (5.37) produces the empirical relation given by Equation (5.35) at higher frequencies. Note that Equation (5.37) can be applied when the DC and AC transport are due to the same hopping mechanism. It is known that the above Dyre equation can be applied to amorphous carbon [54, 55] and amorphous germanium (a-Ge) [56], in which hopping transport of electrons occurs near the Fermi level.

A dispersive AC loss may occur on a macroscopic or mesoscopic scale in inhomogeneous media. This is called a Maxwell–Wagner-type effect (category (iii)). A standard approximation for treating inhomogeneous systems is the effective medium approximation (EMA), although the EMA does not lead to an explicit expression for the conductivity [57, 58]. However, it is known that a classical EMA is useful for inhomogeneous media [44]. The EMA predicts that the effective conductivity σ_m of an inhomogeneous medium in D dimensions follows

$$\left\langle \frac{\sigma_i - \sigma_\mathrm{m}}{\sigma_i + (D-1)\sigma_\mathrm{m}} \right\rangle_i = 0 \tag{5.38}$$

where σ_i is a random value of conductivity. Under the assumption of a random mixture of spherical particles, for example of two components ($i = 1, 2$) with different conductivities, a simple form of the DC conductivity and Hall mobility can be obtained [59]. The EMA has been applied to DC transport in composite amorphous chalcogenides [59–61]. The EMA is also useful for the AC conductivity (and optical conductivity in the optical frequency range), in which the conductivity or dielectric constant is a complex quantity [62]. In fact, the electronic and optical properties of hydrogenated microcrystalline Si (μc-Si:H) have been well interpreted by Shimakawa [63].

The EMA can be connected to a percolation path method (PPM) [64]. Dyre's PPM approach assumes that both the DC and the AC current follow mainly the easiest percolation path, that is, just above the percolation threshold, a critical conducting path can be treated as a one-dimensional series connection of resistors accompanying parallel capacitors. If the local conductivity is thermally activated, the local DC conductivity $\sigma(\Delta E_i)$ is given by

$$\sigma(\Delta E_i) = \sigma_0 \exp(-\Delta E_i / k_B T) \tag{5.39}$$

where ΔE_i is the local activation energy and σ_0 a constant. The effective or overall conductivity $\sigma(\omega)$ in the one-dimensional approximation can be given as

$$\frac{1}{\sigma(\omega)} = \int_0^\infty \frac{p(\Delta E)}{\sigma(\Delta E) + i\omega\varepsilon_0\varepsilon_\infty} d(\Delta E) \tag{5.40}$$

where $p(\Delta E)$ is the distribution function of ΔE and $\varepsilon_0\varepsilon_\infty$ is the local capacitance. For simplicity, the local capacitance is assumed to be the same in all areas and to be equal to the background capacitance.

Assuming a uniform distribution, that is, $p(\Delta E) = 1/E_M$ $(0 < \Delta E < E_M)$, the distribution of the local conductivity is given by [65]

$$p(\sigma) = \frac{p(\Delta E)}{|d\sigma / d(\Delta E)|} = \frac{kT}{E_M} \frac{1}{\sigma} \tag{5.41}$$

It is of interest to note that $p(\sigma)$ is proportional to $1/\sigma$. Equation (5.40) then becomes

$$\frac{1}{\sigma(\omega)} = \frac{kT}{E_M} \int_{\sigma_{min}}^{\sigma_0} \frac{1}{\sigma(\sigma + i\omega\varepsilon_0\varepsilon_\infty)} d\sigma = \frac{kT}{E_M} \frac{1}{s} \int_{\sigma_{min}}^{\sigma_0} \left(\frac{1}{\sigma} - \frac{1}{\sigma + s}\right) d\sigma \tag{5.42}$$

where $\sigma_{min} = \sigma_0 \exp(-E_M/k_B T)$ (the local minimum conductivity) and $s = i\omega\varepsilon_0\varepsilon_\infty$ is called the Laplace frequency. Then the above equation can be written as

$$\frac{1}{\sigma(\omega)} = \frac{kT}{E_M} \frac{1}{s} \log\left(1 + \frac{s}{\sigma_{min}}\right) \tag{5.43}$$

where $\sigma_0 \gg s$ is assumed. As s/σ_{min} is given by $i\omega\tau_{max}$, where τ_{max} is the maximum dielectric relaxation time ($= \varepsilon_0\varepsilon_\infty/\sigma_{min}$), $\sigma(\omega)$ is finally expressed as

$$\sigma(\omega) = \frac{E_M}{kT} \sigma_{min} \frac{i\omega\tau_{max}}{\log(1 + i\omega\tau)} = \sigma_{DC} \frac{i\omega\tau_{max}}{\log(1 + i\omega\tau)} \tag{5.44}$$

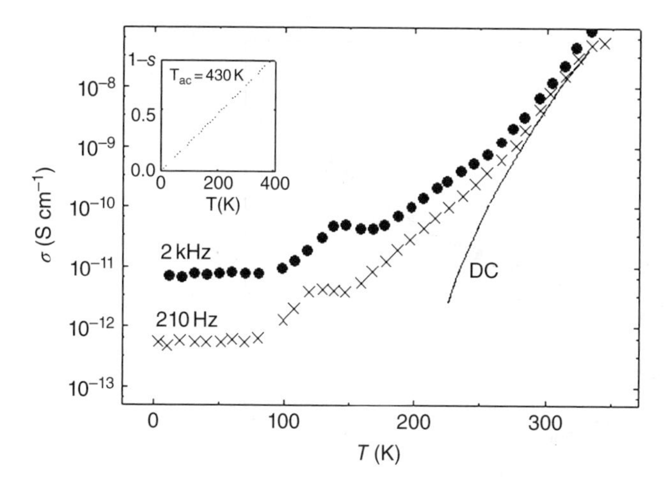

Figure 5.19 Temperature dependence of the AC conductivity of undoped a-Si:H. The inset shows $1-s$ as a function of temperature.

Note that the overall (effective) DC conductivity σ_{DC} is given by $E_{\mathrm{M}}\sigma_{\mathrm{min}}/k_{\mathrm{B}}T$, under the assumption of a uniform distribution of activation energy. It is of interest that the functional form of Equation (5.44) is the same as that of Equation (5.37) for the hopping case, even though the physical meaning of the relaxation time is completely different.

We know that, mathematically, the same equation (Equation (5.37) or (5.44)) can be applied to both hopping and Maxwell–Wagner-type relaxation. It is therefore not easy to distinguish which mechanism dominates the AC loss. In a-Si:H, there is considerable evidence that the electronic properties are dominated by long-range potential fluctuations which induce inhomogeneity in electronic processes [66, 67]. It has been found that the AC loss is dominated by long-range potential fluctuations, which may be due to a small density of localized gap states, whereas initially the AC loss in a-Si:H was thought to be induced by electronic hopping or hydrogen-related atomic relaxation [68, 69].

Let us now discuss the AC loss in a-Si:H. Figure 5.19 shows the temperature dependence of the AC conductivity of undoped a-Si:H films, down to 2 K [70]. Two types of loss were found: one is a small temperature-independent loss, occurring dominantly at low temperatures (2–100 K), and other is a strongly temperature-dependent loss at high temperature (100–300 K), which merges into the DC conductivity. The low-temperature loss can be interpreted in terms of a classical hopping mechanism, that is, hopping of localized electrons over a barrier between defect states; the high-frequency loss, on the other hand, can be attributed to Maxwell–Wagner-type inhomogeneities [71]. As shown in the inset, the temperature-dependent parameter s in the empirical relation expressed by Equation (5.35) follows $s=1-T/T_0$, where T is the

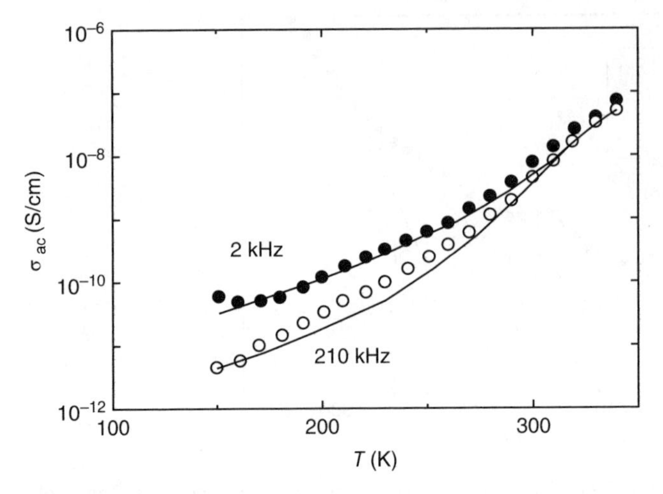

Figure 5.20 Temperature dependence of the AC conductivity replotted from Figure 5.19. The solid curves show the results calculated using Equation (5.38).

ambient temperature (in K) and T_0 is a characteristic temperature (=430 K). In the following, we briefly review the high-frequency loss in terms of the PPM.

The open and filled circles in Figure 5.20 show the AC loss at 210 Hz and 2 kHz, respectively, replotted from Figure 5.19. In the PPM (Equation (5.44)), an important parameter is the relaxation time τ, as already discussed. The solid lines show the results calculated using Equation (5.44) by taking $\varepsilon_\infty = 10$ and using the DC conductivity data, which agree fairly well with the experimental data.

It is of interest to discuss the parameter T_0 which is deduced from the frequency exponent in $\omega^{s(T)}$. A similar temperature dependence is observed in the transient photocurrent $I_p(t)$, which is proportional to t^{-s} with $s = 1 - T/T_0$ [72, 73]. Note that T_0 (=430 K) takes a similar value (450 K) for $I_p(t)$. As will be stated in Sections 5.5.5 and 5.5.6, the transient photocurrent is generally discussed in terms of the multiple trapping of electrons in band tail localized states [72]; alternatively, the behavior of $I_p(t)$ can be interpreted by introducing long-range potential fluctuations as already mentioned [67], in which the parameter T_0 is a measure of the extent of inhomogeneity. It is therefore concluded that the high-temperature AC loss originates from a Maxwell–Wagner type of inhomogeneity.

The AC loss under photoillumination (AC photoconductivity) in the RF [74] and THz [75, 76] ranges has been also reported for a-Si:H. These studies [74, 76] also support the existence of potential fluctuations. It is of interest to note that these potential fluctuations affect carrier relaxation in the THz frequency range. As the details of the dynamics have been stated elsewhere [76, 77], we shall now close the discussion of AC transport in a-Si:H.

5.4.3 Hall Effect

The Hall coefficient R_H of a crystalline semiconductor can be given as follows:

$$R_H = \gamma / nqc \tag{5.45}$$

where n, q, and c are the number of carriers, the charge of the carriers, and the velocity of light, respectively; γ is given by

1) in the case of degenerate semiconductors, $\gamma = 1$;
2) in the case of nondegenerate semiconductors, (i) if the scattering mechanism is phonon scattering, then $\gamma = 3\pi/8 = 1.18$; (ii) if the scattering mechanism is ionized impurity scattering, then $\gamma = 315\pi/512 = 1.93$; (iii) if the scattering mechanism is neutral impurity scattering, then $\gamma = 1$.

In the case of amorphous semiconductors, the Hall coefficient has been obtained in terms of the random-phase model around the mobility edge, E_C. In this model, the Hall effect arises from rotation of an electron (s orbital) through three sites, as shown in Figure 5.21. As a result, the calculated Hall mobility is given by [78]

$$\mu_H = \left(2\pi\eta z'/z^2\right)\left(ea^2/\hbar\right)a^3 BN\left(E_C\right) \tag{5.46}$$

where η is the area of the three-site path projected perpendicular to the magnetic field, z and z' are the coordination number and the average number of three-site paths around a certain site, and B is the band width in the case without disorder.

In the case $z' = z = 6$, μ_H at the mobility edge is given by

$$\mu_H = ea^2/7\hbar \tag{5.47}$$

The value of μ_H, for $a = 3$ Å, for example, is $0.23\,\text{cm}^2\,\text{V}^{-1}\,\text{s}^{-1}$. This can be compared with the drift mobility at the mobility edge E_C, given by

$$\mu = \frac{0.05e/\hbar a}{N\left(E_C\right)k_B T} \tag{5.48}$$

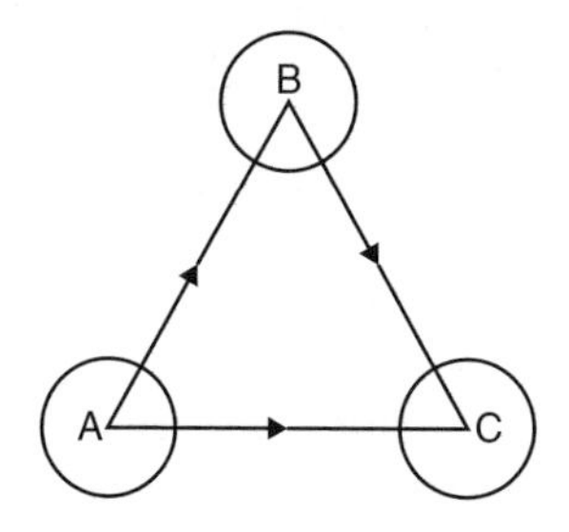

Figure 5.21 Three-site model for the Hall effect.

For $a = 3$ Å, the drift mobility is one or two orders of magnitudes greater than the Hall mobility. Furthermore, we note that the Hall mobility does not depend on the sign of the carriers and that the sign is negative.

An anomaly in the Hall coefficient has been observed in a-Si:H: when electrons are the carriers, the Hall coefficient becomes positive, while when holes are the carriers, it becomes negative. This phenomenon is called double reversal of the Hall coefficient. Emin [79] has proposed a model to account for this anomaly in terms of the three-site model, taking into account antibonding orbitals for electrons and bonding orbitals for holes.

5.4.4 Thermoelectric Power

The thermoelectric power, S, associated with band conduction of electrons is given by

$$S = -\frac{k_B}{e}\left(\frac{E_C - E_F}{k_B T} + A\right) \tag{5.49}$$

where A is a quantity depending on the energy dependence of the relaxation time associated with electrical conduction. When holes are the carriers, S is given by Equation (5.49) except that $E_C - E_F$ is replaced by $E_F - E_V$. In this case, the sign of S is positive. The sign of S coincides with that of the carriers.

The thermoelectric power is related to the electrical conductivity σ as follows:

$$S\sigma = -\frac{k_B}{e}\int \sigma_E \frac{E - E_F}{k_B T}\frac{\partial f}{\partial E}\,dE \tag{5.50}$$

where σ_E is the energy-dependent conductivity given by Equation (5.12). The relationship between the electrical conductivity σ and the thermoelectric power S associated with the band conduction of electrons is given by

$$\ln\sigma + \left|\frac{e}{k_B}S\right| = \ln\sigma_0 + A \equiv Q \tag{5.51}$$

The temperature-dependent quantity Q is defined as above. Q can also be described as follows, using Q_0 and E_Q:

$$Q = Q_0 - E_Q/k_B T,\quad Q_0 = \ln\sigma_0 + A,\quad E_Q = E_\sigma - E_S \tag{5.52}$$

where σ_0, E_σ, and E_S are the prefactor of σ (Equation (5.16)) and the activation energies of σ and S, respectively.

In amorphous semiconductors, the temperature dependence of the thermoelectric power is given by Equation (5.49) at high temperatures, when electrons

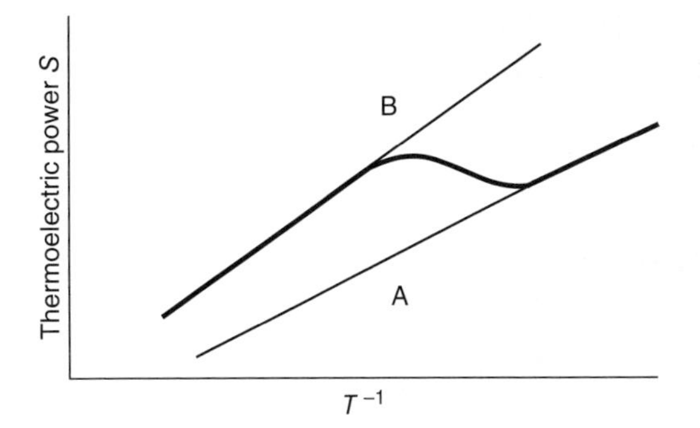

Figure 5.22 Schematic illustration of the temperature dependence of the thermoelectric power *S* of amorphous semiconductors. The lines A and B represent the temperature variation of *S* associated with conduction around the conduction band tail and the conduction band edge, respectively.

are excited into the conduction band, while it is expressed by the following equation at low temperatures, when conduction occurs around the mobility edge E_A:

$$S = -\frac{k_B}{e}\left(\frac{E_A - E_F}{k_B T} + A'\right) \tag{5.53}$$

Thus, the thermoelectric power has the temperature variation shown in Figure 5.22. In Figure 5.22, a turnover from B to A occurs at a certain temperature. Such a variation can be seen in Figure 5.23 [80], where the temperature variation of the thermoelectric power is shown for various samples of a-Ge:H. Here, $E_C - E_F = 0.43\,\text{eV}$ and $E_A - E_F = 0.17\,\text{eV}$. This observation is considered to be evidence for the existence of a mobility edge in amorphous semiconductors.

5.4.5 Doping Effect

In amorphous semiconductors, each constituent atom is coordinated with neighboring atoms, following the so-called $8-N$ rule [42], in which N is the number of valence electrons. Hence, when phosphorus (a group V element) and boron (a group III element) are doped into amorphous silicon, they are expected to be threefold coordinated with neighboring Si atoms. However, n-type and p-type doping were done by Spear and Le Comber [81], in which phosphorus and boron were successfully doped as group V and III elements to supply an electron and a hole, respectively. Nevertheless, as shown in Figure 5.24, the doping efficiency is low, because most of the phosphorus and boron are threefold coordinated as expected from the $8-N$ rule, and only small amounts of them are fourfold coordinated so that they can act as donors and acceptors, respectively. A modified $8-N$ rule

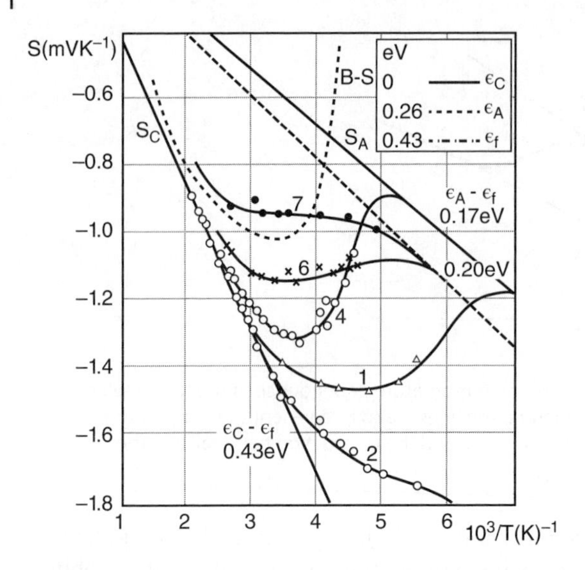

Figure 5.23 Thermoelectric power as a function of the inverse temperature for various samples of a-Ge:H. *Source:* Jones *et al.* 1976 [80]. Reproduced with permission of Elsevier.

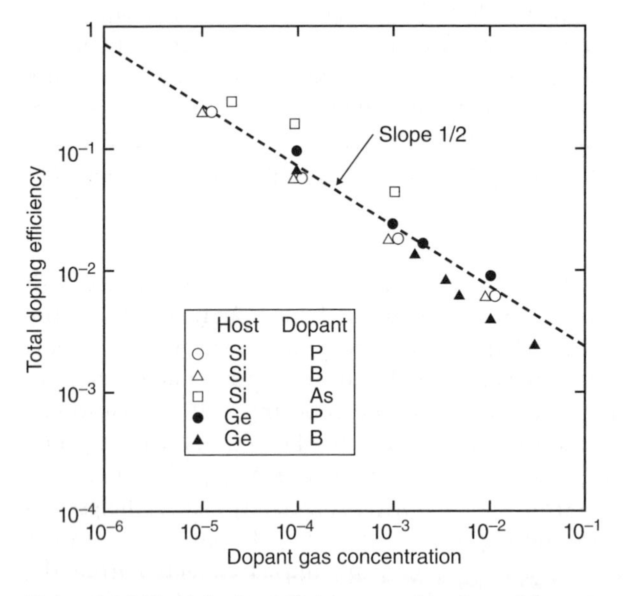

Figure 5.24 Total doping efficiency as a function of dopant gas concentration. *Source:* Stutzmann *et al.* 1987 [83]. Reproduced with permission from the American Physical Society.

was proposed by Street [82] to account for the doping mechanism, as follows. The following reaction is assumed:

$$Si_4{}^0 + P_3{}^0 \rightleftharpoons P_4{}^+ + D^- \tag{5.54}$$

Here, P_4^+ has four valence electrons, so that it is fourfold coordinated with neighboring Si atoms. D^- represents negatively charged threefold-coordinated Si, so that it has five valence electrons. This reaction occurs, following the law of mass action, during film deposition in thermal equilibrium, as follows:

$$N_{P4+} = N_{D-} = \left(N_0 N_P\right)^{1/2} \exp\left(-E_i/2k_B T\right) \tag{5.55}$$

where N_{P4+}, N_{D-}, N_0, and N_P designate the densities of P_4^+, D^-, fourfold-coordinated silicon atoms, and incorporated phosphorus atoms, respectively, and E_i is the formation energy for isolated P_4^+ and D^-. From Equation (5.55), the doping efficiency N_{P4+}/N_P is proportional to $N_P^{-1/2}$, as seen in Figure 5.24.

This model suggests that the Fermi level lies almost at the midpoint between the P_4^+ level and the D^- level, that is, the Fermi level is pinned in the region of the lowest density of states, as shown in Figure 5.25. The P_4^+–D^- pair has been considered

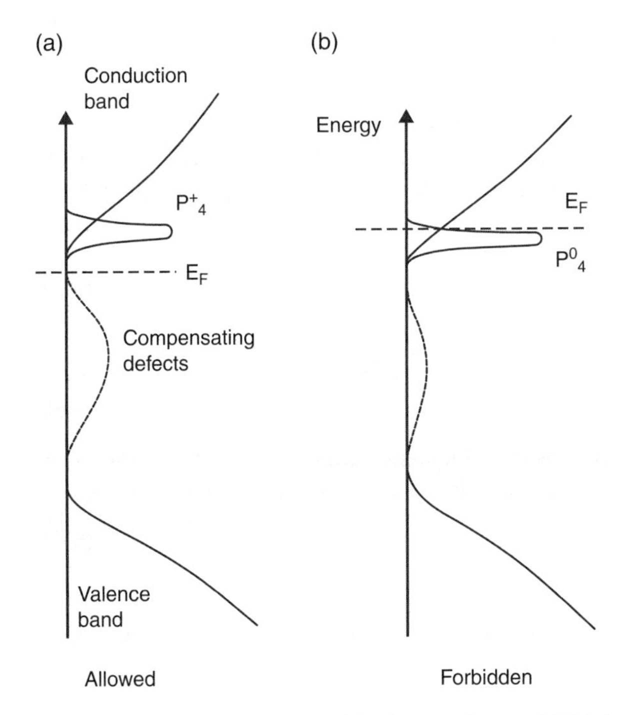

Figure 5.25 Schematic diagram of the density of states: (a) E_F is located between P^{4+} and compensated defects D^-; (b) forbidden fourfold coordination with occupied donor levels (P_4^0). *Source:* Street 1982 [82]. Reproduced with permission from the American Physical Society.

when the doping level is high, that is, when there is a high phosphorus concentration in a-Si:H. Thus, there are isolated P_4^+ and D^-, and P_4^+–D^- pairs. The densities and levels have been discussed on the basis of photoinduced absorption (PA) [84, 85], PA-detected ESR [84, 85], and photo-ICTS measurements [86].

5.5 Optical Properties

5.5.1 Fundamental Optical Absorption

The fundamental optical absorption due to the band-to-band transition in a crystal is treated in the following way. The Bloch functions of the conduction band and the valence band, $|c, \boldsymbol{k}\rangle$ and $|v, \boldsymbol{k}\rangle$, are written as follows:

$$|c,\boldsymbol{k}\rangle = u_{c,\boldsymbol{k}}(r)\exp(i\boldsymbol{k}r) \tag{5.56}$$

$$|v,\boldsymbol{k}'\rangle = u_{v,k'}(r)\exp(i\boldsymbol{k}'r) \tag{5.57}$$

Thus, the transition probability W_{vc} is given by

$$\begin{aligned} W_{vc} &= \frac{2\pi}{\hbar}|\langle c,k|\frac{e}{m}Ap|v,k'|^2\,\delta\left[E_C(k)-E_V(k')-\hbar\omega\right] \\ &= \frac{\pi e^2}{2\hbar m^2}A_0^2\,|M_{vc}|^2\,\delta\left[E_C(k)-E_V(k')-\hbar\omega\right] \end{aligned} \tag{5.58}$$

where A, A_0, p, and $\hbar\omega$ are the vector potential associated with the light, its magnitude, the electronic momentum, and the photon energy, respectively. Now, we consider only the direct transition, where the matrix element M_{vc} is expressed by

$$M_{vc} = \frac{1}{\Omega}\int_{\Omega} u_{ck}^{*}(\mathbf{r})\,ep\,u_{vk}(\mathbf{r})\,d\mathbf{r} \tag{5.59}$$

where Ω is the volume of a unit cell. The absorption coefficient is defined by

$$I = I_0\exp(-\alpha z) \tag{5.60}$$

where I and z are the flux density of incident light and the distance measured from the incident surface of the sample, respectively. Since α is defined by

$$\alpha = -(1/I)(dI/dz) \tag{5.61}$$

α is given by

$$\alpha = \hbar\omega W_{vc}/I \tag{5.62}$$

Taking into account

$$I = (1/2)\bar{n}\varepsilon_0\omega^2 c A_0^2 \tag{5.63}$$

where $\bar{n}$ is the refractive index, we obtain

$$\alpha = \frac{\pi e^2}{\bar{n}\varepsilon_0 c m^2 \omega} \sum_{k,k'} |M_{vc}|^2 \delta\left[E_C(k) - E_V(k') - \hbar\omega\right]\delta_{k,k'} \tag{5.64}$$

$|M_{vc}|^2$ is normally assumed to be a slowly varying function of energy and the summation is replaced by an integral, using the density-of-states spectrum, that is, $N_c(E)$ and $N_v(E)$. Then, we obtain

$$\alpha = \frac{2\pi e^2 V}{\bar{n}\varepsilon_0 c m^2 \omega} \int N_V(E) N_C(E + \hbar\omega) dE \tag{5.65}$$

In the case of amorphous semiconductors, we can make the following assumptions:

1) The k selection rule is relaxed.
2) The energy dependence of the transition probability can be neglected. This seems to be reasonable over a limited range of photon energy. If $N(E)$ is assumed to be a parabolic density of states, α is given by

$$\alpha = B\frac{\left(\hbar\omega - E_{og}\right)^2}{\hbar\omega} \tag{5.66}$$

where B is a factor independent of ω. This relationship can be rewritten as

$$\left(\alpha\hbar\omega\right)^{1/2} = B^{1/2}\left(\hbar\omega - E_{og}\right) \tag{5.67}$$

This relationship is plotted as shown in Figure 5.26, which is called a Tauc plot [87]. As seen in the figure, the optical gap E_{og} can be estimated by extrapolating the straight portion. Experimental examples of Tauc plots are shown in Figure 5.27. The optical gaps estimated from such plots are listed in Table 5.1, as well as other results.

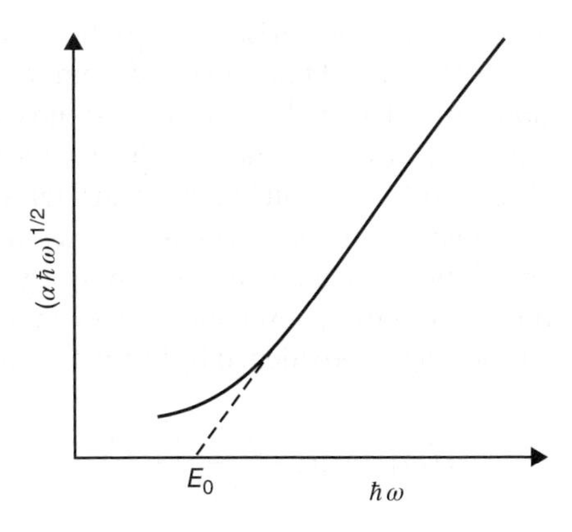

Figure 5.26 Schematic illustration of a Tauc plot.

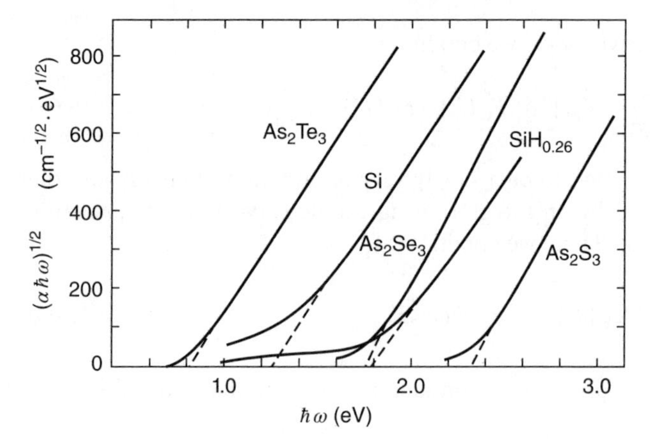

Figure 5.27 Examples of Tauc plots for some amorphous semiconductors (data taken from [88] and [89]).

Table 5.1 Optical gap energy E_{og} and B for amorphous semiconductors, taken from [88].

Material	E_{og} (eV)	B ($cm^{-1} eV^{-1}$)
Si	1.26	5.2×10^5
$SiH_{0.26}$	1.82	4.6×10^5
As_2S_3	2.32	4×10^5
As_2Se_3	1.76	8.3×10^5
As_2Te_3	0.83	4.7×10^5

5.5.2 Weak Absorption

The weak-absorption portion of the optical absorption consists mainly of tail absorption, which arises from optical transitions from the valence band to the conduction band tail states and from the valence band tail states to the conduction band. This tail absorption extends from the absorption edge towards low energy with an exponential shape as a function of photon energy. Thus, this absorption is also called exponential tail absorption. Such an exponential tail was first observed in ionic crystals such as alkali halides and silver halides [90] and is called the Urbach tail. In these ionic crystals, interaction between optical phonons and electronic excitations plays an important role in the optical absorption, whose edge is broadened by this interaction. Thus, the Urbach tail is given by

$$\alpha(\omega) = \exp\left\{\frac{\gamma\left(\hbar\omega - E_{1g}\right)}{k_B T^*}\right\} \tag{5.68}$$

$$T^* = \left(\frac{\hbar\Omega_0}{2k_B}\right)\coth\left(\frac{\hbar\Omega_0}{2k_B T}\right) \tag{5.69}$$

for $\hbar\omega < E_{1g}$, where E_{1g} is called the Urbach focus, and $\hbar\Omega_0$ and γ are the optical phonon energy and a constant, respectively.

The exponential tail absorption in a-Si:H observed by Cody *et al.* [91] is shown in Figure 5.28, in which the following expression for $\alpha(\omega, T)$ was used to fit the experimental points:

$$\alpha(\omega,T) = \alpha_0 \exp\left\{\frac{\left(\hbar\omega - E_{1g}\right)}{E_0(T,X)}\right\} \tag{5.70}$$

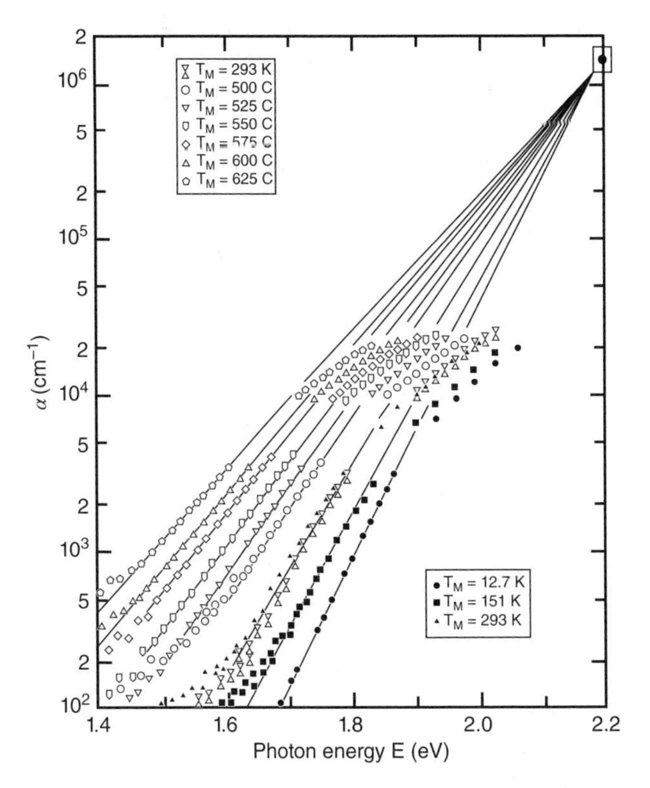

Figure 5.28 Optical absorption coefficient of a-Si:H, α, as a function of photon energy, *E*. The solid symbols refer to data obtained at different measurement temperatures: solid circles, 12.7 K; solid squares, 151 K; solid triangles, 293 K. The open symbols show data measured at 293 K on a similar film, from which hydrogen was evolved in a stepwise manner during isochronal heating in a vacuum. *Source:* Cody *et al.* 1981 [91]. Reproduced with permission from the American Physical Society.

$$E_0(T,X) = K\left[\left\langle U^2 \right\rangle_T + \left\langle U^2 \right\rangle_X\right] \tag{5.71}$$

where E_0 defines the width of the exponential tail as a function of temperature T and disorder X, and U denotes the displacement of the atoms from their equilibrium positions. $K\langle U^2\rangle_T$ is equal to $k_B T^*/\gamma$. The disorder effect is expressed by using a fictitious temperature T_F, which is a measure of the total site disorder, as follows:

$$E_0(T,X) = E_0(T_F,0) \tag{5.72}$$

For a-Si:H, for example, when the annealing temperature T_A is 600 K, T_F is nearly 560 K, but T_F increases with T_A; for example, $T_F \fallingdotseq 930$ K for $T_A \fallingdotseq 900$ K [92].

5.5.3 Photoluminescence

The PL spectrum of a-Si:H consists of a principal PL and a low-energy PL (also called defect PL), as schematically shown in Figure 5.29. The principal PL arises from radiative recombination between tail electrons and self-trapped holes at specific weak Si–Si bonds, that is, weak Si–Si bonds adjacent to an Si–H bond, as explained in Section 5.2.2, at low temperatures below ~100 K, and between tail electrons and tail holes above ~100 K. The low-energy PL arises from radiative recombination between electrons at specific defects, as explained in Section 5.3.4, and tail holes.

5.5.4 Frequency-Resolved Spectroscopy (FRS)

5.5.4.1 Principle of FRS

As shown in Figure 5.30, the PL spectrum of a-Si:H consists of featureless, broad lines. One important method of obtaining useful information from such a PL spectrum is to find the lifetime distribution from it. The method of frequency-resolved spectroscopy (FRS) [93–97] is such a technique, as

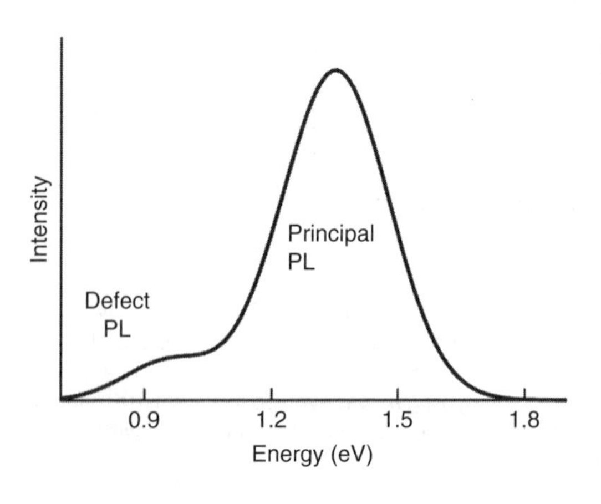

Figure 5.29 Schematic illustration of PL spectrum of a-Si:H films.

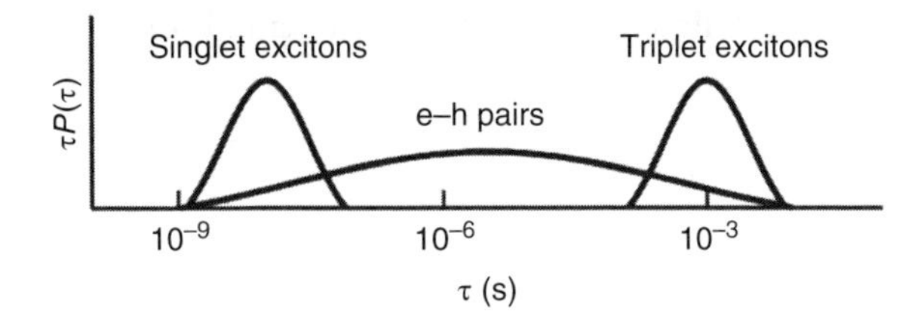

Figure 5.30 Schematic illustration of lifetime distribution of PL in a-Si:H.

described below. The intensity of the excitation light is modulated with an angular frequency ω, and then the output signal of the emission light is detected in phase and out of phase with respect to the excitation-modulated light. In particular, the out-of-phase signal contains information about the lifetime distribution, $P(\tau)$ (where τ is the lifetime), as follows. If the excitation intensity is given by

$$G(t) = G_0 + G_1 \exp(i\omega t) \tag{5.73}$$

the emission signal is given by

$$S(t) = S_0 + \int_0^{\infty} I(t') G_1 \exp[i\omega(t - t')] dt' \tag{5.74}$$

where S_0 and $I(t)$ represent the emission signal corresponding to a constant excitation intensity G_0 and the emission decay signal at t when a sample is excited by a short pulse of light at $t = 0$, respectively.

Fourier transformation of $S(t)$ results in $S(\omega)$, that is, the emission signal as a function of ω, as follows:

$$\begin{aligned} S(\omega) &= T^{-1} \int_0^T \exp(-i\omega t) \left\{ G_1 \int_0^{\infty} I(t') \exp[i\omega(t - t')] dt' \right\} dt \\ &= T^{-1} G_1 \int_0^T \int_0^{\infty} \exp(-i\omega t') I(t') dt' dt = GI(\omega) \end{aligned} \tag{5.75}$$

where $I(\omega)$ is the Fourier transform of $I(t)$, as follows:

$$\begin{aligned} I(\omega) &= \int_0^{\infty} \exp(-i\omega t) I(t) dt \\ &= \int_{-\infty}^{\infty} \exp(-i\omega t) I(t) dt, \quad I(t) = 0 \quad \text{for} \quad t < 0 \end{aligned} \tag{5.76}$$

As shown above, $S(\omega)$ is proportional to $I(\omega)$, that is, the Fourier transform of $I(t)$.

When the PL signal decays, following a simple exponential function with $\tau = \tau_0$,

$$I(t) = J_0 \tau_0^{-1} \exp(-t/\tau_0) \quad \text{for} \quad t \ge 0,$$
$$= 0 \quad \text{for} \quad t < 0 \tag{5.77}$$

where $J_0 \tau_0^{-1} = I(0)$, we have

$$I(\omega) = \int_{-\infty}^{\infty} I(t) \exp(-i\omega t) dt = \int_{0}^{\infty} J \tau_0^{-1} \exp(-t/\tau_0) \exp(-i\omega t) dt = J \frac{1}{1 + i\omega\tau_0} \tag{5.78}$$

From Equation (5.78), the out-of-phase signal (called the quadrature signal) is given by

$$\mathrm{Im} S(\omega) \propto \mathrm{Im} I(\omega) \propto \frac{\omega\tau_0}{1 + \omega^2 \tau_0^2} \tag{5.79}$$

This is illustrated in Figure 5.31, where a single peak appears at $\omega = \tau_0^{-1}$. When the lifetime has a distribution given by $P(\tau)$, we obtain

$$I(t) \propto \int_{0}^{\infty} P(\tau) \tau^{-1} \exp(-t/\tau) dt \quad \text{for} \quad t \ge 0$$
$$= 0 \quad \text{for} \quad t < 0 \tag{5.80}$$

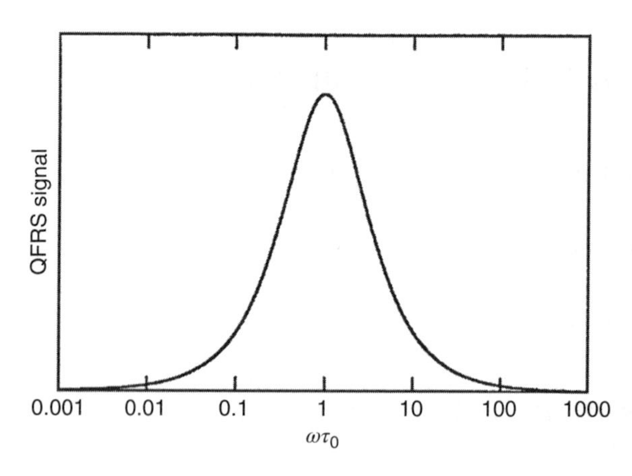

Figure 5.31 Quadrature frequency-resolved spectroscopy (QFRS) signal for a single lifetime τ_0.

$$-\mathrm{Im}S(\omega) \propto \int_0^\infty P(\tau)\frac{\omega\tau}{1+\omega^2\tau^2}d\tau \tag{5.81}$$

This represents a convolution of $P(\tau)$ and $\omega\tau/(1+\omega^2/\tau^2)$, which is peaked at $\tau=\omega^{-1}$. If $\omega\tau/(1+\omega^2/\tau^2)$ has a negligibly narrow width compared with the width of $P(\tau)$, $-\mathrm{Im}\ S(\omega)$ is approximately equal to $P(\omega^{-1})$, that is, the lifetime distribution function. However, the width of $\omega\tau/(1+\omega^2/\tau^2)\omega^{-1}$ increases as ω^{-1} increases, so that $-\mathrm{Im}\,S(\omega)$ is not equal to $P(\omega^{-1})$, but has a shape that is emphasized in the long-lifetime region compared with $P(\tau)$. In order to resolve this problem, we put $y=\ln\tau$, $x=\ln(\omega^{-1})$, $-\mathrm{Im}\,S(\omega)=S'(x)$, and $P'(y)=xP(x)$, and then we obtain

$$S'(x) \propto \int_{-\infty}^{\infty} P'(y)\frac{\exp(y-x)}{1+\exp\left[2(y-x)\right]} \tag{5.82}$$

This shows that $S'(x)$ is a convolution of $P'(y)$ ($=\tau P(\tau)$) and a function of $\exp(y-x)/\{1+\exp[2(y-x)]$, which is peaked at $y=x$. The latter is a function of $y-x$, so that its shape (width, peak height, etc.) does not depend on x. Thus, when this function's width is negligibly narrow, $S'(x)$ is approximately equal to $P'(x)$, that is, $-\mathrm{Im}\,S(\omega)$ expresses $\tau P(\tau)$ ($\tau=\omega^{-1}$). Thus, measurement of the out-of-phase signal enables us to obtain $\tau P(\tau)$. In the case of a-Si:H, the lifetime is distributed over several orders of magnitude, as shown in Figure 5.30, where the lifetime distributions of singlet and triplet excitons and distant electron–hole pairs are shown. Thus, a plot of $\tau P(\tau)$ versus $\ln\tau$ is useful because the following relationship can be obtained:

$$P(\tau)d\tau = \tau P(\tau)d(\ln\tau) \tag{5.83}$$

5.5.4.2 FRS Measurements

In the following, two examples of FRS measurements are presented.

5.5.4.2.1 Geminate and Nongeminate Recombination

A geminate pair is a pair consisting of an electron and a hole created at a certain site by optical excitation and mutually coupled by a Coulomb attractive interaction. Such pairs are photocreated at low optical excitation intensity, for example, for a-Si:H, at a photocarrier generation rate $G \leq 3\times10^{16}\,\mathrm{cm^{-3}\,s^{-1}}$. On the other hand, as the optical excitation intensity increases, the electron–hole pair density becomes large, so that these electrons and holes are coupled with closer counterparts to form nongeminate pairs (also called distant pairs). The critical value of G is approximately equal to $3\times10^{16}\,\mathrm{cm^{-3}\,s^{-1}}$ for a-Si:H, as shown above. Thus, the lifetime of nongeminate pairs shifts towards short

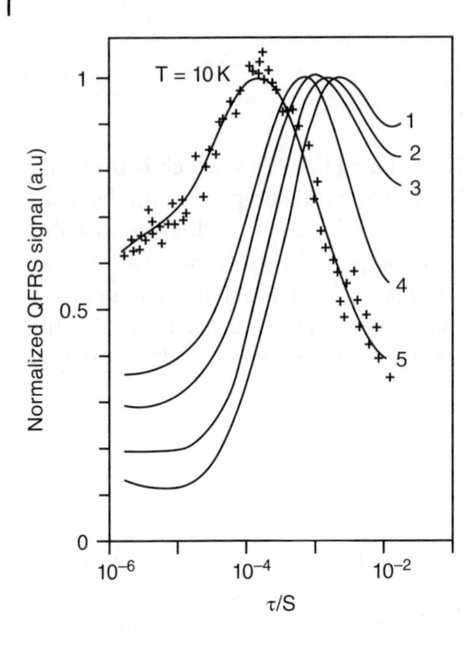

Figure 5.32 Normalized QFRS signals (peak value = 1) as a function of τ in a-Si:H. G ($cm^{-3}s^{-1}$): (1) 3×10^{18}, (2) 2×10^{20}, (3) 5×10^{20}, (4) 2×10^{21}, (5) 5×10^{21}. *Source:* Stachowitz *et al.* 1991 [94]. Reproduced with permission of Elsevier.

times with increasing generation rate. This is shown in Figure 5.32, where the QFRS signal intensities are shown for various generation rates.

5.5.4.2.2 *Localized Triplet Excitons*

Oheda [98] observed a peak at 1 ms at 13 K, as shown in Figure 5.33, but another peak at 10 μs at 50 K, in plots of QFRS versus υ ($=2\pi/\tau$) under conditions of geminate recombination, that is, weak optical excitation. These peaks are almost independent of temperature, as seen in Figure 5.33. From these observations, Oheda concluded that there are two distinct emission centers, one of which is stable at lower temperatures and is replaced by another one with increasing temperature.

The lifetime of the PL, $\tau\approx10\,\mu s$, in the geminate-recombination regime agrees with the decay time of the ODMR signal of triplet excitons [25], that is, their radiative lifetime. Such a triplet exciton is a localized exciton composed of an electron coupled with a hole that is self-trapped at a weak Si–Si bond adjacent to an Si–H bond. This localized nature of triplet excitons coincides with Oheda's proposal. The PL component with $\tau\approx10\,\mu s$ observed in the QFRS spectra by Oheda arises from geminate electron–hole pairs.

Furthermore, as mentioned above, the self-trapped exciton is localized around an Si–H bond. The evidence for its localized nature comes from the following analysis [8]. Electron–phonon interaction generally gives rise to a Stokes shift and broadening of a PL band. The Si–H bond-stretching mode at 2000 cm^{-1} ($\hbar\omega_0=0.248$ eV) participates in this interaction. There is a relationship between the Stokes shift, δE, and the broadening, FWHM, namely FWHM $=1.67(\delta E\,\hbar\omega_0)^{1/2}$.

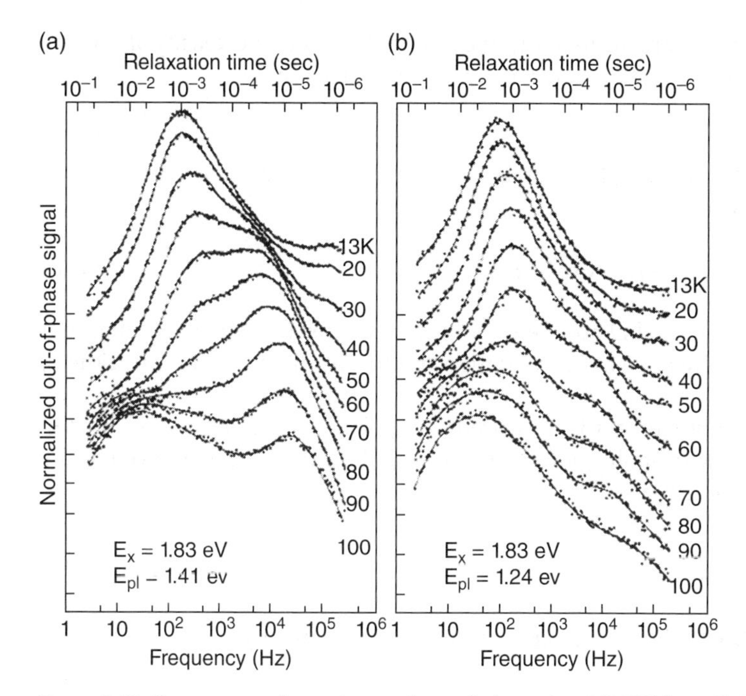

Figure 5.33 Temperature dependence of out-of-phase signal (QFRS) for a-Si:H. Photoluminescence photon energy E_{pl}: (a) 1.41 eV, (b) 1.24 eV. Excitation photon energy $E_x = 1.83$ eV. *Source:* Oheda 1993 [98]. Reproduced with permission of Elsevier.

The PL spectrum due to the triplet exciton can be estimated from the spectral dependence of the ODMR signal due to triplet excitons at $g = 4$ [25]. Thus, FWHM is estimated to be 0.31 eV, which agrees well with the 0.34 eV estimated from the PL spectrum under weak optical excitation. From these values of FWHM, the Stokes shift is obtained from the above relationship as 0.14–0.17 eV, using $\hbar\omega_0 = 0.248$ eV, that is, $\delta E = 0.31$–0.34 eV.

The self-trapped hole level is located at 0.3 eV above the edge of the valence band. This value of the depth corresponds almost to the ionization energy of self-trapped localized excitons. Hence, the absorption energy of self-trapped localized excitons is about 1.5 eV, assuming that the band gap energy is 1.8 eV. The PL peak energy of 1.36 eV leads us to a Stokes shift of 0.14 eV, which is consistent with the estimated range of 0.14–0.17 eV. From these results, it is concluded that the PL component with $\tau = 1$ ms is due to self-trapped triplet excitons.

5.5.5 Photoconductivity

5.5.5.1 Photocurrent

The photocurrent, i_p, is defined by the difference between the current under optical excitation and the dark current as follows:

$$i_p = e\mu F \Delta n \tag{5.84}$$

where μ, F, and Δn are the mobility, the electric field, and the density of carriers created by optical excitation, respectively. Δn is given by

$$\Delta n = G\tau \tag{5.85}$$

where G and τ are the photocarrier generation rate and the carrier lifetime, respectively. G is given by

$$G = N_0\eta(1-R)\{1-\exp(-\alpha D)\} \tag{5.86}$$

where N_0, η, R, α, and D are the photon number per unit time, the quantum efficiency, the reflectivity, the absorption coefficient, and the sample thickness, respectively.

When the photocurrent is measured using plane-type electrodes separated by a distance d with application of a voltage V_0, the photocurrent is given by the following relationship:

$$\frac{i_p d}{eN_0V_0} = \eta\tau\mu(1-R)\{1-\exp(-\alpha D)\} \tag{5.87}$$

For R and $\alpha D \ll 1$, Equation (5.87) can be approximated by

$$\frac{i_p d}{eN_0V_0} \cong \eta\tau\mu\alpha D \tag{5.88}$$

A measurement of i_p leads us to the product of η, τ, μ, α, and D.

5.5.5.2 Fundamental Processes of Photoconduction

The fundamental processes of photoconduction are trapping and recombination. As shown in Figure 5.34, carriers are captured by traps and deexcited into the conduction band, that is, trapping of carriers occurs. Recombination occurs between electrons and holes at some centers. The temperature dependence of photoconduction in semiconductors has been generally understood in terms of the Rose model (see, for example, [99]), in which a demarcation level is defined to separate the roles of a localized level into trapping and recombination.

In the following, we consider the rate equations governing these processes. The density of photogenerated carriers (electrons), n, and density of trapped electrons, n_e, follow the following equation:

$$\frac{dn}{dt} = G - \frac{n}{\tau_n} - \frac{dn_t}{dt} \tag{5.89}$$

$$\frac{dn_t}{dt} = n(N_t - n_t)v\sigma_n - n_t N_C v\sigma_n \exp\left(-\frac{\varepsilon}{k_B T}\right) \tag{5.90}$$

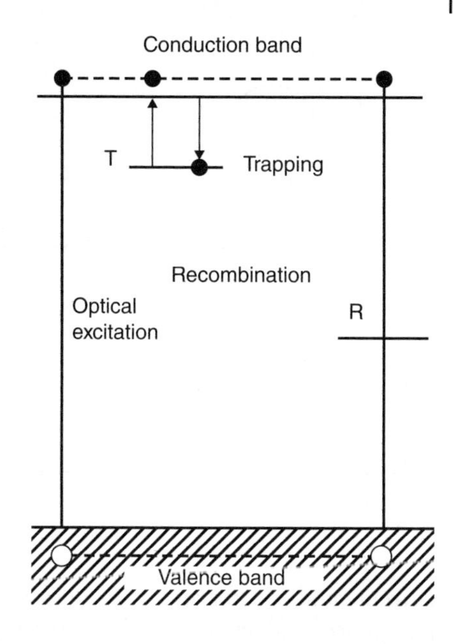

Figure 5.34 Schematic diagram of optical excitation, trapping, and recombination at defects.

where τ_n, N_t, ν, σ_n, N_C, and ε are the carrier lifetime, the number of trapping levels, the carrier velocity, the carrier trapping cross section, the effective density of states at the conduction band edge, and the depth of the trapping level, respectively.

Equation (5.85) can be obtained from the steady-state solution, neglecting the third term in the right-hand side of Equation (5.89). In Equation (5.89), trapping of carriers is taken into account. In the right-hand side of Equation (5.90), the first and second terms represent the rate of carrier trapping and the rate of deexcitation of trapped carriers into the conduction band, respectively.

In the steady state, that is, $dn/dt = dn_t/dt = 0$, we obtain

$$n = G\tau_n \tag{5.91}$$

$$\frac{n}{n_t} = \left(\frac{N_C}{N_t - n_t}\right)\exp\left(-\frac{\varepsilon}{k_B T}\right) \tag{5.92}$$

Equation (5.91) is of the same form as Equation (5.85).

In the following, we consider the temperature dependence of the photocurrent under continuous optical excitation. In a-Si:H and chalcogenide glasses, the temperature dependence of the photocurrent shown in Figure 5.35 is generally observed. In the i_p versus $1/T$ curve, there are three temperature ranges. In range I, the photocurrent is smaller than the dark current and increases with decreasing temperature. In range II, the photocurrent is greater than the dark

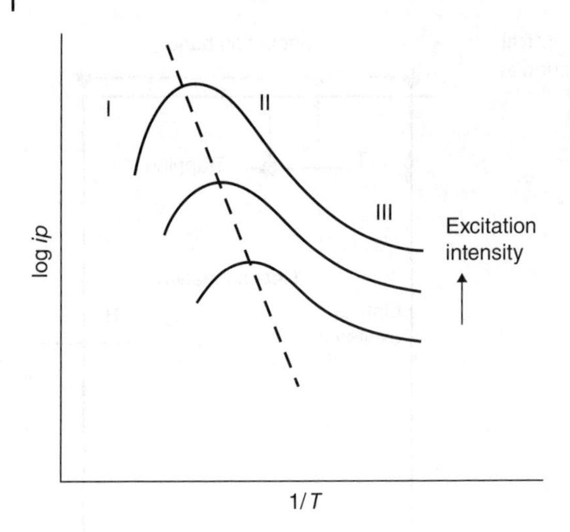

Figure 5.35 Schematic illustration of log i_p versus $1/T$ curves. The dashed line shows the temperature dependence of the dark current.

current and decreases with decreasing temperature. In range III, the photocurrent decreases with decreasing temperature, but it decreases only slowly and tends to saturate. Such temperature dependences of the photocurrent have been observed in a-Si:H; for example, some experimental results [100] are shown in Figure 5.36, where three ranges I, II, and III are clearly seen. The result may be interpreted in terms of the density-of-states model in the band gap region, as shown in Figure 5.37.

5.5.5.2.1 Photoconduction Characteristics in Range I

1) *Recombination of conduction electrons with trapped holes in the level E_Y.* In this temperature range, the photocurrent is smaller than the dark current, so the Fermi level is taken to be E_{F0} (the Fermi level in the dark). The density of trapped holes in the level E_Y, p_Y, is given by

$$P_Y = N_Y \exp\{-(E_{F0} - E_Y)/k_B T\} \tag{5.93}$$

where N_Y is the density of the level E_Y. Then,

$$\sigma_p = ne\mu = \frac{eG\mu}{N_Y R_1} \exp\{(E_{F0} - E_Y)/k_B T\} \tag{5.94}$$

where we take

$$\tau_n^{-1} = R_1 p_Y \tag{5.95}$$

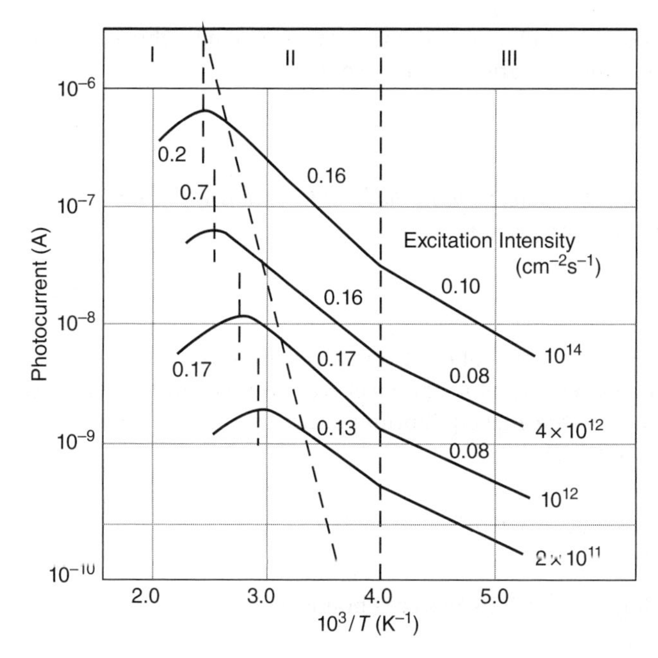

Figure 5.36 Temperature dependence of photocurrent for various excitation intensities in a-Si:H. The numbers alongside the curves are the activation energies in eV. i_d denotes the dark current. *Source:* Spear *et al.* 1974 [100]. Reproduced with permission of Elsevier.

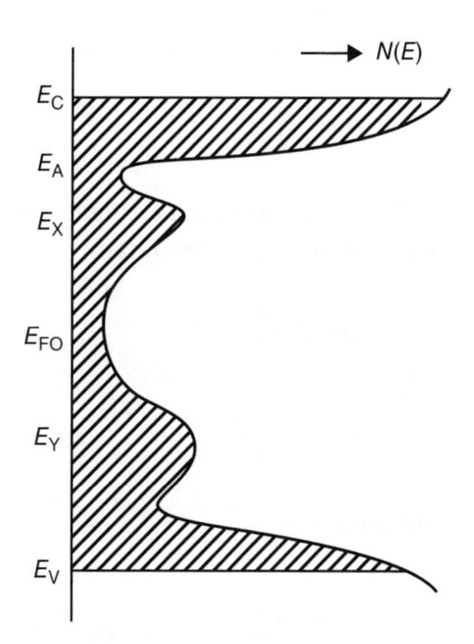

Figure 5.37 Schematic diagram of density-of-states distribution.

where R_1 is the recombination rate in the second term in the right-hand side of Equation (5.89). From Equation (5.94) we obtain

$$\sigma_p \propto G \tag{5.96}$$

2) *Recombination of band tail electrons in the level E_A with trapped holes in the level E_Y.* In the steady state, we take

$$G = R_2 n_A P_Y \tag{5.97}$$

where n_A and R_2 are the density of electrons in the level E_A and the recombination rate between electrons in the level E_A and holes in the level E_Y, respectively. As there is thermal equilibrium between n_A and n, the following relationship is obtained:

$$\frac{n}{n_A} = \frac{N_C}{N_A} \exp\left\{-\left(E_C - E_A\right)/k_B T\right\} \tag{5.98}$$

Furthermore, the following relationships are obtained:

$$N_A \cong N\left(E_A\right) k_B T \tag{5.99}$$

$$\sigma_p = \frac{eGN_C\mu}{R_2 N_Y N_A} \exp \frac{\left(E_{F0} - E_Y\right) - \left(E_C - E_A\right)}{k_B T} \tag{5.100}$$

From Equation (5.100), we obtain

$$\sigma_p \propto G \tag{5.101}$$

5.5.5.2.2 Photoconduction Characteristics in Range II

In the steady state, we obtain

$$G = R_2 n_T^2 \tag{5.102}$$

$$n_T = \left(\frac{G}{R_2}\right)^{1/2} \tag{5.103}$$

and then

$$\sigma_p = \frac{e\mu N_C}{N_T} \left(\frac{G}{R_2}\right)^{1/2} \exp\left\{-\left(E_C - E_T\right)/k_B T\right\} \tag{5.104}$$

As shown in Equation (5.104), we obtain

$$\sigma_p \propto G^{1/2} \tag{5.105}$$

The difference in G dependence between ranges I and II is that the former is as G and the latter as $G^{1/2}$, as shown in Equations (5.96), (5.101), and (5.105).

5.5.5.2.3 Photoconduction Characteristics in Range III

In this temperature range, photoconduction takes place by hopping conduction of electrons and/or holes in their respective band tails. However, even in the low-temperature range, it has been proposed by Murayama *et al.* [101] that photoconduction takes place by the conduction of carriers thermally excited from band tails into a spatially fluctuating band edge.

The results shown in Figure 5.36 in range I have been interpreted in terms of case 2 in Section 5.5.5.2.1 above, from a comparison between the theoretical and observed activation energies, taking into account the difference in the activation energy between Equations (5.94) and (5.100).

In the following, we attempt to interpret the results, including our own measurements [102], in terms of self-trapping of holes at low temperatures ranging from 50 to 200 K. The results of our measurements on an a-Si:H sample prepared at $T_S = 200\,°C$, using plasma-enhanced decomposition of pure silane, namely, σ_{ph} and σ_d versus T^{-1} in the temperature range of 11–300 K, are shown in Figure 5.38. We shall discuss the results in range II. Our model of photoconduction is based on the schematic diagram of recombination and trapping processes involved in a-Si:H shown in Figure 5.39:

$$n = n_T \frac{N_C}{N_T} \exp\{-(E_C - E_T)/k_B T\} \tag{5.106}$$

$$n_T = \tau_T G \tag{5.107}$$

where τ_T is the recombination time of trapped electrons. Below 200 K, it is assumed that most of the holes are self-trapped. Thus, the recombination time of trapped electrons is determined by either their radiative recombination with self-trapped holes or their nonradiative recombination with holes at neutral dangling bonds, as shown in Figure 5.39. In the latter process, holes should escape from their self-trapped sites to recombine with electrons at neutral dangling bonds after hopping through valence band tail states. As the barrier for the hopping process is low compared with the barrier for escape of holes from the self-trapped sites, the recombination time of trapped electrons is given by the following equation:

$$\tau_T = \tau_0 \exp(W/k_B T) \tag{5.108}$$

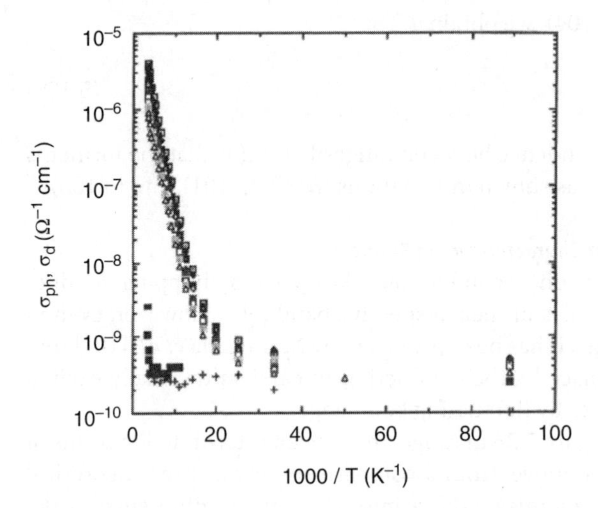

Figure 5.38 Photoconductivity σ_{ph} and dark conductivity σ_d as functions of the inverse of the temperature T for an a-Si:H sample. The symbols used in the figure are as follows: under optical excitation at 1 mW/cm^2, open squares with dots; 0.81 mW/cm^2, filled diamonds; 0.55 mW/cm^2, filled squares with white dots; 0.45 mW/cm^2, open diamonds; 0.31 mW/cm^2, gray squares; 0.25 mW/cm^2, open squares; 0.17 mW/cm^2, cross-hatched squares; 0.14 mW/cm^2, open triangles; and in the dark, filed squares, plus signs, and squares with horizontal strokes. *Source:* Morigaki *et al.* 1998 [102].

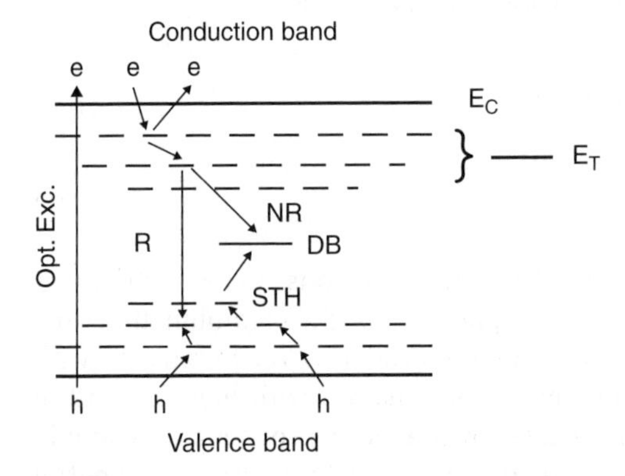

Figure 5.39 Schematic diagram of recombination and trapping processes in a-Si:H. The symbols used in the figure are as follows: e, electrons; h, holes; Opt. Exc., optical excitation; R, radiative recombination; NR, nonradiative recombination; DB, dangling bonds; STH, self-trapped holes; E_c, energy position of the conduction band edge; E_T, energy position of a single trapping level representing the conduction band tail states. *Source:* Morigaki *et al.* 1998 [102].

where W is the barrier height for escape of self-trapped holes from their trapping sites, that is, the relaxation energy of self-trapped holes.

Then, σ_p is given by Vommas and Fritzsche [103] as follows:

$$\begin{aligned}\sigma_p &= e\mu\tau_T G \frac{N_C}{N_T} \exp\{-(E_C - E_T)/k_B T \\ &= e\mu\tau_0 G \frac{N_C}{N_T} \exp \frac{-(E_C - E_T) - W}{k_B T}\end{aligned} \tag{5.109}$$

Thus, the activation energy E_A is given by

$$E_A = E_C - E_T - W \tag{5.110}$$

The value of $E_C - E_T$ is estimated to be 0.13–0.17 eV [103]. The value of W has been estimated to be 100 meV for the a-Si:H sample measured [104]. Thus, if we take the value of $E_C - E_T$ to be 0.15 eV, an average value of those mentioned above, then we obtain 50 meV for the value of the activation energy, which is almost consistent with the observed value. It has been observed from photoinduced absorption measurements [105] that self-trapping of holes occurs below 210 K. This is consistent with the present photoconduction measurement. The present model is an alternative approach to the Rose model mentioned earlier. This alternative model may be applied to the case of device quality a-Si:H samples, in which the activation energy is about 0.11 eV [104]. For device quality samples, we have estimated W to be about 50 meV [105]. If we take the value of $E_C - E_T$ to be 0.15 eV, then we obtain E_A as 0.10 eV, which agrees with the observed value, that is, 0.11 eV. The assumption of $E_C - E_T = 0.15$ eV is not unreasonable, and so the model can account for the temperature dependence of photoconduction in device quality a-Si:H samples.

5.5.6 Dispersive Photoconduction

Measurements of transient photoconductivity are useful for elucidating the conduction mechanism and also the localized levels within the band gap. These measurements are used to measure the drift mobility of carriers in semiconductors; the technique is called time-of-flight measurement [106, 107], the apparatus for which is schematically shown in Figure 5.40. The transient photoconductivity can be measured using blocking electrodes. In this case, pulsed light is shone on the left-hand electrode, and then carriers, that is, electron–hole pairs, are generated and these carriers move towards the right-hand electrode. When the left-hand electrode is positive and the right-hand electrode is negative, the moving carriers are holes. A sheet of carriers reaches the opposite electrode (negative), and then a transient current flows in the external circuit and an output voltage is obtained. The pulsed photocurrent flowing in the

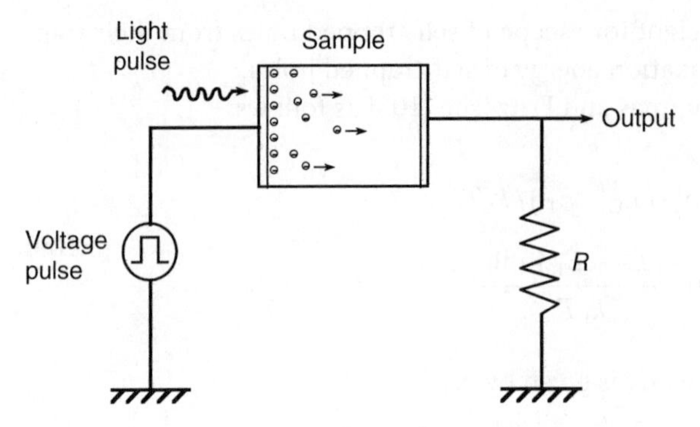

Figure 5.40 Schematic illustration of time-of-flight measurement.

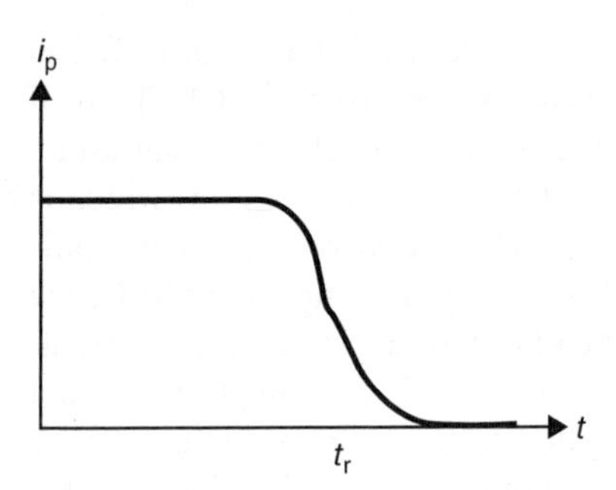

Figure 5.41 Schematic illustration of Gaussian conduction: time variation of a pulsed current flowing in the external circuit.

external circuit is shown as a function of time in Figure 5.41. When the transient photocurrent decays exponentially, this type of conduction is called Gaussian conduction.

When the flight time is distributed, that is, it has a dispersion, the conduction is said to be non-Gaussian. The pulsed photocurrent flowing in the external circuit in this case is shown as a function of time in Figure 5.42.

In the following, we consider non-Gaussian conduction, that is, dispersive photoconduction. In this case, the transient photocurrent i_p flowing in the external circuit is described by

$$i_p \propto t^{-(1-\alpha)} \quad \text{for} \quad t < t_T \tag{5.111}$$

$$i_p \propto t^{-(1+\alpha)} \quad \text{for} \quad t > t_T \tag{5.112}$$

The sum of the two exponents is −2. These transient photocurrents are schematically shown in Figure 5.42. Scher and Montroll [108] developed a general theory of dispersive photoconduction, taking into account hopping conduction with a dispersive flight time. Their theory was also applied to the

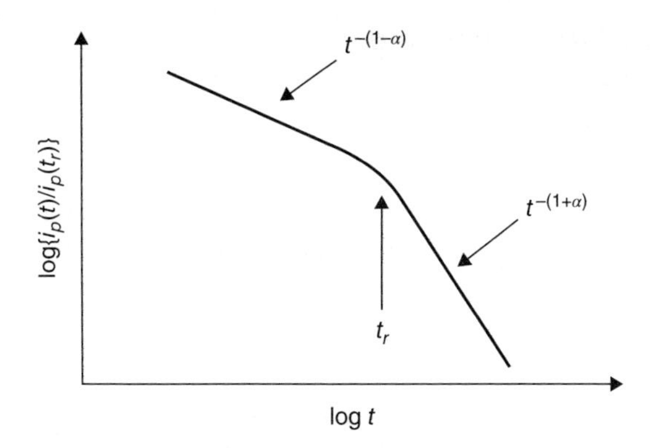

Figure 5.42 Schematic illustration of non-Gaussian conduction: time variation of a pulsed current flowing in the external circuit.

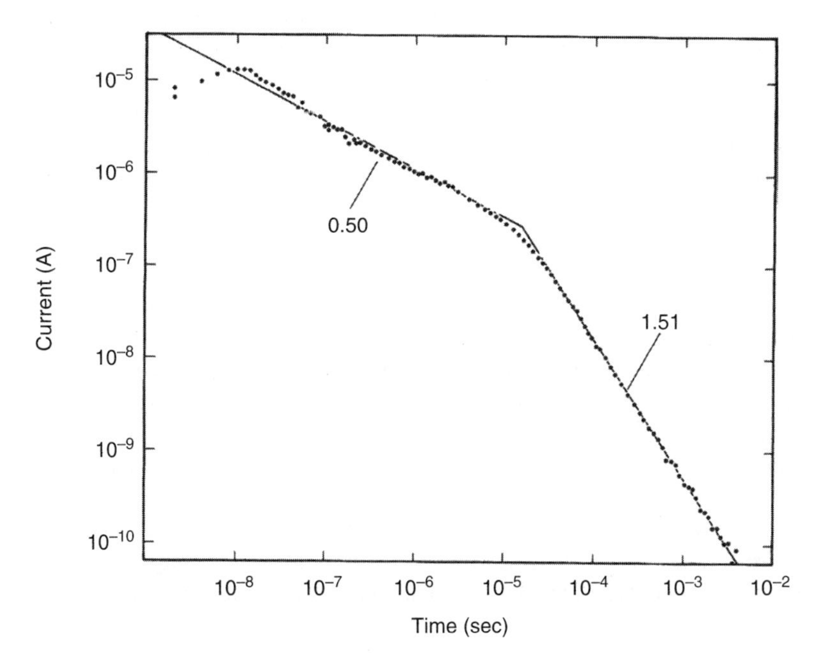

Figure 5.43 Time-of-flight electron photocurrent decay measured at 160 K for an a-Si:H sample 2.8 μm thick. The numbers indicate the slope of the curve. *Source:* Tiedje 1984 [107]. Reproduced with permission of Elsevier.

case in which carriers are captured by traps and thermally excited carriers contribute to the photoconduction. This is the case observed in chalcogenide glasses and amorphous silicon. A typical example in the case of a-Si:H is shown in Figure 5.43.

5.6 Electron Magnetic Resonance and Spin-Dependent Properties

5.6.1 Introduction

In this section, we treat electron magnetic resonance, particularly pulsed electron magnetic resonance and spin-dependent properties, to elucidate the nature of magnetic centers in a-Si:H.

5.6.2 Electron Magnetic Resonance

Electron magnetic resonance, also called electron spin resonance or electron paramagnetic resonance (EPR), is a powerful tool for elucidating the electronic states of magnetic centers as well as for identifying the magnetic centers themselves. A combination of ESR and nuclear magnetic resonance (NMR) called ENDOR is also a powerful tool for elucidating electronic states of magnetic centers. These techniques have been widely used in crystalline semiconductors. Some important defects in a-Si:H, namely, silicon dangling bonds, were first identified with this tool [109]. In this section, pulsed electron magnetic resonance [110, 111] is treated so as to present recent developments in the investigation of defects in a-Si:H using spin echo phenomena and Rabi oscillations.

5.6.2.1 Spin Echo Phenomena and Electron Spin Echo Envelope Modulation (ESEEM)

First, the principle of the spin echo phenomenon will be briefly explained. Here, a magnetic field $\boldsymbol{B}_0$ and an oscillating magnetic field $\boldsymbol{B}_1$ with an angular frequency ω are applied along the z-direction and the x-direction, respectively. The magnetic moment rotates around the z-axis (an effect called the Larmor precession) at the Larmor frequency, that is, $\gamma = g\mu_B/\hbar$, where γ, g, and μ_B are the gyromagnetic ratio, g-factor, and Bohr magneton, respectively.

Now, we consider a spin echo in a rotating coordinate system. The first pulse, with a width corresponding to a $\pi/2$ rotation around the x-axis, is applied along the x-direction, as shown in Figure 5.44. The magnetic moment then rotates by

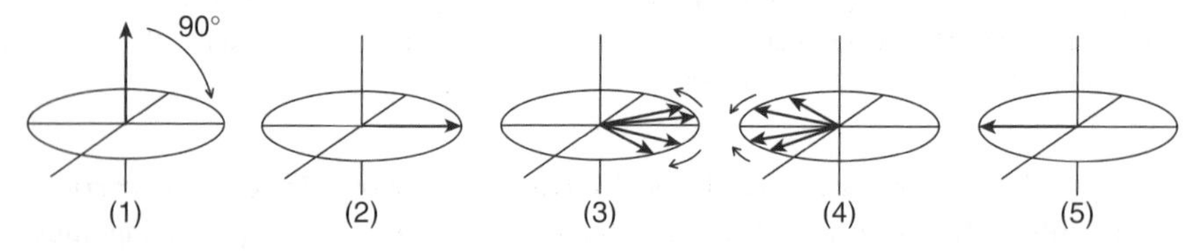

Figure 5.44 Diagrams describing the motion of the magnetization vectors in spin echo measurements. See text for more detail.

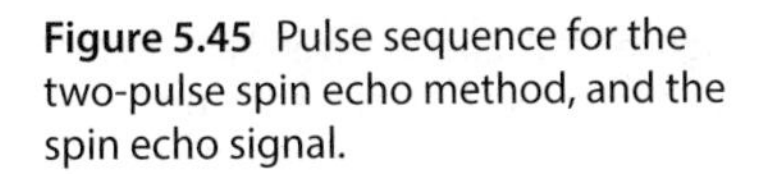

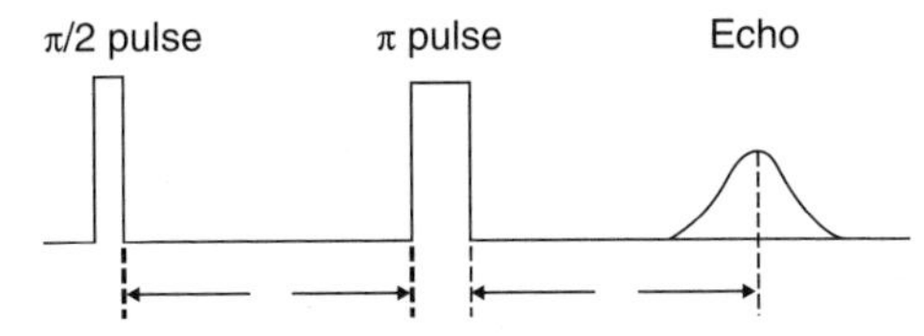

Figure 5.45 Pulse sequence for the two-pulse spin echo method, and the spin echo signal.

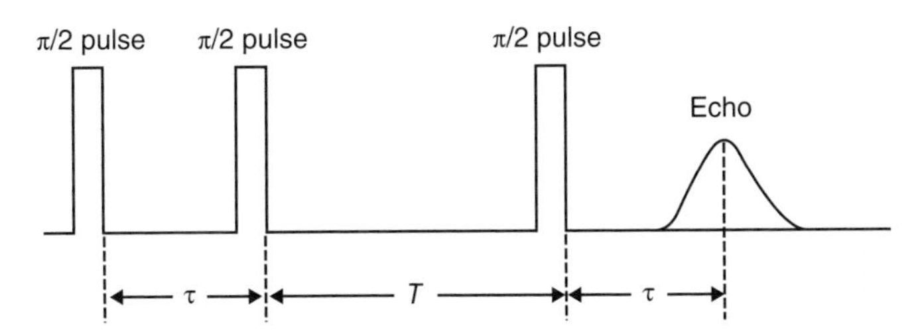

Figure 5.46 Pulse sequence for the three-pulse spin echo method, and the spin echo signal.

π/2 from the z-axis to the y-axis. However, although the magnetic moment is oriented towards the y-direction, each magnetic moment rotates at its own Larmor frequency, and these are different from each other owing to spin–spin interaction. At a time τ after application of the first π/2 pulse, a second pulse, with a width corresponding to a rotation of π around the x-axis, is applied, and then a spin echo signal appears at τ after the second π pulse, as shown in Figure 5.45. A three-pulse spin echo method is also used, as shown in Figure 5.46.

The decay of the electron spin echo signal with τ is influenced by the electron–nucleus interaction and the nuclear quadrupolar interaction. As an example, we take a system with $S=1/2$ and $I=1/2$, in which the g-value is isotropic and the hyperfine interaction is axially symmetric. The Hamiltonian is given as follows [111]:

$$\mathrm{H}/\hbar = \omega_{\mathrm{e}} S_z + \omega_{\mathrm{n}} I_z + A S_z I_z + B S_z I_x \tag{5.113}$$

$$\omega_{\mathrm{e}} = g \mu_{\mathrm{B}} B/\hbar \tag{5.114}$$

$$\omega_{\mathrm{n}} = g_{\mathrm{n}} \mu_{\mathrm{n}} B/\hbar \tag{5.115}$$

$$A = \hbar^{-1} r^{-3} g_{\mathrm{n}} \mu_{\mathrm{n}} g \mu_{\mathrm{B}} \left(\cos^2\theta - 1\right) + A_0 \tag{5.116}$$

$$B = \hbar^{-1} r^{-3} g_{\mathrm{n}} \mu_{\mathrm{n}} g \mu_{\mathrm{B}} \left(3\cos\theta \sin\theta\right) \tag{5.117}$$

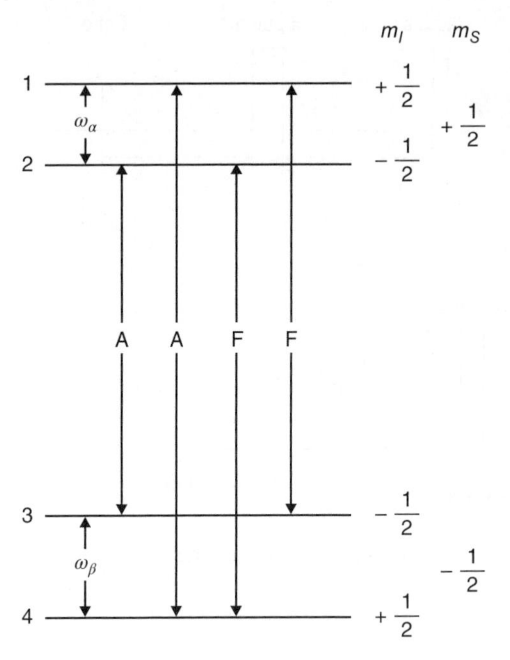

Figure 5.47 Energy level diagram for an interacting electron-spin–nuclear-spin pair. m_S and m_I are the magnetic quantum numbers of the electron spin and nuclear spin, respectively. A and F designate allowed and forbidden transitions, respectively. ω_α and ω_β are the ENDOR frequencies for $m_I = 1/2$ and $-1/2$, respectively.

where g_n and μ_n are the nuclear g-factor and the nuclear magneton, respectively, θ is the angle between the z-direction and the line connecting the electron spin and the nuclear spin, and $\hbar A_0$ is a Fermi-type hyperfine interaction.

The decay of the electron spin echo signal with τ is influenced by the electron–nucleus interaction and the nuclear quadrupolar interaction. In the case of the electron–nucleus interaction, the energy levels of the electron spin S and the nuclear spin I are illustrated in Figure 5.47. The electron spin echo intensity $V(\tau)$ is given by [111]

$$V(\tau) = 1 - 2(\omega_n B/\omega_\alpha \omega_\beta)^2 \sin^2(\omega_\alpha \tau/2) \sin^2(\omega_\beta \tau/2) \tag{5.118}$$

$$= 1 - \frac{1}{4}\left(\frac{\omega_n B}{\omega_\alpha \omega_\beta}\right)^2 \left[2 - 2\cos(\omega_\alpha \tau) - 2\cos(\omega_\beta \tau) + \cos\{(\omega_\alpha + \omega_\beta)\tau\} + \cos\{(\omega_\alpha - \omega_\beta)\tau\}\right] \tag{5.119}$$

$$\omega_\alpha = \left[\{(A/2) + \omega_n\}^2 + (B/2)^2\right]^{1/2} \tag{5.120}$$

$$\omega_\beta = \left[\{(A/2) - \omega_n\}^2 + (B/2)^2\right]^{1/2} \tag{5.121}$$

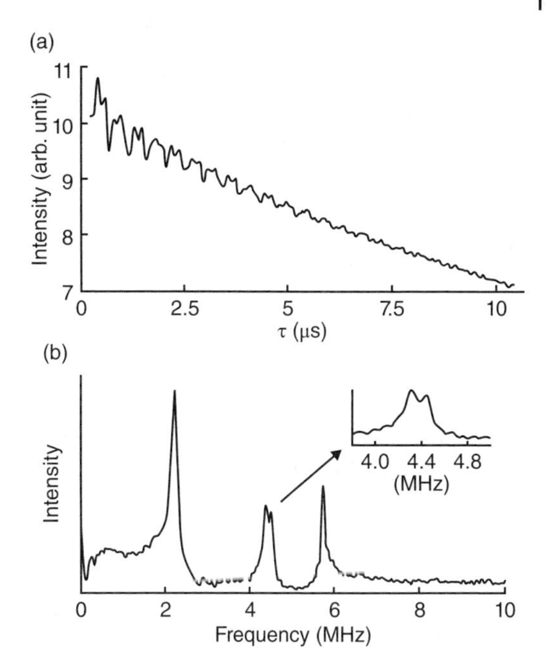

Figure 5.48 (a) Two-pulse echo decay and time-domain ESEEM spectrum (9.54 GHz, 56 K) of a photocreated dangling bond ($g = 2.0055$) in a-Si:D. (b) FT-ESEEM spectrum showing ν_n(D), $2\nu_n$(D), which has a doublet structure, and $2\nu_n$(^{29}Si) peaks. The ν_n(^{29}Si) peak is not observed, since the modulation is damped out within the dead time (~150 ns) of the pulsed ESR spectrometer. *Source:* Isoya *et al.* 1993 [112]. Reproduced with permission from the American Physical Society.

where $\hbar\omega_\alpha$ and $\hbar\omega_\beta$ are the energy difference between levels 1 and 2 and the energy difference between levels 3 and 4, respectively, as shown in Figure 5.47.

The electron spin echo intensity for the three-pulse spin echo method is given by

$$V(\tau,T) = 1 - \left(\frac{\omega_n B}{\omega_\alpha \omega_\beta}\right)^2 \left[\sin^2(\omega_\alpha \tau/2)\sin^2\{\omega_\beta(\tau+T)/2\} + \sin^2(\omega_\beta \tau/2)\sin^2\{\omega_\alpha(\tau+T)/2\}\right] \tag{5.122}$$

These time dependences of the electron spin echo intensity are Fourier transformed to obtain the frequency spectrum.

In the following, the observation of ESEEM in a-Si:H is described in detail. Two-pulse and three-pulse ESEEM have been observed by Isoya *et al.* [112]. Figures 5.48(a) and (b) show the decay curve of the two-pulse electron spin echo signal and the frequency spectrum of its Fourier transform, respectively, for light-soaked a-Si:H. The decay curve is expressed by Equation (5.118); ESEEM can be seen in it, as shown in Figure 5.48(a). In Figure 5.49, only the ESEEM parts of the decay curve are shown, as follows: curves a, the observed spectrum of light-soaked a-Si:D; curves b, the sum of the spectra shown in curves c and d; curves c, a simulated spectrum; and curves d, the observed spectrum of light-soaked a-Si:H. The simulated spectrum in curves c was

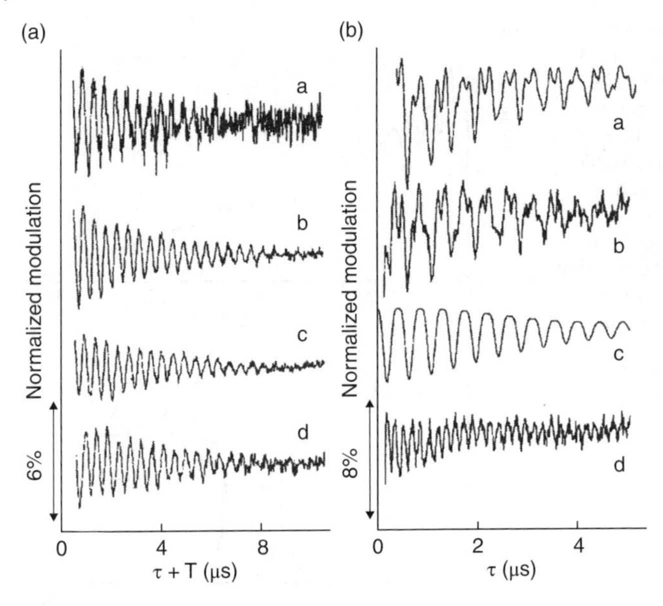

Figure 5.49 (a) Three-pulse ESEEM spectra plotted against $\tau + T$ ($\tau = 220$ ns) for (curve a) annealed a-Si:H sample and (curve b) light-soaked a-Si:H sample. Curve c, LESR signal of $g = 2.004$ for annealed a-Si:H sample; curve d, LESR signal of $g = 2.013$ for annealed a-Si:H sample. (b) Two-pulse spin-echo modulation for (curve a) the observed spectrum of light-soaked a-Si:D, (curve b) a spectrum made from curve c plus curve d, (curve c) a simulated spectrum (see text for more detail), and (curve d) the observed spectrum of light-soaked a-Si:H. *Source:* Isoya *et al.* 1993 [112]. Reproduced with permission from the American Physical Society.

obtained using the following values of parameters: $A_0 = 0.05$ MHz for one D nucleus with $r = 4.2$ Å, and $A_0 = 0$, $B \neq 0$ for one D nucleus with $r = 6$ Å, two D nuclei with $r = 7$ Å, and eight D nuclei with $r = 8$ Å. For all deuterium nuclei, $e^2qQ/h = 87$ kHz and an asymmetry parameter $\eta = 0$ were assumed. The spectrum in curves b was made from a superposition of the spectra in curves c and d, because the spectrum that is constructed from the ^{29}Si nuclei is the spectrum in curves d, which is composed of hyperfine interactions with ^{29}Si nuclei. Thus, the spectrum in curves b should be compared with the observed spectrum in curves a.

The observed three-pulse ESEEM spectrum of light-soaked a-Si:D is shown in Figure 5.50 (curve a). Simulated spectra obtained from Equation (5.122), using the following values of parameters, are shown in curves b–e: curve b, one D nucleus with $r = 4.2$ Å and distant D nuclei; curve c, two D nuclei with $r = 4.8$ Å and distant D nuclei; curve d, one D nucleus with $r = 4.2$ Å; and curve e, one D nucleus with $r = 6$ Å, two D nuclei with $r = 7$ Å, and eight D nuclei with $r = 8$ Å.

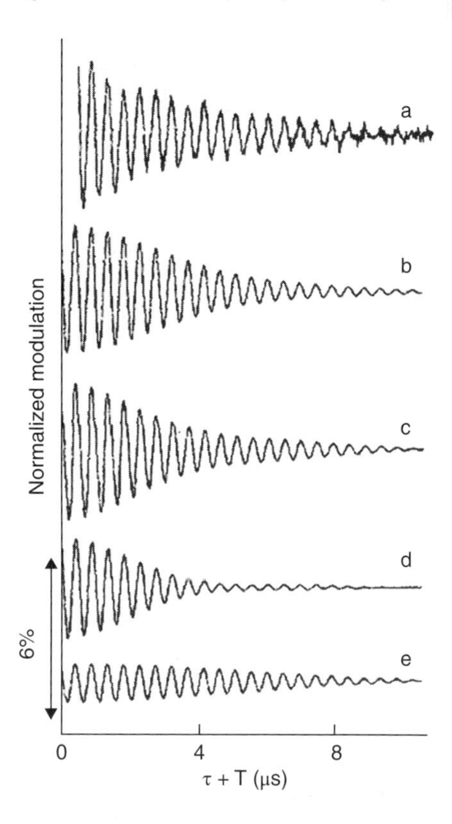

Figure 5.50 Three-pulse ESEEM spectra. Curve a, observed three-pulse ESEEM spectrum of light-soaked a-Si:D. Simulated spectra obtained from Equation (5.122), using the following values of parameters: curve b, one D nucleus with $r = 4.2$ Å and distant D nuclei; curve c, two D nuclei with $r = 4.8$ Å and distant D nuclei; curve d, one D nucleus with $r = 4.2$ Å; and curve e, one D nucleus with $r = 6$ Å, two D nuclei with $r = 7$ Å, and eight D nuclei with $r = 8$ Å. In all cases, $A_0 = 0$ was assumed. *Source:* Isoya *et al.* 1993 [112]. Reproduced with permission from the American Physical Society.

In all cases, $A_0 = 0$ was assumed. The dipolar interaction between the electron spins and nuclear spins was estimated by a computer simulation of ESEEM using a point-dipole approximation.

From these two-pulse and three-pulse electron spin echo measurements, Isoya *et al.* [112] concluded that the distance from the closest deuterium (or hydrogen) atom was 4.2 Å for both native and metastable dangling bonds, and 4.8 Å for the LESR signals. When there were two closest deuterium (or hydrogen) atoms, the distance was 4.8 Å for both native and metastable dangling bonds and 5.3 Å for the LESR signals.

5.6.2.2 Pulsed Electron–Nuclear Double Resonance

The electron–nuclear double resonance method was invented by Feher [113, 114], by which he evaluated a map of the wave function of the unpaired electrons in F-centers in alkali halides [115], as well as of donor electrons in doped crystalline silicon [116]. The energy levels of the electron spin and nuclear spin involved in an ENDOR measurement are illustrated in Figure 5.51.

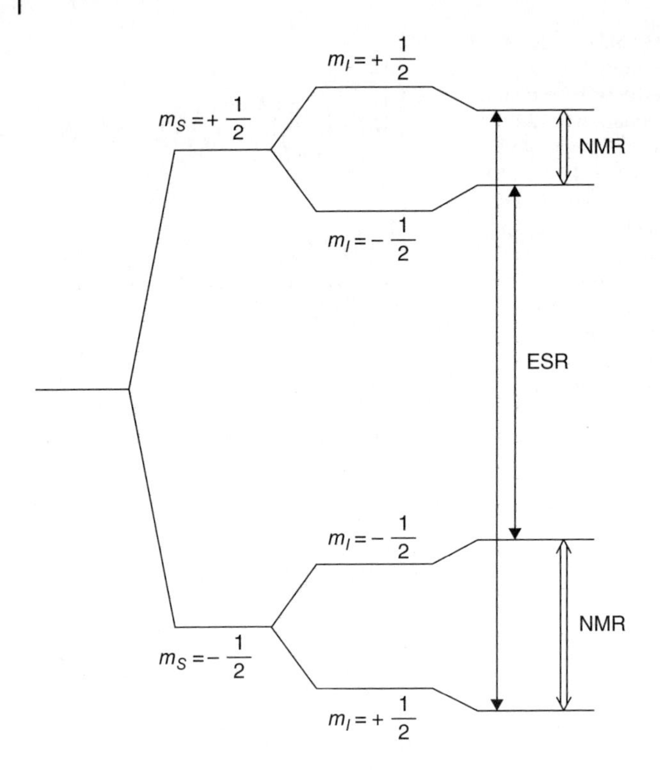

Figure 5.51 Energy level diagram for an $S = 1/2$, $I = 1/2$ spin system with hyperfine interaction and in the presence of a static magnetic field. ESR and NMR are indicated in the figure.

The ENDOR frequency is given by

$$\nu_1 = |(A_0/2) - \nu_0|, \qquad \nu_2 = |(A_0/2) + \nu_0| \tag{5.123}$$

where A_0 and ν_0 are the isotropic hyperfine interaction constant (in units of MHz) and the natural nuclear frequency, respectively.

In the following, the Davies method [117] for the pulse sequence used in the pulsed ENDOR measurement technique is described. The pulse sequence is illustrated in Figure 5.52. The RF pulse is applied between the first and second pulses (a π pulse and a π/2 pulse). Then, the ENDOR signal is detected through the electron spin echo that appears after the third pulse (a π pulse). The amplitude of the electron spin echo signal is increased when the NMR transition occurs during the RF pulse. The details of the mechanism are described in [117].

In the following, we describe the pulsed ENDOR measurements made by Fehr *et al.* [118] on a-Si:H, which were concerned with the location of hydrogen atoms. Fehr *et al.* [118] made pulsed ENDOR measurements at 34.5 GHz

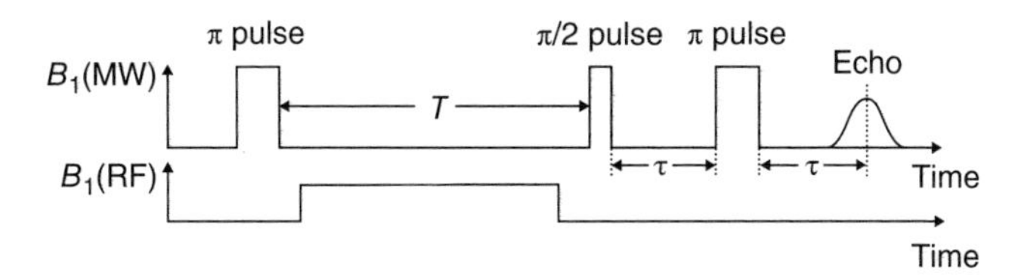

Figure 5.52 Three-pulse sequence (with microwave (MW) and radio-frequency (RF) pulses) and spin echo signal in the Davies method.

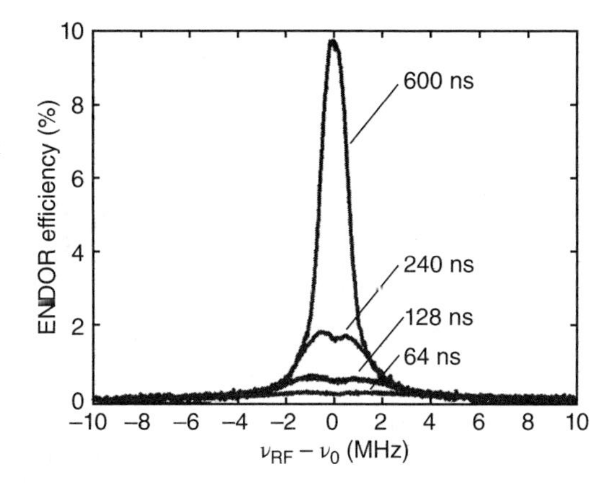

Figure 5.53 ENDOR spectra taken at $g = 2.0055$ with different widths of the initial microwave π pulse (see Figure 5.52), indicated by the number beside each curve; ν_{RF} and ν_0 designate the RF frequency and the natural resonance frequency of the H nucleus, respectively. *Source:* Fehr *et al.* 2010 [118]. Reproduced with permission from John Wiley & Sons.

and 80 K on an a-Si:H sample prepared at 160 °C ($N_S = 4 \times 10^{16}\,\text{cm}^{-3}$, [H] = 10 at.%), using the Davies method ($T = 22\,\mu\text{s}$, $\tau = 400\,\text{ns}$, and RF pulse width = 19 μs). The ENDOR signal was obtained by monitoring the ESR signal due to the dangling bonds. ENDOR spectra were obtained for various widths of the first pulse, as shown in Figure 5.53, where ν_{RF} and ν_0 are the RF frequency and the proton natural resonance frequency, respectively.

The ENDOR efficiency is defined by

$$\text{ENDOR efficiency} = \frac{1}{2}\left[I_{\text{echo}}(\text{RF on}) - I_{\text{echo}}(\text{RF off})\right]/I_{\text{echo}}(\text{RF off}) \tag{5.124}$$

As seen in Figure 5.53, a strong ENDOR signal is observed at $\nu_{RF} = \nu_0$, that is, the natural nuclear resonance frequency. When the width of the first pulse is decreased, the ENDOR efficiency decreases and the signal is broadened with a dip in the central part. The peak observed at a width of 600 ns seems to be a distant or matrix ENDOR signal. The broad peak observed at shorter widths of the first pulse seems to be an ENDOR signal due to a ^{1}H nucleus being located near a dangling bond. The dip may be due to a doublet structure following Equation (5.123). The ENDOR spectrum observed at a pulse width of 40 ns is

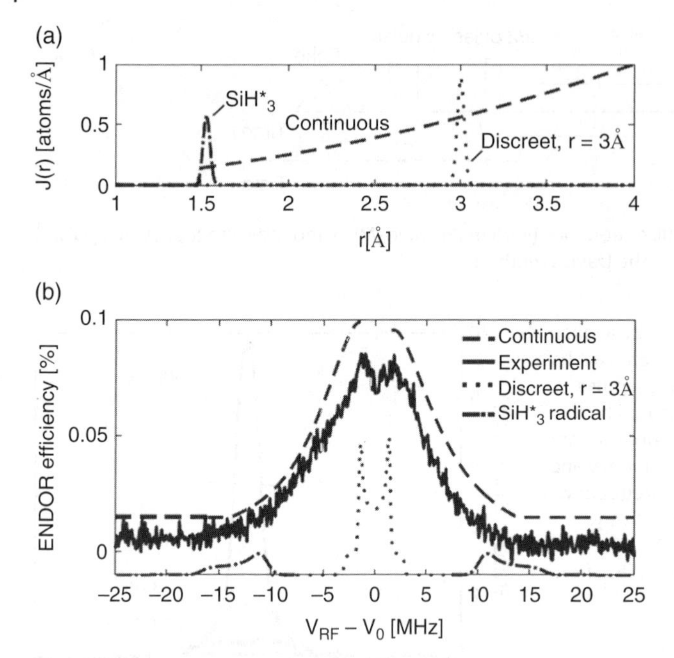

Figure 5.54 (a) Radial distribution function *J(r)* for a continuous (dashed line) and discrete (dotted line) distribution of H atoms around dangling bonds. The position of H atoms in free SiH_3* radicals is shown by the dash–dotted line. (b) ENDOR spectrum for the distribution in (a) and comparison with the observed spectrum for an initial π pulse width of 40 ns. The dashed, dotted, and dash–dotted spectra are offset vertically with respect to the observed spectrum for a better overview. The ENDOR efficiency is given by Equation (5.124). *Source:* Fehr *et al.* 2010 [118]. Reproduced with permission from John Wiley & Sons.

shown in Figure 5.54(b), where the ENDOR spectra due to dipolar interaction with ^{1}H nuclei distributed continuously and with ^{1}H at $r = 3$ Å, as well as the spectrum due to SiH_3^+ radicals, are shown for comparison. In Figure 5.54(a), the radial distribution function *J*(*r*) for a continuous distribution (dashed line) and a discrete distribution (dotted line, $r = 3$ Å) of hydrogen atoms around a dangling bond are shown, along with that for a discrete distribution of SiH_3^+ radicals, as a function of distance (in Å) from the dangling bond site. As ca be seen in Figure 5.54(b), the ENDOR signal is observed over a wide range of ν_{RF}.It is obvious that an ENDOR signal is observed for $r < 3$ Å. This is inconsistent with Isoya *et al.*'s conclusion mentioned above [112]. Fehr *et al.* [118] concluded that there is no correlation between the dangling bonds and the hydrogen distribution, that is, hydrogen is continuously distributed around a dangling bond. However, it should be noted that there is a shoulder around $\nu_{RF} - \nu_0 = -5$ MHz in Figure 5.54(b). This suggests that hydrogen with $r \approx 2.2$ Å exists more commonly than statistically distributed hydrogen. This distance of 2.2 Å corresponds to the distance between a dangling bond site and its nearby hydrogen in a hydrogen-related dangling bond (see Section 5.2.3). The ENDOR signal at the shoulder should be symmetrical with respect to ν_0, but this is not the case in the observed ENDOR spectra shown in Figure 5.54(b). This is often seen in continuous wave (CW) ENDOR spectra observed at low temperatures. This may be related to the relaxation mechanism, particularly at low temperatures. The anisotropic hyperfine interaction also contributes to the asymmetric pattern through line broadening. The existence of hydrogen-related dangling bonds has been confirmed by CW ENDOR measurements in a-Si:H and a-Si:D [17], as mentioned below.

ENDOR spectra of a-Si:H and a-Si:D are shown in Figures 5.55(a) and (b), respectively, where their microwave power dependences are shown. In Figure 5.55(a), the peak at 10.7 MHz is the local ENDOR signal due to the ^{1}H nucleus, but the expected peak at higher frequency is hidden by the local ENDOR signal due to the ^{29}Si nucleus. In Figure 5.55(b), the local ENDOR signal due to the ^{2}D nucleus is seen at 3.3 MHz. The ENDOR frequencies of

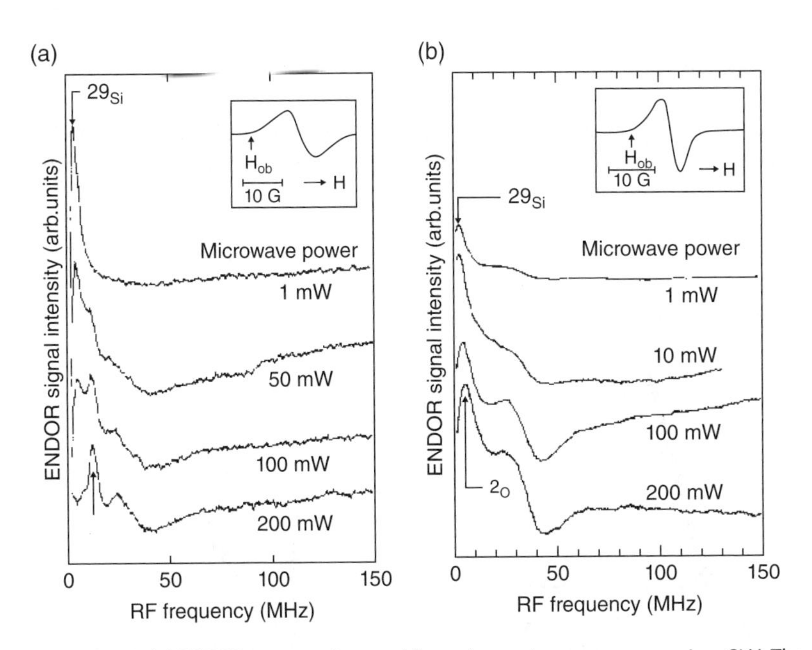

Figure 5.55 (a) ENDOR spectra observed for various microwave powers in a-Si:H. The magnetic field was set at H_{ob} as shown in the inset. A local ENDOR signal due to the ^{1}H nucleus is indicated by an arrow labeled ^{1}H. A peak in the distant ENDOR signal due to the ^{29}Si nucleus is seen in the spectra for microwave powers of 1, 50, and 100 mW. Source: Yokomichi and Morigaki 1987 [15]. Reproduced with permission of Elsevier. (b) ENDOR spectra observed for various microwave powers in a-Si:D. The magnetic field was set at H_{ob} as shown in the inset. A local ENDOR signal due to the ^{2}D nucleus and a distant ENDOR signal due to the ^{29}Si nucleus are indicated by arrows. *Source:* Yokomichi and Morigaki 1996 [17]. Reproduced with permission of Elsevier.

the ^{2}D nucleus expected from Equation (5.123), using $A_0 = 6$ MHz for the ^{1}H nucleus, are 1.64 and 2.56 MHz. The above frequency of 3.3 MHz seems to correspond to the latter frequency.

5.6.2.3 Rabi Oscillations

As shown in Figure 5.56(a), the magnetic moment $\boldsymbol{\mu}$ precesses about an external magnetic field $\boldsymbol{B}_0$ directed parallel to the z-axis in the laboratory frame with coordinates x, y, and z; that is, a Larmor procession at a frequency ω_0 occurs. Furthermore, it rotates about an oscillating magnetic field, and this rotating motion is called a Rabi oscillation [119]. Here, we use a frame rotating at an angular velocity of $\omega_0 - \omega$ with coordinates x', y', and z', where ω is the angular frequency of the circularly polarized oscillating magnetic field $\boldsymbol{B}_1$, as illustrated in Figure 5.56(b).

The motion of the magnetic moment $\boldsymbol{\mu}$ is described as follows:

$$d\boldsymbol{\mu}/dt = \gamma\left(\boldsymbol{\mu}\times\boldsymbol{B}\right) \tag{5.125}$$

$$\boldsymbol{B} = \boldsymbol{B}_0 + \boldsymbol{B}_1 \tag{5.126}$$

where γ is the gyromagnetic ratio. When the rotating oscillatory magnetic field is zero, the magnetic moment rotates around the z-axis at the Larmor

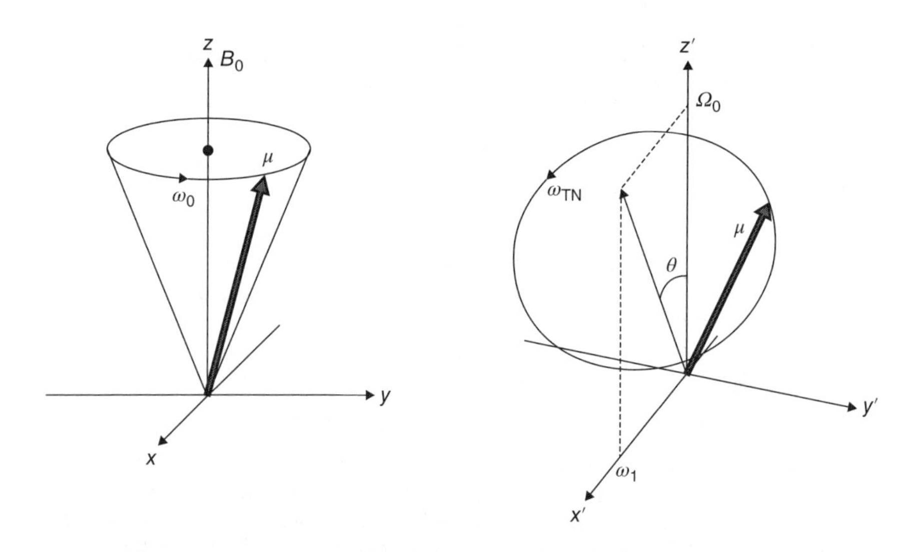

Figure 5.56 (a) Motion of the moment $\boldsymbol{\mu}$ around a static magnetic field $\boldsymbol{B}_0$ parallel to the z-direction in the laboratory coordinate system. (b) Nutation motion of the moment $\boldsymbol{\mu}$ and the composite field ($= \boldsymbol{B}_0 + \boldsymbol{B}_1$) in a rotating coordinate system, where $\boldsymbol{B}_1$ is a circularly polarized microwave magnetic field. $\Omega_0 = \omega_0 - \omega$ (Ω_0, Larmor angular frequency; ω_0, microwave angular frequency); θ is given by Equation (5.127).

angular frequency ω_0 (= $g_B B_0/\hbar$). When the rotating oscillatory magnetic field is applied, the magnetic moment nutates around the composite magnetic field $\boldsymbol{B}$ arising from $\boldsymbol{B}_0$ and $\boldsymbol{B}_1$. Now, we consider the magnetic moment in terms of coordinates rotating at a speed of $\omega_0 - \omega$. The axis of the composite magnetic field is inclined at θ from the z-axis, as shown in Figure 5.56(b), where

$$\theta = \tan^{-1}\left(\omega_1/\Omega_0\right) \tag{5.127}$$

$$\Omega_0 = \omega_0 - \omega \tag{5.128}$$

Then, the nutation frequency ω_{TN} is given by

$$\omega_{\mathrm{TN}} = \left\{\omega_1^{\,2} + \left(\omega_0 - \omega\right)^2\right\}^{1/2} \tag{5.129}$$

When the microwave angular frequency coincides with the Larmor angular frequency, that is, $\omega = \omega_0$, the nutation frequency for $S = 1/2$ is given by

$$\omega_{\mathrm{TN}} = \omega_1 = g_1 \mu_B B_1/h \tag{5.130}$$

where g_1 is the g-factor along the x-direction. For a transition from the kth level to the lth level in a multilevel system, we obtain

$$\omega_{\mathrm{TN}} = \left(a_{kl}/a_0\right)\omega_1 \tag{5.131}$$

where a_{kl} is the transition amplitude for this transition and a_0 is the transition amplitude for an isotropic system with $S = 1/2$.

In the case where $S > 1/2$ with isotropic g-values, the nutation frequency for the transition between two levels m_S and $m_S - 1$ (where m_S is the spin magnetic quantum number) is obtained as

$$\omega_{\mathrm{TN}} = \left\{S\left(S+1\right) - m_S\left(m_S - 1\right)\right\}^{1/2}\omega_1 \tag{5.132}$$

using the value of $a_{m-1,m}/a_0$. From Equation (5.132), the spin multiplicity can be determined.

When a dipolar interaction operates between spin a and spin b, as shown in the following equation,

$$\mathcal{H}_{\mathrm{d}} = -D^{\mathrm{d}}\left(3S_{az}S_{bz} - \boldsymbol{S}_a\boldsymbol{S}_b\right) \tag{5.133}$$

the Rabi frequency ω_{TN} is given by [120]

$$\omega_{\mathrm{TN}} = \{(\gamma B_1)^2 + \left(3D_d/4\hbar\right)^2\}^{1/2} \tag{5.134}$$

for $\omega_a - \omega_b \ll \gamma B_1$, where ω_a and ω_b are the Larmor angular frequencies of spins a and b. From Equation (5.134), we note that ω_{TN} deviates from γB_1 ($\equiv \omega_1$)

owing to the dipolar interaction. Thus, a measurement of ω_{TN} allows us to know of the existence of a dipolar interaction between two spins a and b.

There are two types of observation of Rabi oscillations, namely, direct and indirect observations. In the former type, one can observe Rabi oscillations by detecting the transverse magnetization [121]. This was performed by Torrey [122] for NMR. In the latter type, Rabi oscillations can be observed in the free induction decay after the oscillating pulse is switched off, and also in the electron spin echo signal.

The technique has been applied to elucidate the recombination processes and the nature of recombination centers in a-Si:H by Boehme *et al.* [123] and Herring *et al.* [124], using the detection of ODMR. PL excited by light from an Ar^+ ion laser at 514 nm was observed in a magnetic field corresponding to the peak of the ODMR signal ($g = 2.008$) after a microwave pulse of width 350 ns was applied, as shown in Figure 5.57(a), where the relative change in the PL intensity at resonance for $g = 2.008$, that is, $\Delta I_{PL}/I_{PL}$, is shown. The photon number

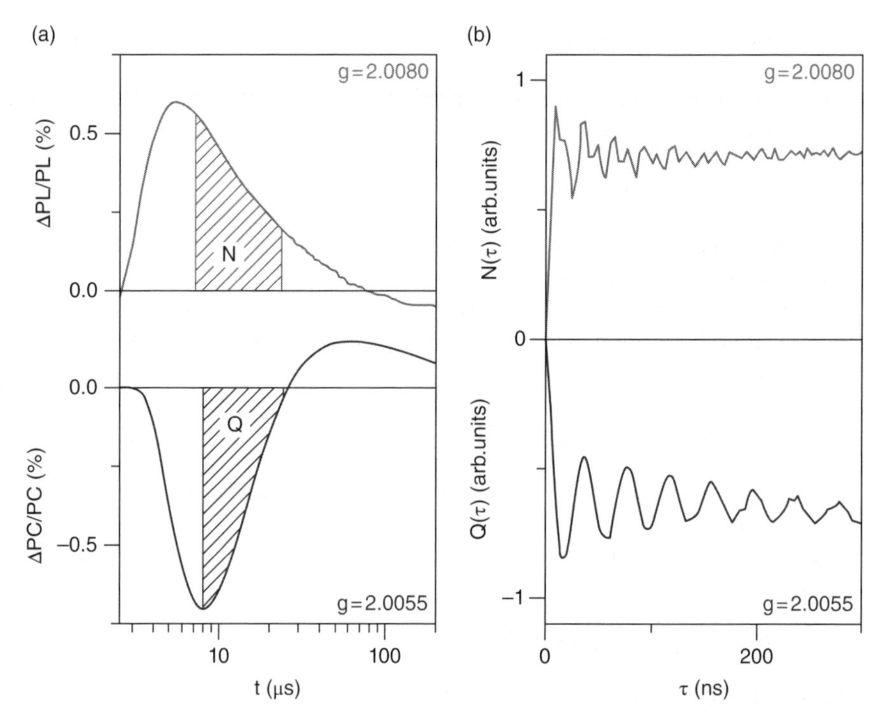

Figure 5.57 (a) Plot of the changes in PL and photoconductivity (PC) as functions of time after a short microwave pulse during ESR with $g = 2.008$ and $g = 2.0055$, respectively. (b) Plot of the photon number $N(\tau)$ and the charge $Q(\tau)$ obtained by integration of the PL and PC, respectively (see shaded areas in (a)), as functions of the pulse length τ. *Source:* Herring *et al.* 2009 [124]. Reproduced with permission from the American Physical Society.

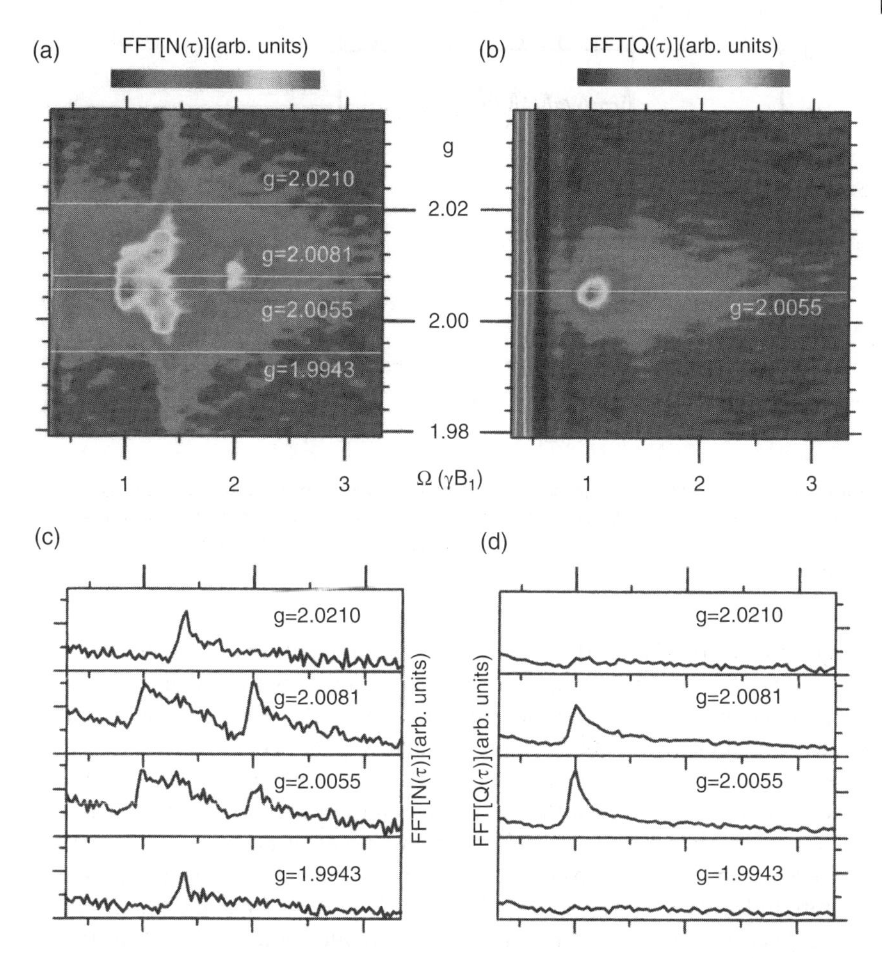

Figure 5.58 Three-dimensional plot of the Fourier transform of the Rabi nutation signal as a function of the *g*-factor measured with (a) pulsed ODMR (pODMR) and (b) pulsed EDMR (pEDMR) under identical conditions. Plots for four selected *g*-factors measured with (c) pODMR and (d) pEDMR under the same conditions as in (a) and (b). The pODMR measurements reveal a number of recombination signals which are not detectable with pEDMR. Only one signal ($g = 2.0055$, $\Omega = \gamma_{B1}$) is visible with both pEDMR and pODMR. *Source:* Herring *et al.* 2009 [124]. Reproduced with permission from the American Physical Society.

corresponding to the area shown by N in Figure 5.57(a) is plotted as a function of the microwave pulse width τ in Figure 5.57(b), where the microwave pulse width changes in steps of 2 ns. The signal shown in Figure 5.57(b) represents the Rabi oscillation. The TN spectrum is obtained from the Fourier transform of the signal shown in Figure 5.57(b), as shown in Figure 5.58(a), as a three-dimensional diagram. In this figure, the TN frequency and the g-value corresponding to the

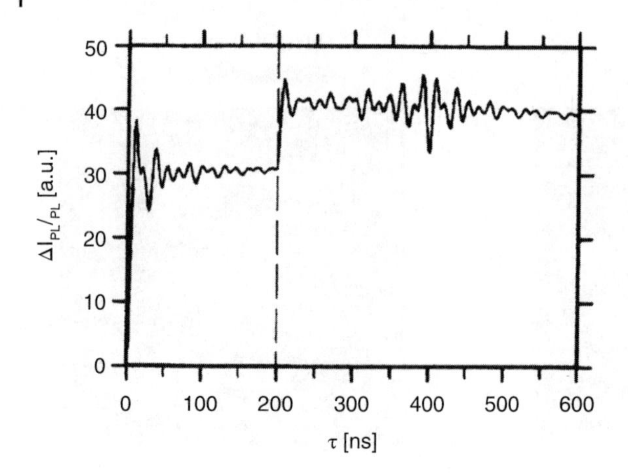

Figure 5.59 PL-detected recombination echo in a-Si:H with a phase change at 200 ns. One can recognize a step-like dephasing after the phase change and an echo at $\tau_{180} = 400$ ns. Note that many different Rabi frequency components determine the signal. *Source:* Boehme *et al.* 2005 [123]. Reproduced with permission of Elsevier.

magnetic field are taken as the abscissa (i.e., the angular frequency is normalized by $\omega_1 = \Omega = \gamma B_1$) and the ordinate, respectively. As shown in Figures 5.57 and 5.58, the relative change in photocurrent (PC), ΔPC/PC, was also observed, and was used to obtain a TN spectrum in a similar way to ΔPL.

Next, a measurement at $g = 2.005$ similar to that shown in Figure 5.57(b) was performed, but the phase of the microwave pulse was changed by 180° above $\tau = 200$ ns, and then a transient echo signal was seen at $\tau = 400$ ns, as shown in Figure 5.59. The Fourier transform of the signal is shown as a three-dimensional diagram in Figure 5.60. Figure 5.61 shows the Fourier transform of the $\Delta I_{PL}/I_{PL}$ versus g-value diagram for three cases of TN frequencies, namely, (1) a strong exchange interaction, $\omega_{TN} = 2\omega_1 = 36$ MHz; (2) a strong dipolar interaction, $\omega_{TN} = 25.5$ MHz; and (3) a weak dipolar interaction, $\omega_{TN} = \omega_1 = 18$ MHz. That is, it shows a cross section of the three-dimensional diagram shown in Figure 5.60. From Equation (5.134) we obtain $\omega_{TN} = 2\omega_1 = 25.5$ MHz in the case of $3D_d/4\hbar \approx \gamma B_1$. This value coincides with ω_{TN} in case (2) above. Boehme *et al.* interpreted these results in terms of recombination between tail electrons with $g_e = 2.004$ and tail holes with $g_h = 2.012$, that is, cases (1) and (2) correspond to the exchange between a pair of tail electrons and tail holes and to distant pairs, respectively, because they thought that the peak of the enhancing signal shown in the top spectrum in Figure 5.61 corresponded to $g = (g_e + g_h)/2 = (2.004 + 2.012)/2 = 2.008$ and that the middle spectrum shown in Figure 5.61 involved two enhancing signals at $g = 2.004$ and 2.012.

In the following, these results are interpreted in terms of our model. Our model of recombination centers is described in Section 5.3. Our ODMR

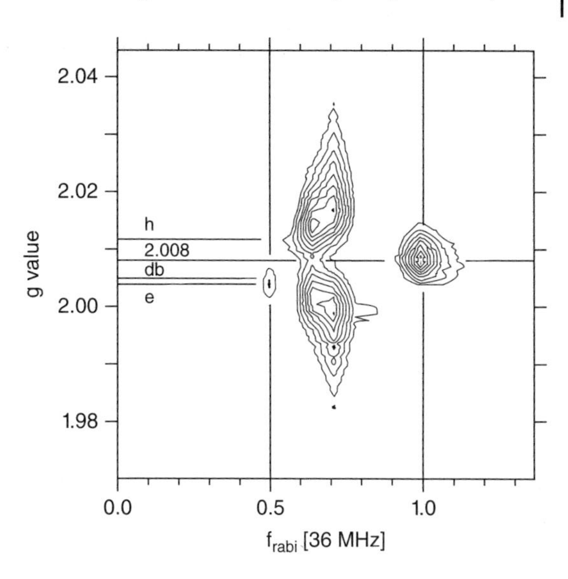

Figure 5.60 Three-dimensional contour plot of TN signal as g versus f_{rabi} ($= \omega_{TN}$). *Source:* Lips *et al.* 2005 [125]. Reproduced with permission.

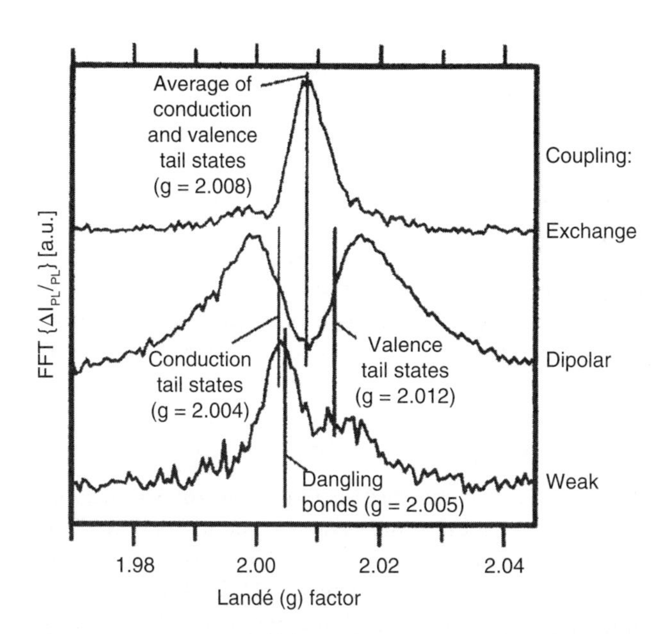

Figure 5.61 Magnetic field dependence (expressed in g factors) of the Rabi frequency components corresponding to weakly coupled, dipolar coupled, and exchange-coupled spin pairs. The graphs are shifted vertically by an arbitrary offset. The dangling bond state participates only in electronic transitions between weakly coupled states. *Source:* Boehme *et al.* 2005 [123]. Reproduced with permission of Elsevier.

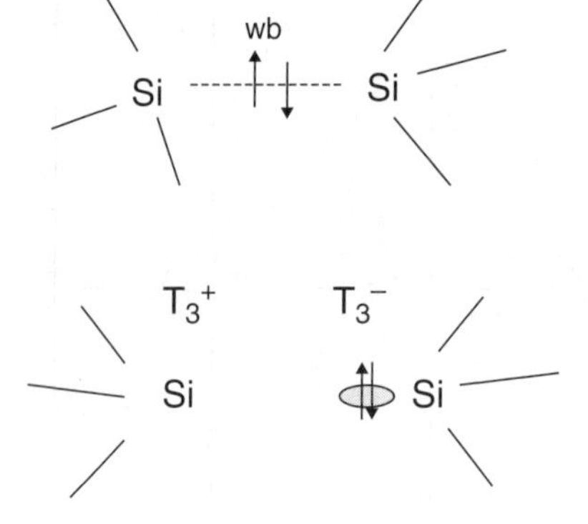

Figure 5.62 Top: before breaking of a weak Si–Si bond (wb). Bottom: after breaking of a weak Si–Si bond. Two centers, T^{3+} and T^{3-}, are formed.

measurements showed that the two signals at $g = 2.004$ and 2.012 were quenching signals, as mentioned before. The middle spectrum in Figure 5.61 shows a dip containing two signals at $g = 2.004$ and 2.012. Furthermore, the dangling bond ODMR signal contributes to this dip as a quenching signal. The two signals originate from T_3^+ + e and T_3^- + h, where T_3^+, T_3^-, e, and h designate positively and negatively charged threefold silicon centers, an electron, and a hole, respectively. Our recombination process is described in Section 5.3. A close T_3^+–T_3^- pair has a strong dipolar interaction, as shown in Figure 5.62. Such pairs contribute to the middle spectrum in Figure 5.61. The enhancing signal at $g = 2.008$ shown in the top spectrum in Figure 5.61 may also possibly be due to triplet excitons, which have been observed at $g = 4$ and 2 in a-Si:H by ODMR measurements. The observation of Rabi oscillations through ODMR at $g = 4$ may allow us to identify the triplet exciton in a-Si:H.

As mentioned above, the Rabi frequency ω_{TN} takes some value relative to the Larmor frequency ω_1 between 1 and 2, and thus the observation of Rabi oscillations is useful for identifying the existence of a dipolar interaction between two carrier spins, that is, two carriers involved in the recombination process.

5.6.3 Spin-Dependent Properties

The electronic properties of semiconductors are related to electron spins: spin-dependent recombination is a typical example, as mentioned in Section 5.3.5. Spin-dependent recombination provides a highly sensitive detection method for ESR and ENDOR signals. This was applied to a-Si:H for the first time by Biegelsen *et al.* [126] and Morigaki *et al.* [127] independently, by monitoring the intensity of PL. In Section 5.3.5, spin-dependent photoconductivity was treated. This was applied to a-Si:H for the first time by Solomon *et al.* [128]. Thus, in this section, we consider other examples, namely, spin-dependent transport and spin-dependent photoinduced absorption.

5.6.3.1 Spin-Dependent Transport

Here, we are concerned with the spin-dependent dark conductivity. This was studied in doped n-type crystalline InSb for the first time by Guéron and Solomon [129]. In the case of doped n-type crystalline silicon (c-Si) and crystalline germanium (c-Ge), spin-dependent conductivity measurements were performed by Toyoda and Hayashi [130] and by Toyotomi and Morigaki [131] independently. We consider two electron systems in n-type c-Si and c-Ge, namely, conduction electrons and neutral donors. When a donor spin is reversed during an ESR transition, the observed ESR energy of the donor spin is transferred to the conduction electrons, so that their mobility increases and, as a result, the conductivity increases, as observed in n-type c-Si and c-Ge [132, 133]. This technique has also been applied to a-Si, in which the hopping conductivity is spin-dependent, by Kishimoto *et al.* [134]. In the case of hopping conductivity, hopping conduction is enhanced during ESR.

5.6.3.2 Spin-Dependent Photoinduced Absorption

The energy levels and optical transitions involved in PA in a-Si:H are shown in Figure 5.63, where the levels of self trapped holes (A centers), neutral Si dangling bonds, and negatively charged Si dangling bonds are shown. When the trapped carriers have spins, their ESR affects their recombination and relaxation processes, so that the PA intensity changes at their ESR when the PA intensity is monitored, since their recombination rate is enhanced at their ESR. For example, in Figure 5.63, when a self-trapped hole at an A center is excited into the valence band during PA, the PA intensity decreases at ESR. This is called PA-detected ESR (PADESR) [135]; the transmitted light intensity is increased, as shown in Figure 5.64, in which, for comparison, the ODMR

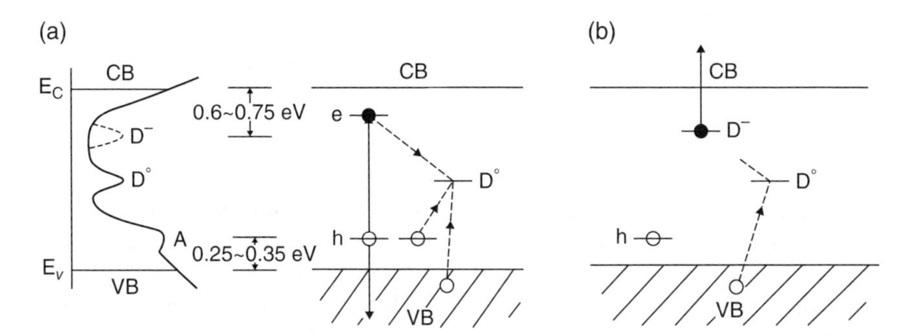

Figure 5.63 Schematic diagram of PA processes in a-Si:H: (a) hole excitation from an A center into the valence band and (b) electron excitation from D^- states into the conduction band. The solid and dashed lines show the radiative and nonradiative recombination channels, respectively. The energy positions of the A and D^- states were obtained from ODMR and PA line shape analyses. *Source:* Hirabayashi and Morigaki 1983 [135]. Reproduced with permission of Elsevier.

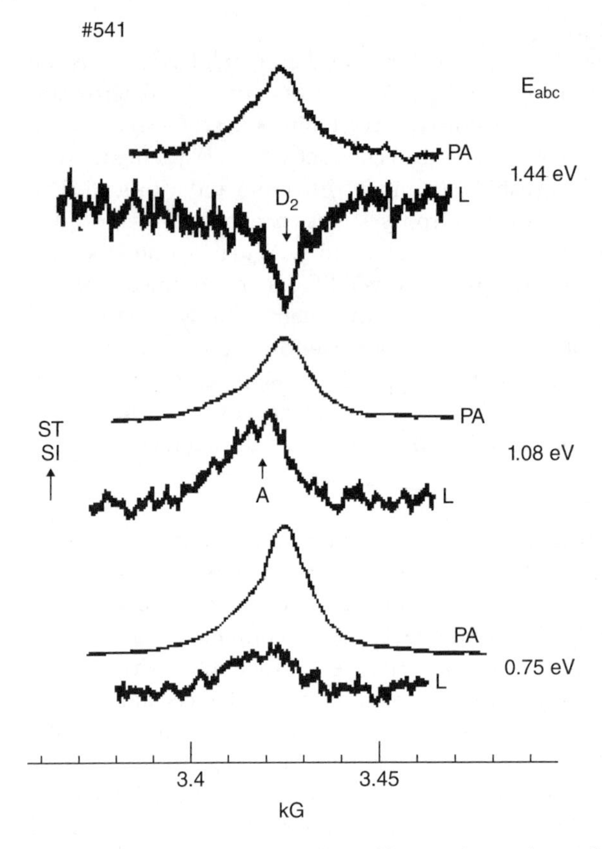

Figure 5.64 PADESR spectra (indicated by PA) observed at 2 K by monitoring the intensity of transmitted probe light with photon energies (E_{obs}) of 1.44, 1.08, and 0.75 eV for a-Si:H sample No. 541. The optical excitation by argon ion laser light was about 3 W/cm^2 at 514.5 nm. ODMR spectra (indicated by L) are also shown for comparison with the luminescence energy under the same excitation conditions as for PADESR. *Source:* Hirabayashi and Morigaki 1983 [135]. Reproduced with permission of Elsevier.

spectrum is also shown, where the PL intensity is monitored. The ESR spectrum is due to dangling bond electrons and self-trapped holes (A centers). The PADESR spectrum can be deconvoluted into spectra of the dangling bond electrons and the self-trapped holes. The D^- level is spin-independent, because the D^- center is the spin-singlet state, so that the PA due to the D^- level is spin-independent. This is shown in Figure 5.65, which shows the spectral dependences of the PA and PADESR signals. These are drawn as a Lucovsky plot [137], that is, $(\Delta T/T)^{2/3}(\hbar\omega)^2$ versus $\hbar\omega$, where ΔT is the change in the transmitted light intensity associated with ESR and $\hbar\omega$ is the photon energy. Such spectral dependences are useful for distinguishing the two contributions

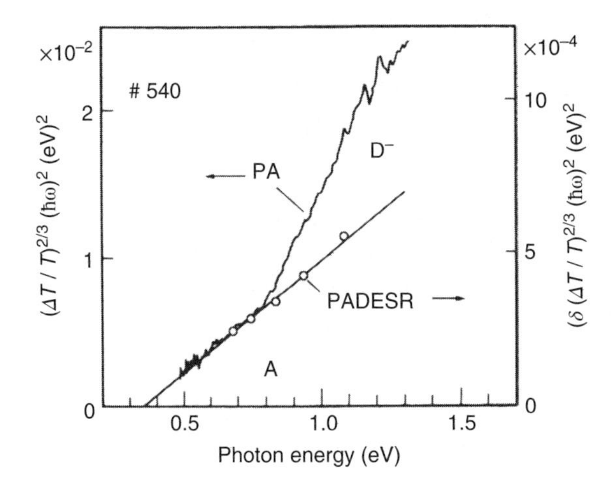

Figure 5.65 Lucovsky plots of the PA spectrum and the spectral dependence of the PADESR signal for a-Si:H. The arrow against the PA spectrum indicates the threshold energy for the bleaching effect in PA (see text). *Source:* Hirabayashi and Morigaki 1986 [136]. Reproduced with permission from Taylor & Francis Ltd.

to PA spectra, by monitoring the PA intensity for ESR detection, as seen from Figure 5.65. PADESR is a powerful detection tool for the ESR signal when the PA is rather strong but the ESR signal is weak for normal detection. Schultz *et al.* [138] have observed ESR spectra of triplet excitons in a-Si:H using the PADESR technique.

5.7 Light-Induced Phenomena and Light-Induced Defect Creation

5.7.1 Introduction

After prolonged illumination, the dark conductivity and photoconductivity of a-Si:H decrease compared with their values before illumination [139]. These light-induced effects are called the Staebler–Wronski effect. Staebler and Wronski [140] suggested that this may be due to light-induced creation of Si dangling bonds. This suggestion has been confirmed by ESR measurements by Hirabayashi *et al.* [141] and Dersch *et al.* [142]. In the following, we survey light-induced phenomena in a-Si:H, and then consider the mechanism of light-induced creation of Si dangling bonds and specific properties of light-induced dangling bonds, particularly the correlation between hydrogen atoms and dangling bonds, and specific sites of dangling bonds.

5.7.2 Light-Induced Phenomena

5.7.2.1 Light-Induced Effects in Conductivity and Photoconductivity

Prolonged illumination causes the dark conductivity to decrease, as shown in Figure 5.66. The photoconductivity also decreases during illumination, as also shown in Figure 5.66. This change in the dark conductivity is due to a shift of the Fermi level towards the valence band edge as a result of light-induced creation of Si dangling bonds [139, 140]. The decrease in the photoconductivity is due to a decrease in the lifetime of carriers caused by recombination at light-induced defects, that is, Si dangling bonds.

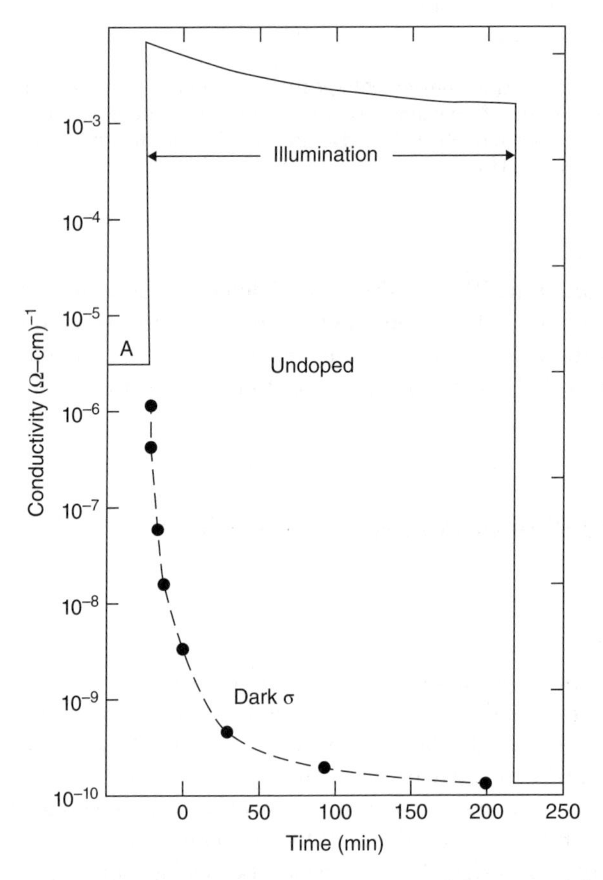

Figure 5.66 Decrease in photoconductivity (solid curve) and dark conductivity during illumination with 200 mW/cm^{-2} of filtered tungsten light (600–900 nm) for undoped a-Si:H. *Source:* Staebler and Wronski 1980 [140]. Reproduced with permission from AIP Publishing LLC.

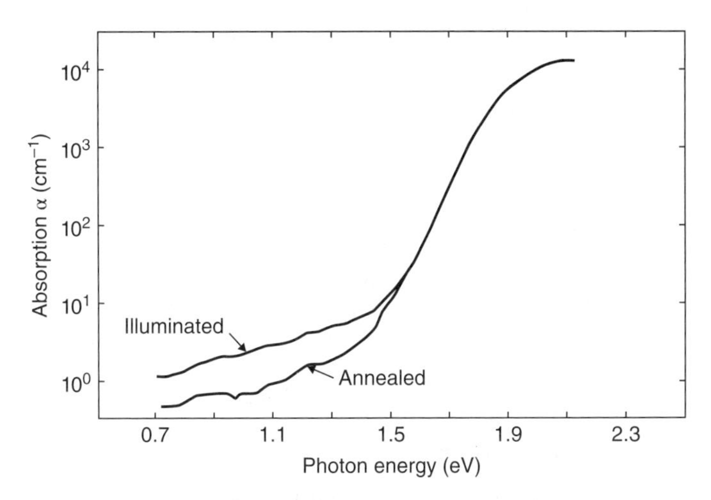

Figure 5.67 Effect of illumination on the absorption spectrum of undoped a-Si:H. *Source:* Amer *et al.* 1983 [143]. Reproduced with permission of Elsevier.

5.7.2.2 Optical Absorption

The low-energy absorption band due to optical transitions from the neutral Si dangling bond level to the conduction band and from the valence band to the neutral Si dangling bonds in a-Si:H is increased by prolonged illumination [143], as shown in Figure 5.67. Optical transitions involving ionized Si dangling bonds have also been observed in a-Si:H. These measurements were performed using more sensitive methods such as photothermal deflection spectroscopy [144] and the constant photocurrent method [145].

5.7.2.3 Photoluminescence

The principal PL in a-Si:H is decreased in intensity by prolonged illumination [146, 147]; that is, so-called PL fatigue occurs, like that in chalcogenide glasses [148]. This is due to the enhancement of nonradiative recombination between electrons and holes associated with light-induced creation of Si dangling bonds, as shown in Figure 5.68. This fatigue effect is recovered by annealing of the sample at higher temperature, as also shown in Figure 5.68. On the other hand, the defect PL is increased by prolonged illumination; this is associated with light-induced creation of radiative defects [147]. Detailed investigations of the defect PL have been performed by Ogihara *et al.* [149–151]. PL fatigue in a-Si:H has been examined for low-temperature illumination in more detail [152] and compared with the case of illumination at room temperature [153].

5.7.2.4 ESR, ODMR

The light-induced creation of Si dangling bonds in a-Si:H was first confirmed by ESR measurements, as mentioned in Section 5.7.1. The variation of the ESR

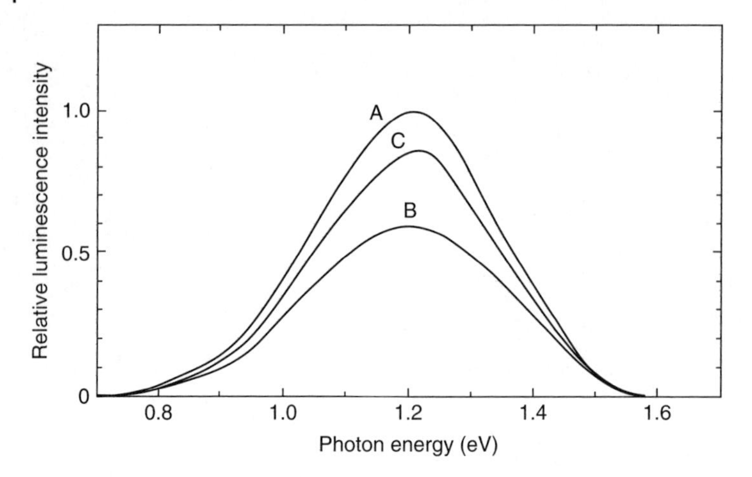

Figure 5.68 Luminescence spectra of a-Si:H sample No. 49 at 4.2 K under optical excitation with $P_{EX} = 9$ mW: (A) initial spectrum before fatigue occurs; (B) after prolonged illumination with 514.5 nm light at 540 mW; (c) after annealing the sample at 270 K. *Source:* Morigaki *et al.* 1980 [146]. Reproduced with permission of Elsevier.

signal with illumination time due to Si dangling bonds exhibits a nonexponential growth, that is, stretched exponential relaxation, as will be mentioned again in Section 5.7.3.

ODMR measurements at low temperature provide evidence for light-induced creation of Si dangling bonds at low temperature [154], as shown in Figure 5.69. Furthermore, ODENDOR measurements have shown the self-trapping of holes at a weak Si–Si bond adjacent to an Si–H bond at low temperature. A detailed description is given in Section 5.2.2.

5.7.3 Light-Induced Defect Creation

5.7.3.1 Mechanism of Light-Induced Defect Creation in a-Si:H

The following several models have been proposed for light-induced creation of Si dangling bonds in a-Si:H: (1) the weak Si–Si bond-breaking model, (2) the hydrogen-related bond-breaking (Si–H, Si–H–Si) model, (3) the hydrogen-mediated model, (4) the charged-center-trapping model, and (5) the impurity-involved model. These models have been overviewed by Morigaki *et al.* [13].

Here, we present the first model in more detail. Hirabayashi *et al.* [141] proposed a model of breaking of a weak Si–Si bond adjacent to an Si–H bond to create an Si dangling bond in a-Si:H. Furthermore, Dersch *et al.* [142] suggested that Si–H bond switching and a continuous exchange of hydrogen atoms stabilized the dangling bond configuration. The kinetics of light-induced creation of Si dangling bonds has been considered in detail by Stutzmann *et al.* [155].

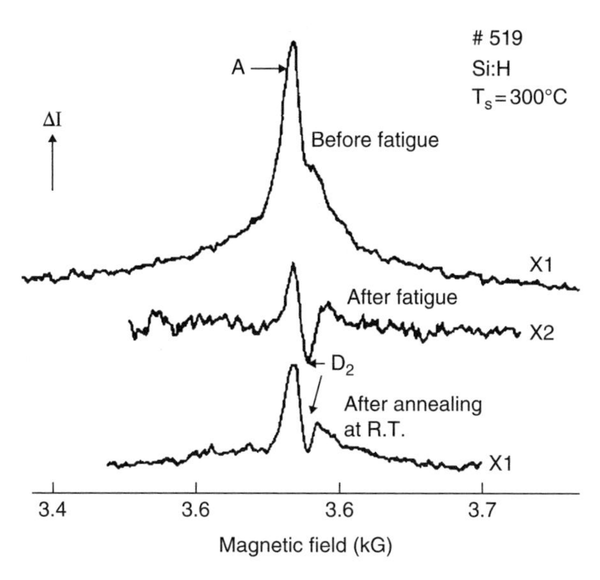

Figure 5.69 ODMR spectra taken at 2 K before and after fatigue (light soaking) and also after annealing at room temperature for an a-Si:H sample prepared at 300 °C, where the intensity of emitted light at a photon energy of 1.16 eV was monitored. *Source:* Morigaki *et al.* 1982 [154]. Reproduced with permission from the Physical Society of Japan.

In the following, we consider light-induced defect creation on the basis of our model [156]. The processes involved in light-induced defect creation under illumination in a-Si:H consist of the following. A hole is self-trapped on a specific weak Si–Si bond, that is, a weak Si–Si bond adjacent to an Si–H bond (Figure 5.70(a)), and then recombines with an electron mostly nonradiatively (Figure 5.70(b)), and eventually the weak bond is broken. Using the recombination energy associated with nonradiative recombination between the electron and the hole, the Si–H bond is switched towards the weak Si–Si bond (Figure 5.70(c)). Following switching of the Si–H bond (Figure 5.70(c)), movement of hydrogen due to hopping and/or tunneling (Figure 5.70(d)) and then breaking of the weak Si–Si bond occur, and eventually two separate dangling bonds (Figure 5.70(d)), that is, a normal dangling bond and a hydrogen-related dangling bond (a dangling bond having hydrogen at a nearby site), are created under illumination. Furthermore, hydrogen is dissociated from an Si–H bond located near a hydrogen-related dangling bond as a result of nonradiative recombination between an electron and a hole at the hydrogen-related dangling bond. A dissociated (or metastable) hydrogen atom is inserted into a nearby weak Si–Si bond to form a hydrogen-related dangling bond, or terminates either a normal dangling bond or a hydrogen-related dangling bond, or collides with another hydrogen atom to form a hydrogen molecule.

(a)

H
Si wb Si
h

(b)

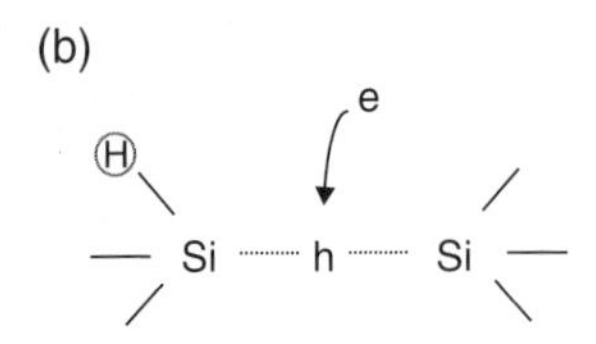

(c)

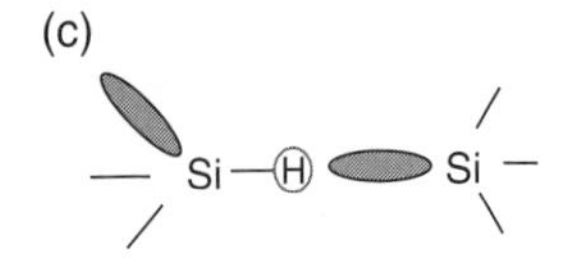

(d)

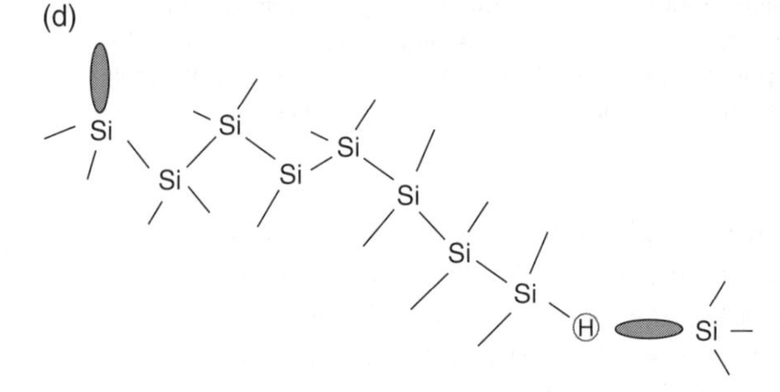

Figure 5.70 Atomic configurations involved in the formation of two types of dangling bonds, namely, normal dangling bonds and hydrogen-related dangling bonds, under illumination: (a) self-trapping of a hole on a weak Si–Si bond (wb) adjacent to an Si–H bond; (b) electron–hole recombination at a weak Si–Si bond; (c) switching of the Si–H bond towards the weak Si–Si bond, leaving a dangling bond behind; (d) formation of two separate dangling bonds through hydrogen movement after repeating the processes shown in (a)–(c).

The light-induced defect creation can be considered in terms of a rate equation model, taking into account the processes mentioned above. The rate equations are as follows:

$$dN_a / dt = C_d np - C_2 N_m N_a \tag{5.135}$$

$$dN_b / dt = C_d np - C_1 npN_b + C_3 N_m N_{Si} - C_4 N_m N_b \tag{5.136}$$

$$dN_m/dt = C_1 npN_b - C_2 N_m N_a - C_3 N_m N_{Si} - C_4 N_m N_b - C_5 N_m{}^2 \quad (5.137)$$

where N_a, N_b, N_m, N_{Si}, n, and p are the densities of normal dangling bonds, hydrogen-related dangling bonds, metastable hydrogen atoms, Si–Si bonds, free electrons, and free holes, including band tail electrons and band tail holes, respectively, and C_d, C_1, C_2, C_3, C_4, and C_5 are reaction coefficients for the following processes: C_d, for the light-induced creation of two separate dangling bonds; C_1, for the dissociation of a hydrogen atom from an Si–H bond located near a hydrogen-related dangling bond; C_2, for the termination of a normal dangling bond by a metastable hydrogen atom; C_3, for the insertion of a metastable hydrogen atom into an Si–Si bond; C_4, for the termination of a hydrogen-related dangling bond by a metastable hydrogen atom; and C_5, for the formation of a hydrogen molecule by two metastable hydrogen atoms. In the case of high-quality a-Si:H, the C_3 term can be neglected, because the density of weak Si–Si bonds is relatively small compared with low-quality a-Si:H, which contains a large amount of hydrogen. In our numerical calculations, the C_5 term was also neglected for simplicity. We consider the case of relatively weak illumination such as continuous illumination, so that the density of weak Si–Si bonds under illumination is assumed not to be changed. We have also considered this issue for strong illumination such as intense pulsed illumination [158].

Since the lifetimes of the carriers are short compared with the timescale of light-induced creation of dangling bonds, the carrier densities n and p are approximately given by their steady-state values,

$$n = p \cong G/(aN_d) \quad (5.138)$$

where it is assumed that n and p are determined by trapping of carriers (electrons) at neutral dangling bonds followed by rapid recombination with holes, and G, α, and N_d are the generation rate of free carriers, the trapping coefficient of free carriers by neutral dangling bonds, and the total density of neutral dangling bonds, respectively. The value of α was assumed to be $1 \times 10^{-8}\,\text{cm}^3\,\text{s}^{-1}$ in the following calculations. The rate equations, Equations (5.135)–(5.137), were solved numerically, using normalized densities r, q, and s defined by

$$r = N_a/N_{d0} \quad (5.139)$$

$$q = N_b/N_{d0} \quad (5.140)$$

$$rs = N_s/N_{d0} \quad (5.141)$$

The details of the numerical calculations are described in [156]. The results for two examples for a-Si:H are shown in Figures 5.71(a) and (b). The values of the parameters used in the calculations are listed in Table 5.2. Figures 5.71(a) and (b) show the calculated curves for r, q, and s as a function of illumination time in the cases of a-Si:H sample No. 601211 with $G = 1.9 \times 10^{22}\,\text{cm}^{-3}\,\text{s}^{-1}$ and

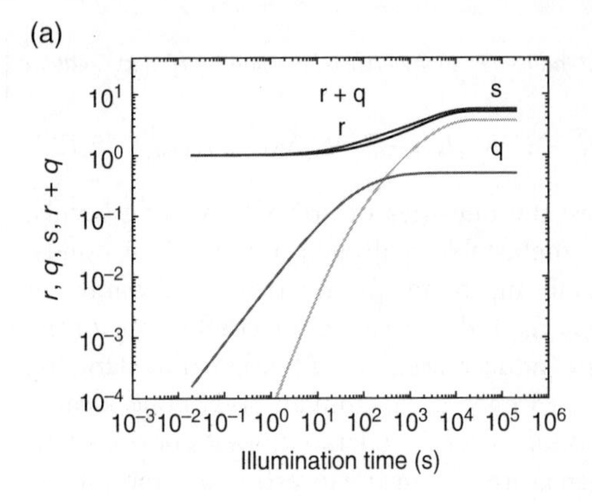

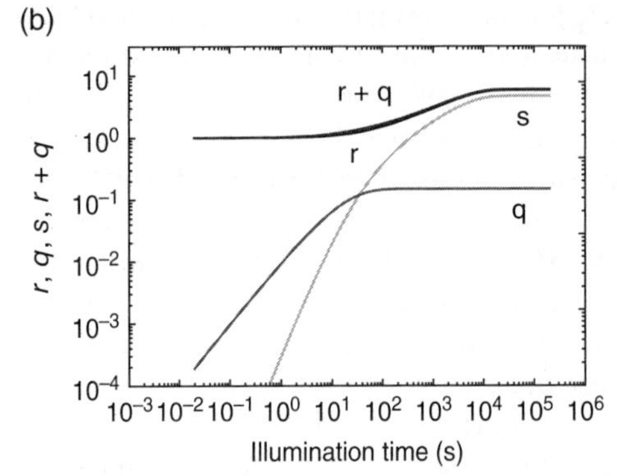

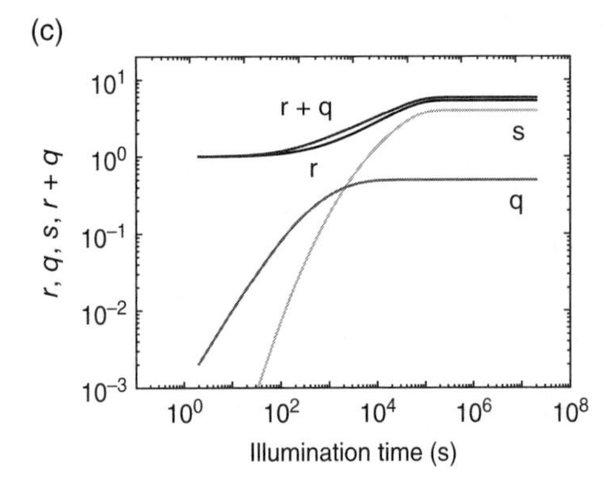

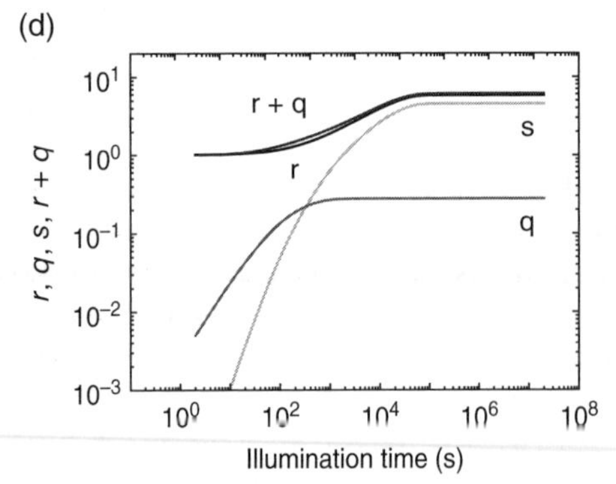

Figure 5.71 Densities of total dangling bonds, $r+q$, normal dangling bonds, r, hydrogen-related dangling bonds, q, and metastable hydrogen atoms, s, relative to N_{d0} as a function of illumination time under illumination with G.
(a) a-Si:H ($T_s = 250\,°C$), No. 601211, $N_{d0} = 1.0 \times 10^{16}\,cm^{-3}$, $G = 1.9 \times 10^{22}\,cm^{-3}\,s^{-1}$;
(b) a-Si:H ($T_s = 250\,°C$), No. 15y17, $N_{d0} = 1.0 \times 10^{16}\,cm^{-3}$, $G = 4.0 \times 10^{22}\,cm^{-3}\,s^{-1}$;
(c) pm-Si:H ($T_s = 250\,°C$), No. 810091, $N_{d0} = 1.0 \times 10^{16}\,cm^{-3}$, $G = 7.2 \times 10^{21}\,cm^{-3}\,s^{-1}$;
(d) pm-Si:H ($T_s = 200\,°C$), No. 601201, $N_{d0} = 1.8 \times 10^{16}\,cm^{-3}$, $G = 2.7 \times 10^{22}\,cm^{-3}\,s^{-1}$. See text for the values of the parameters. T_s, substrate temperature.

Table 5.2 Values of parameters used in the calculations in Section 5.7.3.1.

Case	C_d (cm^3 s^{-1})	C_1 (cm^6 s^{-1})	C_2 (cm^3 s^{-1})	C_3 (cm^3 s^{-1})	C_4 (cm^3 s^{-1})	N_{SS} (cm^{-3}) ($G \rightarrow 0$)
I	2×10^{-15}	4×10^{-31}	1.28×10^{-21}	0	0	2.0×10^{16}
II	6×10^{-15}	4×10^{-31}	1.00×10^{-21}	0	0	1.3×10^{16}
III	2×10^{-15}	4×10^{-31}	1.43×10^{-22}	0	0	2.0×10^{16}
IV	2×10^{-15}	4×10^{-31}	1.43×10^{-22}	0	0	2.8×10^{16}
V	6×10^{-15}	4×10^{-31}	1.28×10^{-21}	0	0	4.8×10^{16}

a-Si:H sample No. 15y17 with $G = 4.0 \times 10^{22}$ cm^{-3} s^{-1}. In Figures 5.71(c) and (d), the calculated curves correspond to cases of hydrogenated polymorphous silicon (pm-Si:H), as mentioned below.

As shown in Figure 5.72, the $r+q$ versus t curves can be fitted well by stretched exponential functions with a dispersive parameter β and a characteristic time τ, as given in the following equation [159, 160]:

$$N_d(t) = N_{SS} - \{N_{SS} - N_d(0)\}\exp\{-(t/\tau)^{\beta}\} \tag{5.142}$$

where N_{SS} is the saturated (steady-state) dangling bond density. The values of β, τ, and N_{SS} for a-Si:H are listed in Table 5.3, in which the errors in β and τ are those associated with the fitting. The values of β and τ are also shown as a function of N_{SS} in Figures 5.73 and 5.74, respectively, as well as values for pm-Si:H. In these figures, the values of β and τ, both calculated and experimental, are also included for pm-Si:H consisting of a relaxed amorphous network containing a small fraction of small crystallites 2–4 nm in size (for details, see [161, 162]). In this material, Si dangling bonds are created in the amorphous network under illumination. We note that the parameters β and τ characterize the features of the growth curve of N_d versus t in the following way: β mainly determines the growth curve in the initial stage, that is, when β is small, N_d initially grows rapidly, whereas τ determines the characteristics of the growth curve in the long term, that is, when τ is large, N_d grows slowly in the long term. In our model, when G is large, β and τ become small, that is, N_d increases quickly during the initial stage of illumination and reaches the steady-state value N_{SS} in a short time.

For each illumination condition, the dangling bond density was measured as a function of illumination time and the observed dangling bond density was fitted to the stretched exponential function given in Equation (5.142) as a function of illumination time. As examples, Figure 5.75 shows the results for a-Si:H samples Nos. 009151 and 601211, prepared at 250 °C. The values of β and τ obtained from the fitting are shown as functions of N_{SS} in Figures 5.73 and 5.74

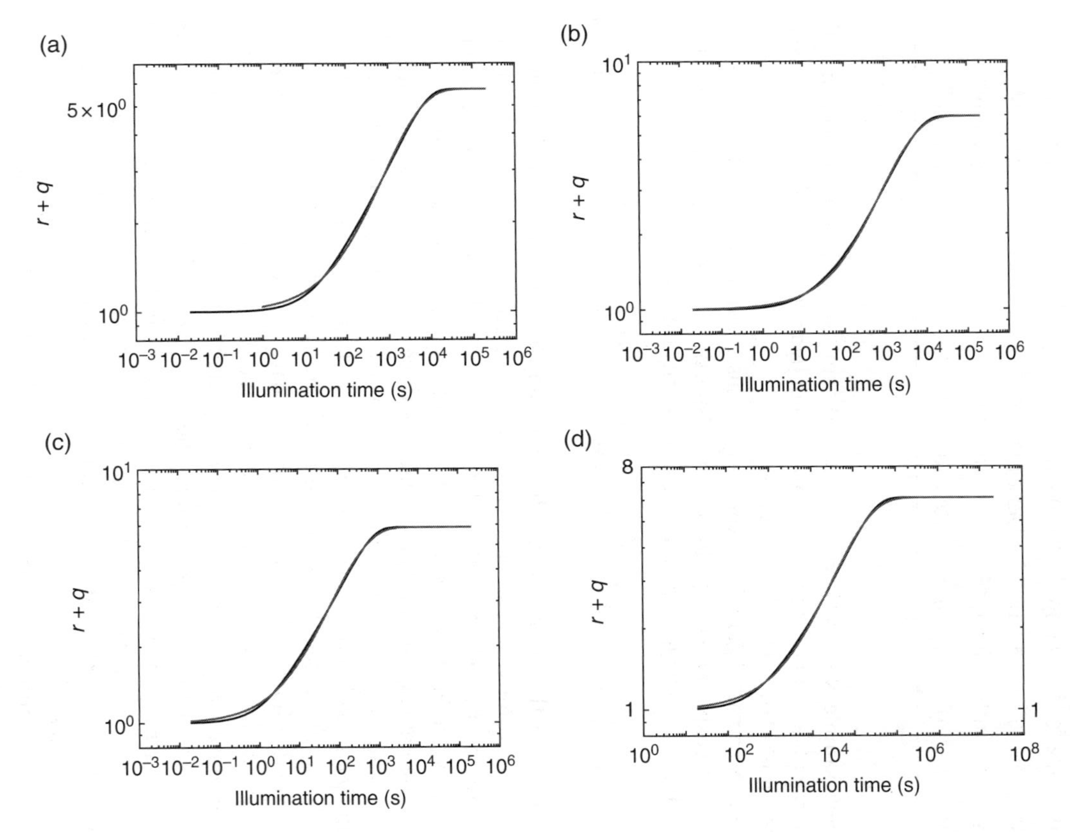

Figure 5.72 Fitting of the calculated curves (thick lines) of $r+q$ versus illumination time shown in Figure 5.72(a)–(d) by the stretched exponential function (thin lines) given in Equation (5.142).

Table 5.3 Estimated values of β and τ for two a-Si:H samples.

Sample	G ($cm^{-3}s^{-1}$)	N_{SS} ($10^{16}cm^{-3}$)	β	τ (s)
a-Si:H No. 601211	1.9×10^{22}	5.7	0.610×0.002	$(2.34 \pm 0.01) \times 10^3$
a-Si:H No. 15y17	4.0×10^{22}	6.0	0.635×0.002	$(2.47 \pm 0.01) \times 10^3$

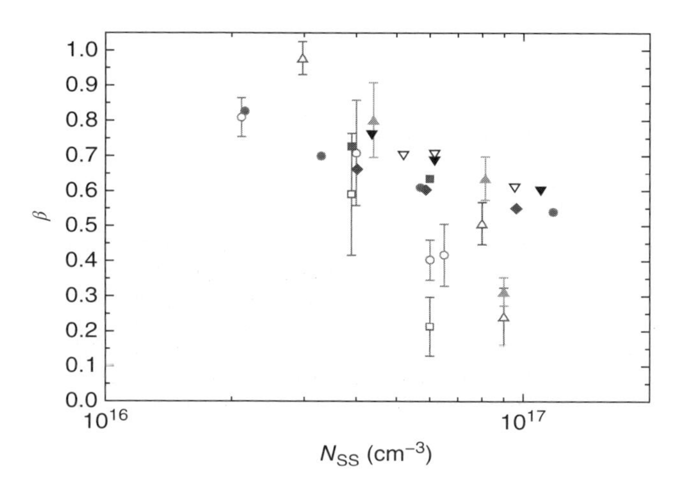

Figure 5.73 Plots of the experimental and calculated values of β as a function of saturated dangling bond density N_{SS} in pm-Si:H and a-Si:H. Experimental results: open triangles, pm-Si:H ($T_s = 250\,°C$); filled triangles, pm-Si:H ($T_s = 200\,°C$); open circles, a-Si:H ($T_s = 250\,°C$, Ecole Polytechnique); open squares, a-Si:H ($T_s = 250\,°C$, Yamaguchi University). Calculated results: filled diamonds, case III for pm-Si:H ($T_s = 250\,°C$); filled inverted triangles, case IV for pm-Si:H ($T_s = 200\,°C$); open inverted triangles, case V for pm-Si:H ($T_s = 200\,°C$); filled circles, case I for a-Si:H ($T_s = 250\,°C$, Ecole Polytechnique); filled squares, case II for a-Si:H ($T_s = 250\,°C$, Yamaguchi University).

along with the values obtained from calculation, as shown above. The value of N_{SS} depends on the illumination intensity, that is, on the generation rate of free carriers, as shown in Figure 5.76, in which the results obtained for a-Si:H are shown along with the calculated curves. The calculated curves were obtained from Equation (24) of [156] for the case of monomolecular recombination. The values of the parameters used for the calculated curves I and II in Figure 5.76 are those for cases I and II, respectively, in Table 5.2.

From Figure 5.71, it is found that the density of hydrogen-related dangling bonds is much lower than that of normal dangling bonds, similarly to the high-quality a-Si:H sample.

In the following, we give some details of how the calculation was done so as to fit the experimental results. First, we attempted to determine the value of C_d by fitting the experimental points in Figure 5.76 to the calculated curve of N_{SS} versus generation rate. The expected value of C_d for high-quality a-Si:H samples

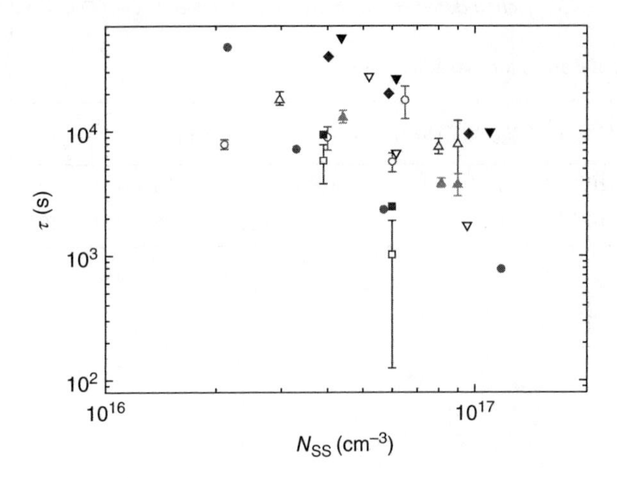

Figure 5.74 Plots of the experimental and calculated values of τ as a function of saturated dangling bond density N_{SS} in pm-Si:H and a-Si:H. Experimental results: open triangles, pm-Si:H (T_s = 250 °C); filled triangles, pm-Si:H (T_s = 200 °C); open circles, a-Si:H (T_s = 250 °C, Ecole Polytechnique); open squares, a-Si:H (T_s = 250 °C, Yamaguchi University). Calculated results: filled diamonds, case III for pm-Si:H (T_s = 250 °C); filled inverted triangles, case IV for pm-Si:H (T_s = 200 °C); open inverted triangles, case V for pm-Si:H (T_s = 200 °C); filled circles, case I for a-Si:H (T_s = 250 °C, Ecole Polytechnique); filled squares, case II for a-Si:H (T_s = 250 °C, Yamaguchi University).

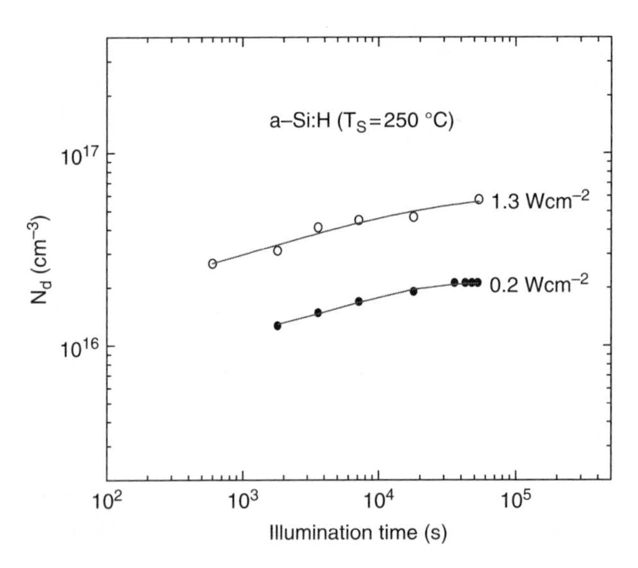

Figure 5.75 Plots of dangling bond density N_d versus illumination time for illumination intensities of 0.2 and 1.3 W cm^{-2} in a-Si:H (T_s = 250 °C) samples Nos. 009151 and 601211. The curves of N_d versus illumination time were fitted by the stretched exponential function given in Equation (5.142) with $\beta = 0.809 \pm 0.055$ and $\tau = (7.84 \pm 0.46) \times 10^3$ s for 0.2 W cm^{-2} and with $\beta = 0.403 \pm 0.057$ and $\tau = (5.69 \pm 1.00) \times 10^3$ s for 1.3 W cm^{-2}. *Source:* Morigaki *et al.* 2007 [157]. Reproduced with permission from Elsevier.

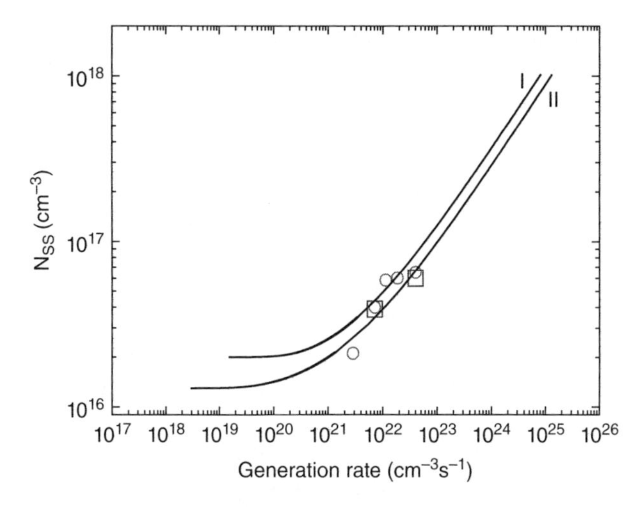

Figure 5.76 Plots of saturated dangling bond density N_{SS} versus generation rate of free carriers. I, curve calculated using the values of parameters for case I in Table 5.2; II, curve calculated using the values of parameters for case II in Table 5.2. Open circles, samples prepared at Ecole Polytechnique; open squares, samples prepared at Yamaguchi University. The values of N_{SS} in the zero-generation-rate limit are 2×10^{16} cm^{-3} for curve I and 1.3×10^{16} cm^{-3} for curve II. The values of N_{SS} have an error of ±25%. For details, see text. *Source:* Morigaki *et al.* 2008 [159]. Reproduced with permission from Elsevier.

was taken from the experimental results obtained by Vignoli *et al.* [163]. The values of the other parameters were also estimated, as described in detail in [156]. The total density given by N_a and N_b was calculated as a function of illumination time, using Equations (5.135)–(5.137) with appropriate values of parameters. Then, the calculated results for β and τ were obtained by fitting the calculated curve to a stretched exponential function, as mentioned before.

Before discussing the results for β and τ, we shall discuss the experimental results for the N_{SS} versus G curves, as well as the calculated results. The calculated curve I for N_{SS} versus G shown in Figure 5.76 was fitted to the experimental points for a-Si:H sample No. 601211, except for an experimental point measured at weak G for a-Si:H sample No. 009151. For this point, we attempted to use the other calculated curve to fit it in [159]. However, overall agreement with curve I was obtained for these samples. For a-Si:H samples Nos. 15y17 and 15y18, curve II was fitted to them, as seen in Figure 5.76.

We now discuss β and τ, which are correlated with the amorphous network. The value of β in particular determines the extent of the distribution function of the lifetime of light-induced dangling bonds. This distribution function is the Fourier transform of a stretched exponential function. A higher value of β indicates a sharply defined lifetime distribution of light-induced dangling bonds with a well-defined lifetime close to the characteristic time τ.

As shown in Figure 5.73, β becomes small with increasing N_{SS}, that is, with increasing G, in a-Si:H. We note that β becomes small under strong illumination. This means that the lifetime distribution of normal dangling bonds becomes broad compared with that under weak illumination or in the dark. This is connected to an increase in the dangling bond density. This is also related to the fact that the distance between normal dangling bonds and hydrogen-related dangling bonds becomes shorter with increasing dangling bond density, that is, the annihilation rate of normal dangling bonds with metastable hydrogen atoms becomes high. In our model, the value of β is determined by competition between the light-induced creation of dangling bonds and the annihilation of dangling bonds by metastable hydrogen atoms. The above tendency of β under illumination is consistent with the calculations. From the above consideration, it can be seen that the light-induced defect creation process is an example of a stretched exponential relaxation process in a disordered system.

Light-induced defect creation has been explored by a Monte Carlo computer simulation in a model system, namely, a simple cubic lattice [164]. The estimated light-induced dangling bond density agrees with the observed value [164].

5.7.3.2 Light-Induced Defects

As mentioned above, two types of light-induced defect are created under illumination in our model, that is, a normal dangling bond and a hydrogen-related dangling bond, whose natures were mentioned in Section 5.2.3. Here, first we treat the correlation between a dangling bond and a hydrogen atom. In a hydrogen-related dangling bond, the distance between the dangling bond site and its nearby hydrogen atom is estimated to be 2.1–2.9 Å [13, 165]. In hydrogen-implanted c-Si, Nielsen *et al.* [21] observed the existence of a vacancy–hydrogen complex from the hyperfine structure of its ESR spectra, in which the distance between the hydrogen atom and the dangling bond site is estimated to be 2.8 Å, assuming 1.5 Å for the Si–H bond length in the diamond structure. This distance lies between 2.1 and 2.9 Å. As mentioned in Section 5.62, we have suggested on the basis of the pulsed ENDOR spectra observed by Fehr *et al.* [118] that there exists a hydrogen atom near to a dangling bond at a distance of 2.2 Å [13]. This distance is consistent with that of a hydrogen-related dangling bond. Furthermore, the results of these pulsed ENDOR measurements are remarkable in that there exist hydrogen atoms closer than 3 Å to a dangling bond, for example, at a distance of ~1.5 Å. This result is inconsistent with the conclusion of Isoya *et al.* [112] and Yamasaki and Isoya [166] that there are no hydrogen atoms within a distance of 3 Å from a dangling bond. That conclusion has been criticized from an experimental point of view [17]. However, some models for light-induced defect creation in a-Si:H have been constructed by using that conclusion as a constraint (e.g., [167]).

From electron spin echo relaxation measurements, Fehr *et al.* [168] concluded that two types of light-induced dangling bonds are created: one type consists of dangling bonds that are isolated and uniformly distributed in the bulk of the a-Si:H sample, and the other type consists of dangling bonds located on the internal surfaces of microvoids. Kondo and Morigaki [6] suggested, on the basis of optically detected ENDOR measurements for various types of a-Si:H samples with different hydrogen contents, that the light-induced creation of dangling bonds was related to microvoids whose internal surfaces were covered by Si–H bond clusters.

References

1 Street, R.A. and Biegelsen, D.K. (1980) Luminescence and ESR studies of defects in hydrogenated amorphous silicon. *Solid State Commun.*, **33**, 1159–1162.

2 Yan, B., Schultz, N.A., Efros, A.L., and Taylor, P.C. (2000) Universal distribution of residual carriers in tetrahedrally coordinated amorphous semiconductors. *Phys. Rev. Lett.*, **84**, 4180–4183.

3 Takenaka, H., Ogihara, C., and Morigaki, K. (1985) Conduction band tail electrons and A centres in a-Si:H as elucidated by time-resolved optically detected magnetic resonance measurements. *J. Non-Cryst. Solids*, **77&78**, 655–658.

4 Takenaka, H., Ogihara, C., and Morigaki, K. (1988) Time-resolved optically detected magnetic resonance experiment on conduction band tail electrons and A centres in hydrogenated amorphous silicon. *J. Phys. Soc. Japan*, **57**, 3858–3867.

5 Kondo, M. and Morigaki, K. (1990) Light-induced phenomena in a-Si:H as elucidated by optically detected electron-nuclear double resonance, in Proceedings *of the International Conference on Physics of Semiconductors* (eds E.M. Anastassakis and J.D. Joannopoulos), World Scientific, Singapore, pp. 2083–2086.

6 Kondo, M. and Morigaki, K. (1991) The role of hydrogen clusters in the Staebler–Wronski effect of amorphous silicon as elucidated by optically detected electron-nuclear double resonance. *J. Non-Cryst. Solids*, **137–138**, 247–250.

7 Kondo, M. and Morigaki, K. (1993) Possibility of hydrogen migration in photoinduced defect creation processes of a-Si:H. *J. Non-Cryst. Solids*, **164–166**, 227–230.

8 Morigaki, K., Hikita, H., and Kondo, M. (1995) Self-trapping of holes and related phenomena in a-Si:H. *J. Non-Cryst. Solids*, **190**, 38–47.

9 Hirabayashi, I. and Morigaki, K. (1983) Light-induced metastable effect on the short lived photoinduced midgap absorption in hydrogenated amorphous silicon. *J. Non-Cryst. Solids*, **59&60**, 433–436.

10 Morigaki, K., Takenaka, H., Hirabayashi, I., and Yoshida, M. (1985) Gap states in hydrogenated amorphous silicon: The trapped hole centres (the A centres), in *Tetrahedrally-Bonded Amorphous Semiconductors* (eds D. Adler and H. Fritzsche), Plenum Press, New York and London, pp. 221–232.

11 Vardeny, Z. and Olszakier, M. (1987) Infrared photomodulation spectroscopy of band tail states in a-Si:H and a-Si:F. *J. Non-Cryst. Solids*, **97–98**, 109–112.
12 Oheda, H. (1999) Real-time modulation of Ai-H vibration in hydrogenated amorphous silicon. *Phys. Rev. B*, **60**, 16531–16542.
13 Morigaki, K., Hikita, H., and Ogihara, C. (2014) *Light-Induced Defects in Semiconductors*, Pan Stanford, Singapore.
14 Yokomichi, H., Hirabayashi, I., and Morigaki, K. (1987) Electron-nuclear double resonance of dangling-bond centres in a-Si:H. *Solid State Commun.*, **61**, 697–701.
15 Yokomichi, H. and Morigaki, K. (1987) Electron-nuclear double resonance of dangling bond centres associated with hydrogen incorporation in a-Si:H. *Solid State Commun.*, **63**, 629–632.
16 Yokomichi, H. and Morigaki, K. (1993) Clustered dangling bonds in a-Si:H as elucidated by ENDOR measurements. *Solid State Commun.*, **85**, 759–761.
17 Yokomichi, H. and Morigaki, K. (1996) Evidence for existence of hydrogen-related dangling bonds in hydrogenated amorphous silicon. *Philos. Mag. Lett.*, **73**, 283–287.
18 Hikita, H., Takeda, K., Kimura, Y., Yokomichi, H., and Morigaki, K. (1997) Deconvolution of ESR spectra and their light-induced effect in a-Si:H. *J. Phys. Soc. Japan*, **66**, 1730–1740.
19 Morigaki, K., Hikita, H., Yamaguchi, M., and Fujita, Y. (1998) The structure of dangling bond having hydrogen at a nearby site in a-Si:H. *J. Non-Cryst. Solids*, **227–230**, 338–342.
20 Morigaki, K. (1988) Microscopic mechanism for the photo-creation of dangling bonds in a-Si:H. *Japan. J. Appl. Phys.*, **27**, 163–168.
21 Bech Nielsen, B., Johanneson, P., Stallinga, P., Bonde Nielsen, K., and Byberg, J.R. (1997) Identification of the silicon vacancy containing a single hydrogen atom by ESR. *Phys. Rev. Lett.*, **79**, 1507–1510.
22 Onsager, L. (1938) Initial recombination of ions. *Phys. Rev.*, **54**, 554–557.
23 Street, R.A., Knights, J.C., and Biegelsen, D.K. (1978) Luminescence studies of plasma-deposited hydrogenated silicon. *Phys. Rev. B*, **18**, 1880–1891.
24 Yoshida, M. and Morigaki, K. (1987) Triplet exciton recombination in silicon-based amorphous semiconductors. *J. Non-Cryst. Solids*, **97&98**, 579–582.
25 Yoshida, M. and Morigaki, K. (1989) Triplet exciton states in a-Si:H and its alloys as elucidated by optically magnetic resonance measurements. *J. Phys. Soc. Japan*, **58**, 3371–3382.
26 Morigaki, K. (1985) Recombination mechanisms in amorphous semiconductors deduced from resonance measurements. *J. Non-Cryst. Solids*, **77&78**, 583–592.
27 Englman, R. and Jortner, J. (1970) The energy gap law for radiative transitions in large molecules. *Mol. Phys.*, **18**, 145–164.
28 Morigaki, K. (1983) Optically detected magnetic resonance in amorphous semiconductors. *Japan. J. Appl. Phys.*, **22**, 375–388.

29 Morigaki, K. (1984) Optically detected magnetic resonance, in *Semiconductors and Semimetals*, Vol. **21**, Part C (ed. J.I. Pankove), Academic Press, Orlando, FL, pp. 155–191.
30 Morigaki, K. and Kondo, M. (1995) Optically detected magnetic resonance in a-Si:H. *Solid State Phenomena*, **44–46**, 731–764.
31 Kaplan, D., Solomon, I., and Mott, N.F. (1978) Explanation of the large spin-dependent recombination effect in semiconductors. *J. Phys. Paris*, **19**, L51.
32 Dunstan, D.J. and Davies, J.I. (1979) *J. Phys. C*, **12**, 2927–2944.
33 Cavenett, B.C. (1981) Optically detected magnetic resonance (O.D.M.R.) investigations of recombination processes in semiconductors. *Adv. Phys.*, **30**, 475–538.
34 Solomon, I., Biegelsen, D.K., and Knights, J.C. (1977) Spin-dependent photoconductivity in n-type and p-type amorphous silicon. *Solid State Commun.*, **22**, 505–508.
35 Solomon, I. (1979) Spin effects in amorphous semiconductor, in *Amorphous Semiconductors* (ed. M. H. Brodsky), Springer, Berlin, pp. 189–213.
36 Fuhs, W. and Lips, K. (1993) Recombination in a-Si:H films and pin-structure studied by electrically detected magnetic resonance (EDMR). *J. Non-Cryst. Solids*, **164–166**, 541–546.
37 Lips, K., Lerner, C., and Fuhs, W. (1996) Semiclassical model of electrically detected magnetic resonance in undoped a-Si:H. *J. Non-Cryst. Solids*, **198–200**, 267–270.
38 Hindley, N.K.(1970) Random phase model of amorphous semiconductors I. Transport and optical properties. *J. Non-Cryst. Solids*, **5**, 17–30.
39 Mott, N.F. (1970) Conduction in non-crystalline systems IV. Anderson localization in a disordered lattice. *Philos. Mag.*, **22**, 7–29.
40 Kishimoto, N. and Morigaki, K. (1979) Metal–nonmetal transition in amorphous Si–Au system at low temperatures: Measurement of electrical conductivity and thermoelectric power. *J. Phys. Soc. Japan*, **46**, 846–856.
41 Miller, A. and Abrahams, E. (1960) Impurity conduction at low temperatures. *Phys. Rev.*, **120**, 745–755.
42 Mott, N.F. (1969) Conduction in non-crystalline materials III. Localized states in a pseudogap and near extremities of conduction and valence bands. *Philos. Mag.*, **19**, 835–852.
43 Ambegaokar, V., Halpern, B.I., and Langer, J.S. (1971) Hopping conductivity in disordered systems. *Phys. Rev. B*, **4**, 2612–2620.
44 Kirkpatrick, S. (1973) Hopping conduction: Experiment versus theory, in *Amorphous and Liquid Semiconductors* (eds J. Stuke and W. Brenig), Taylor & Francis, London, pp. 183–187.
45 Nishida, N., Furubayashi, T., Yamaguchi, M., Morigaki, K., Ishimoto, H., and Ono, K. (1982) Superconductivity and metal–insulator transition in amorphous $Si_{1-x}Au_x$ system. *Solid State Commun.*, **44**, 305–309.

46 Nishida, N., Furubayashi, T., Yamaguchi, M., Morigaki, K., and Ishimoto, H. (1985) Metal–insulator transition in the amorphous $Si_{1-x}Au_x$ system with a strong spin–orbit interaction. *Solid State Electron.*, **28**, 81–86.
47 Long, A.R. (1982) Frequency-dependent loss in amorphous semiconductors. *Adv. Phys.*, **31**, 553–637.
48 Elliott, S.R. (1987) AC conduction in amorphous chalcogenide and pnictide semiconductors, *Adv. Phys.*, **36**, 135–218.
49 Pollak, M. and Geballe, T.H. (1961) Low-frequency conductivity due to hopping processes in silicon, *Phys. Rev.*, **122**, 1742–1753.
50 Austin, I.G. and Mott, N.F. (1969) Polarons in crystalline and non-crystalline materials, *Adv. Phys.*, **18**, 41–102.
51 Shimakawa, K. (1982) On the temperature dependence of a.c. conduction in chalcogenide glasses. *Philos. Mag. B*, **46**, 123–135.
52 Dyre, J.C. (1988) The random free-energy barrier model for ac conduction in disordered solids, *J. Appl. Phys.*, **64**, 2456–2468.
53 Scher, H. and Lax, M. (1973) Stochastic transport in disordered solids I. Theory, *Phys. Rev. B*, **7**, 4491–4501.
54 Shimakawa, K. and Miyake, K. (1988) Multiphonon tunnelling conduction of localized π electrons in amorphous carbon films. *Phys. Rev. Lett.*, **61**, 994–996.
55 Shimakawa, K. and Miyake, K. (1989) Hopping transport of localized π-electrons in amorphous carbon films. *Phys. Rev. B*, **39**, 7578–7584.
56 Shimakawa, K. (1989) Multiphonon hopping of electrons on defect clusters in amorphous germanium. *Phys. Rev. B*, **39**, 12933–12936.
57 Bottger, H. and Bryskin, V.V. (1985) *Hopping Conduction in Solids*, Akademie-Verlag, Berlin.
58 Shroder, T.B. and Dyre, J.C. (2008) AC hopping conduction at extreme disorder takes place on the percolating cluster. *Phys. Rev. Lett.*, **101**, 025901-1–4.
59 Cohen, M.H. and Jortner, J. (1973) Inhomogeneous transport regime in disordered materials. *Phys. Rev. Lett.*, **30**, 699–701.
60 Ast, D.G. (1974) Evidence of percolation-controlled conductivity in amorphous As_xTe_{1-x} films. *Phys. Rev. Lett.*, **33**, 1042–1045.
61 Shimakawa, K. and Nitta, S. (1977) Influence of silver additive on electronic and ionic natures in amorphous As_2Se_3, *Phys. Rev. B*, **18**, 4348–4354.
62 Springett, B.E. (1973) Effective medium theory for the ac behavior of a random media. *Phys. Rev. Lett.*, **31**, 1463–1465.
63 Shimakawa, K. (2000) Percolation-controlled electronic properties in microcrystalline silicon. *J. Non-Cryst. Solids*, **266**, 223–226.
64 Dyre, J.C. (1993) Universal low-temperature ac conductivity of macroscopically disordered nonmetals. *Phys. Rev. B*, **48**, 12511–12526.
65 Papaulis, A. (1965) *Probability, Random Variables and Stochastic Processes*, McGraw-Hill Kogakusha, Tokyo.
66 Overhof, H. (1981) The Hall mobility in amorphous semiconductors in the presence of long-range potential fluctuations. *Philos. Mag. B*, **44**, 317–322.

67 Overhof, H. and Thomas, P. (1989) *Electronic Transport in Hydrogenated Amorphous Semiconductors*, Springer, Berlin.

68 Shimakawa, K., Watanabe, A., and Imagawa, O. (1985) AC conduction originated from the atomic two-level systems in hydrogenated amorphous silicon. *Solid State Commun.*, **55**, 245–248.

69 Shimakawa, K., Watanabe, A., Hattori, K., and Imagawa, O. (1986) Frequency-dependent transport in glow-discharge amorphous silicon. *Philos. Mag. B*, **54**, 391–414.

70 Shimakawa, K., Long, A.R., and Imagawa, O. (1987) Frequency-dependent loss in sandwich samples of hydrogenated amorphous silicon. *Philos. Mag. Lett.*, **56**, 79–84.

71 Shimakawa, K., Kondo, A., Goto, M., and Long, A.R. (1996) AC loss originating from mesoscopic and macroscopic inhomogeneities in hydrogenated amorphous silicon. *J. Non-Cryst. Solids*, **198–200**, 157–160.

72 Tiedje, T. and Rose, A. (1980) A physical interpretation of dispersive transport in disordered semiconductors. *Solid State Commun.*, **37**, 49–52.

73 Murayama, K. and Mori, M. (1992) Monte Carlo simulation of dispersive transient transport in percolation clusters. *Philos. Mag. B*, **65**, 501–524.

74 Shimakawa, K. and Ganjoo, K. (2002) AC photoconductivity of hydrogenated amorphous silicon: Influence of long-range potential fluctuations. *Phys. Rev. B*, **65**, 165213-1–5.

75 Fekete, L., Kuzel, P., and Nemec, H. (2009) Ultrafast carrier dynamics in microcrystalline silicon probed by time-resolved terahertz spectroscopy. *Phys. Rev. B*, **79**, 115306-1–13.

76 Shimakawa, K., Wagner, T., and Frumar, M. (2013) THz photoconductivity in a-Si:H. *Phys. Stat. Sol. (b)*, **250**, 1004–1007.

77 Kugler, S. and Shimakawa, K. (2015) *Amorphous Semiconductors*, Cambridge University Press, Cambridge.

78 Friedman, L. (1971) Hall conductivity of amorphous semiconductors in the random phase model. *J. Non-Cryst. Solids*, **6**, 329–341.

79 Emin, D. (1977) The sign of the Hall effect in hopping conduction. *Philos. Mag.*, **35**, 1189–1198.

80 Jones, D.I., Spear, W.E., and Le Comber, P.G. (1976) Transport properties of amorphous germanium prepared by the glow discharge technique. *J. Non-Cryst. Solids*, **20**, 259–270.

81 Spear, W.E. and Le Comber, P.G. (1976) Electronic properties of substitutionally doped amorphous Si and Ge. *Philos. Mag.*, **33**, 935–949.

82 Street, R.A. (1982) Doping and the Fermi energy in amorphous silicon. *Phys. Rev. Lett.*, **49**, 1187–1190.

83 Stutzmann, M., Bigelsen, D.K., and Street, R.A. (1987) Detailed investigation of doping in hydrogenated amorphous silicon and germanium. *Phys. Rev. B*, **35**, 5666–5701.

84 Hirabayashi, I., Morigaki, K., Yamasaki, S., and Tanaka, K. (1984) Photoinduced absorption and photoinduced absorption-detected ESR in

P-doped a-Si:H, in *Optical Effects in Amorphous Semiconductors*, AIP Conference Proceedings, No. 120 (eds P.C. Taylor and S.G. Bishop), AIP, New York, pp. 8–15.

85 Hirabayashi, I., Morigaki, K., and Nitta, S. (1985) Defect creation associated with phosphorus doping as elucidated by photoinduced absorption, photoinduced absorption detected ESR and ODMR. *J. Non-Cryst. Solids*, **77&78**, 519–522.

86 Okushi, H., Banenjee, R., and Tanaka, K. (1989) Thermally and light-induced change in the gap states of P-doped a-Si:H, in *Amorphous Semiconductors and Related Materials* (ed. H. Fritzsche), World Scientific, Singapore, pp. 657–685.

87 Tauc, J. (1968) Optical properties and electronic structure of amorphous Ge and Si. *Mater. Res. Bull.*, **3**, 37–46.

88 Mott, N.F. and Davis, E.A. (1979) *Electronic Processes in Non-Crystalline Materials*, 2nd edn, Clarendon Press, Oxford.

89 Tsai, C.C. and Fritzsche, H. (1979) Effect of annealing on the optical properties of plasma deposited amorphous hydrogenated silicon. *Solar Energy Mater.*, **1**, 29–42.

90 Urbach, F. (1953) The long wavelength edge of photographic sensitivity and of the electronic absorption of solids. *Phys. Rev.*, **92**, 1324–1324.

91 Cody, G.D., Tiedje, T., Abeles, B., Brooks, B., and Goldstein, Y. (1981) Disorder and the optical absorption edge of hydrogenated amorphous silicon. *Phys. Rev. Lett.*, **47**, 1480–1483.

92 Cody, C.D. (1984) The optical absorption edge of a-Si:H, in *Semiconductors and Semimetals*, vol. **21**, Part B (ed. J.I. Pankove), Academic Press, Orlando, FL, pp. 11–82.

93 Depinna, S.P. and Dunstan, D.J. (1984) Frequency-resolved spectroscopy and its application to the analysis of recombination in semiconductors. *Philos. Mag. B*, **50**, 579–597.

94 Stachowitz, R., Bort, M., Carius, R., Fuhs, W., and Liedtke, S. (1991) Geminate recombination in a-Si:H. *J. Non-Cryst. Solids*, **137&138**, 551–554.

95 Stachowitz, R., Schubert, M., and Fuhs, W. (1993) Frequency-resolved spectroscopy and its application to lifetime studies in a-Si:H. *J. Non-Cryst. Solids*, **164–166**, 583–586.

96 Stachowitz, R., Schubert, M., and Fuhs, W. (1994) Frequency-resolved spectroscopy and its low-temperature geminate recombination in a-Si:H. *Philos. Mag. B*, **70**, 1219–1230.

97 Aoki, T. (2006) Photoluminescence, in *Optical Properties of Condensed Matter and Applications* (ed. J. Singh), John Wiley & Sons, Chichester, pp. 75–106.

98 Oheda, H. (1993) Discontinuous change of photoluminescence lifetime with temperature in hydrogenated amorphous silicon. *J. Non-Cryst. Solids*, **164–166**, 559–562.

99 Morigaki, K. (1999) *Physics of Amorphous Semiconductors*, World Scientific, Singapore and Imperial College Press, London.

100 Spear, W.E., Loveland, R.J., and Al-Shartaty, A. (1974) The temperature dependence of photoconductivity in a-Si. *J. Non-Cryst. Solids*, **15**, 410.

101 Murayama, K., Oheda, H., Yamasaki, S., and Matsuda, A. (1992) Electron time of flight experiment under high electric field in a-Si: H. *Solid State Commun.*, **81**, 887.

102 Morigaki, K., Yagi, K., and Yamaguchi, M. (1998) Temperature dependence of photoconductivity and self-trapping of holes in a-Si:H. *Res. Bull. Hiroshima Inst. Tech.*, **32**, 9–15.

103 Vommas, A. and Fritzsche, H. (1987) The temperature dependence of the photoconductivity of n-type a-Si:H and the effect of Staebler–Wronski defects. *J. Non-Cryst. Solids*, **97&98**, 823–826.

104 Yamaguchi, M. and Morigaki, K. (1991) The correlation between hydrogen content and electronic properties in a-Si:H. *J. Non-Cryst. Solids*, **137&138**, 57–60.

105 Morigaki, K., Yamaguchi, M., and Hirabayashi, I. (1993) Temperature dependence of photoluminescence spectra and model of self-trapping of holes in a-Si:H. *J. Non-Cryst. Solids*, **164–166**, 571–574.

106 Pfister, G. and Scher, H. (1978) Anomalous transit-time dispersion in amorphous solids. *Adv. Phys.*, **27**, 747–798.

107 Tiedje, T. (1984) Information about band-tail states from time-of-flight experiments, in *Semiconductors and Semimetals*, vol. **21**, Part C (ed. J.I. Pankove), Academic Press, Orlando, FL, pp. 207–238.

108 Scher, H. and Montroll, E.W. (1975) Anomalous transit-time dispersion in amorphous solids. *Phys. Rev. B*, **12**, 2455–2477.

109 Brodsky, M.H. and Title, R.S. (1969) Electron spin resonance in amorphous silicon, germanium, and silicon carbide. *Phys. Rev. Lett.*, **23**, 581–584.

110 Schweiger, A. and Jeschke, G. (2001) *Principles of Pulse Electron Paramagnetic Resonance*, Oxford University Press, Oxford.

111 Mims, W.B. (1977) Electron spin echoes, in *Electron Paramagnetic Resonance* (ed. S. Geschwind), Plenum Press, New York, pp. 263–351.

112 Isoya, J., Yamasaki, S., Okushi, H., Matsuda, A., and Tanaka, K. (1993) Electron-spin echo envelope-modulation study of the distance between dangling bonds and hydrogen atoms in hydrogenated amorphous silicon. *Phys. Rev. B*, **47**, 7013–7024.

113 Feher, G. (1956) Method of polarizing nuclei in paramagnetic substances. *Phys. Rev.* **103**, 500.

114 Feher, G. (1956) Observation of nuclear magnetic resonances via the electron spin resonance line. *Phys. Rev.* **103**, 834–835.

115 Feher, G. (1957) Electronic structure of F centers in KCl by the electron spin double resonance technique. *Phys. Rev.* **105**, 1122–1123.

116 Feher, G. (1959) Electron spin resonance experiments on donor in silicon, electronic structure of donor by the electron nuclear double resonance technique. *Phys. Rev.* **114**, 1219–1244.

117 Davies, E.R. (1974) A new pulse ENDOR technique. *Phys. Lett. A*, **47**, 1–2.
118 Fehr, M., Schnegg, A., Teutloff, C., Bittl, R., Astakhov, O., Finger, F., Roch, B., and Lips, K. (2010). Hydrogen distribution in the vicinity of dangling bonds in hydrogenated amorphous silicon (a-Si:H). *Phys. Stat. Solidi A*, **207**, 552–555.
119 Rabi, I.I. (1937) Space quantization in a gyrating magnetic field. *Phys. Rev.* **51**, 652–654.
120 Boehme, C. and Lips, K. (2003) Theory of time-domain measurement of spin-dependent recombination with pulsed electrically detected magnetic resonance. *Phys. Rev. B*, **68**, 245105-1–19.
121 Astashkin, A.V. and Schweiger, A. (1990) Electron-spin transient nutation: A new approach to simplify the interpretation of ESR spectra. *Chem. Phys. Lett.*, **174**, 595–602.
122 Torry, H.C. (1949) Transit nutation in nuclear magnetic resonance. *Phys. Rev.*, **76**, 1059–1067.
123 Boehme, C., Friedrich, F., Ehara, T., and Lips, K. (2005) Recombination at silicon dangling bonds. *Thin Solid Films*, **487**, 132–136.
124 Herring, T.W., Lee, S.-Y., McCamey, D.R., Taylor, P.C., Lips, K., Hu, J., Zhu, F., Madan, A., and Boehme, C. (2009) Experimental discrimination of geminate and non-geminate recombination in a-Si:H. *Phys. Rev. B*, **79**, 195205-1–5.
125 Lips, K., Boehme, C., and Ehara, T. (2005) The impact of the electron spin on charge carrier recombination – the example of amorphous silicon. *J. Optoelectron. Adv. Mater.*, **7**, 13–24.
126 Biegelsen, D.K., Knights, J.C., Street, R.A., Tsang, C., and White, R.M. (1978) Spin dependent luminescence in hydrogenated amorphous silicon. *Philos. Mag. B*, **37**, 477–488.
127 Morigaki, K., Dunstan, D.J., Cavenett, B.C., Dawson, P., Nicholls, J.E., Nitta, S., and Shimakawa, K. (1978) Optically detected electron spin resonance in amorphous silicon. *Solid State Commun.*, **26**, 981–985.
128 Solomon, I., Biegelsen, D.K., and Knights, J.C. (1977) Spin-dependent photoconductivity in n-type and p-type amorphous silicon. *Solid State Commun.*, **22**, 505–508.
129 Guéron, M. and Solomon, I. (1965) Effect of spin resonance on hot electrons by spin-orbit coupling in n-type InSb. *Phys. Rev. Lett.*, **15**, 667–669.
130 Toyoda, Y. and Hayashi, Y. (1970) Bolometric detection of ESR in P-doped Si at low temperature. *J. Phys. Soc. Japan*, **29**, 247–248.
131 Toyotomi, S. and Morigaki, K. (1970) Microwave hot electron effect and resistivity decrease due to donor spin resonance in P-doped Si. *Solid State Commun.*, **8**, 1307–1308.
132 Morigaki, K. and Toyotomi, S. (1971) Spin energy transfer from donor spin system to mobile electron system in P-doped Si. *J. Phys. Soc. Japan*, **30**, 1207–1208.

133 Morigaki, K. and Onda, M. (1972) Resistivity decrease due to donor spin resonance in n-type germanium. *J. Phys. Soc. Japan*, **33**, 1031–1046.

134 Kishimoto, N., Morigaki, K., and Murakami, K. (1981) Conductivity change due to electron spin resonance in amorphous Si–Au system. *J. Phys. Soc. Japan*, **50**, 1970–1977.

135 Hirabayashi, I. and Morigaki, K. (1983) Spin-dependent photoinduced absorption in hydrogenated amorphous silicon. *J. Non-Cryst. Solids*, **59&60**, 133–136.

136 Hirabayashi, I. and Morigaki, K. (1986) Level of dangling-bond centres in a-Si:H. *Philos. Mag. B*, **54**, L119–L123.

137 Lucovsky, G. (1965) On the photoionization of deep impurity centers in semiconductors. *Solid State Commun.*, **3**, 299–302.

138 Schultz, N., Vardeny, Z.V., and Taylor, P.C. (1997) Spin-dependent photoinduced absorption in a-Si:H, in *Materials Research Society Symposium*, vol. **467** (eds. S. Wagner, M. Hack, E.A. Schiff, R.R. Schropp, and I. Shimizu), Materials Research Society, Pittsburgh, pp. 179–183.

139 Staebler, D.L. and Wronski, C.R. (1977) Reversible conductivity changes in discharge-produced amorphous Si. *Appl. Phys. Lett.*, **31**, 292–294.

140 Staebler, D.L. and Wronski, C.R. (1980) Optically induced conductivity changes in disordered-produced hydrogenated amorphous silicon. *J. Appl. Phys.*, **51**, 3262–3268.

141 Hirabayashi, I., Morigaki, K., and Nitta, S. (1980) New evidence for defect creation by high optical excitation in glow discharge amorphous silicon. *Japan. J. Appl. Phys.*, **19**, L357–L360.

142 Dersch, H., Stuke, J., and Beichler, J. (1981) Light-induced dangling bonds in hydrogenated amorphous silicon. *Appl. Phys. Lett.*, **38**, 456–458.

143 Amer, N.M., Skumanich, A., and Jackson, W.B. (1983) The contribution of the Staebler–Wronski effect to gap state absorption in hydrogenated amorphous silicon. *Physica*, **117B**, 897–898.

144 Jackson, W.B. and Amer, N.M. (1982) Direct measurement of gap state absorption in hydrogenated amorphous silicon by photothermal deflection spectroscopy. *Phys. Rev. B*, **25**, 5559–5562.

145 Vanecek, M., Kocka, J., Stuchlik, J., Kozisek, Z., Stika, O., and Triska, A. (1983) Density of the gap states in undoped and doped glow discharge a-Si:H. *Solar Energy Mater.*, **8**, 411–423.

146 Morigaki, K., Hirabayashi, I., Nakayama, M., Nitta, S., and Shimakawa, K. (1980) Fatigue effect in luminescence of glow discharge amorphous silicon at low temperatures. *Solid State Commun.*, **33**, 851–856.

147 Pankove, J.I. and Berkeyheiser, J.E. (1980) Light-induced radiative recombination centers in hydrogenated amorphous silicon. *Appl. Phys. Lett.*, **37**, 705–706.

148 Cernogora, J., Mollot, F., and Benoit à la Guillaume, C. (1973) Radiative recombination in amorphous As_2Se_3. *Phys. Stat. Solidi*, **15**, 401–407.

149 Ogihara, C. and Morigaki, K. (2009) Temperature variation of radiative recombination rate of electron–hole pairs responsible for defect photoluminescence. *Phys. Stat. Solidi C*, **6**, 5167–5170.

150 Ogihara, C., Inagaki, Y., Takata, A., and Morigaki, K. (2012) Thermal quenching of defect photoluminescence and recombination rates of electron–hole pairs in a-Si:H. *J. Non-Cryst. Solids*, **358**, 2004–2006.

151 Ogihara, C., Nakayama, A., Yamaguchi, K., and Morigaki, K. (2015) Defects in hydrogenated amorphous silicon created by intense pulsed illumination at low temperature and the decay of their density. *J. Phys. Conf. Ser.*, **619**, 012016–012019.

152 Han, D., Yoshida, M., and Morigaki, K. (1987) Light-induced photoluminescence fatigue in a-Si:H. *J. Non-Cryst. Solids*, **97&98**, 651–654.

153 Yoshida, M. and Morigaki, K. (1990) Difference in light-induced photoluminescence fatigue in a-Si:H between 300 K and 7 K. *J. Phys. Soc. Japan*, **59**, 1733–1739.

154 Morigaki, K., Sano, Y., and Hirabayashi, I. (1982) Defect creation by optical excitation in hydrogenated amorphous silicon as elucidated by optically detected magnetic resonance. *J. Phys. Soc. Japan*, **51**, 147–152.

155 Stutzmann, M., Jackson, W.B., and Tsai, C. (1985) Light-induced metastable defects in hydrogenated amorphous silicon: A systematic study. *Phys. Rev. B*, **32**, 23–47.

156 Morigaki, K. and Hikita, H. (2007) Modeling of light-induced defect creation in hydrogenated amorphous silicon. *Phys. Rev. B*, **76**, 085201-1–17.

157 Morigaki, K., Takeda, K., Hikita, H. and Roca i Cabarrocas, P. (2007) Dispersive processes of light-induced defect creation in hydrogenated amorphous silicon. *Solid State Commun.*, **142**, 232–236.

158 Morigaki, K., Hikita, H., and Ogihara, C. (2009) Light-induced defect creation under intense pulsed illumination in hydrogenated amorphous silicon. *J. Optoelectron. Adv. Mater.*, **11**, 1–14.

159 Morigaki, K., Takeda, K., Hikita, H., Ogihara, C., and Roca i Cabarrocas, P. (2008) The kinetics of light-induced defect creation in hydrogenated amorphous silicon – Stretched exponential relaxation. *J. Non-Cryst. Solids*, **354**, 2131–2134.

160 Morigaki, K., Hikita, H., Takeda, K., and Roca i Cabarrocas, P. (2010) The kinetics of light-induced defect creation in hydrogenated polymorphous silicon – Stretched exponential relaxation. *Phys. Stat. Solidi C*, **7**, 692–695.

161 Roca i Cabarrocas, P., Fontouberta, A., and Poissant, Y. (2002) Growth and optoelectronic properties of polymorphous silicon thin films. *Thin Solid Films*, **403–404**, 39–46.

162 Roca i Cabarrocas, P., Châbane, N., Kharchenko, A.V., and Tchakarov, S. (2004) Polymorphous silicon thin films produced in dusty plasmas: Applications to solar cells. *Plasma Phys. Control. Fusion*, **46**, B235–B243.

163 Vignoli, S., Meaudre, R., and Meaudre, M. (1996) Metastable defect creation and annealing under illumination in intrinsic hydrogenated amorphous silicon deposited from helium–silane mixtures. *Philos. Mag. B*, **73**, 261–276.
164 Morigaki, K. and Hikita, H. (2015) Model-simulation of light-induced defect creation in hydrogenated amorphous silicon. *J. Phys. Conf. Ser.*, **619**, 012013–012016.
165 Morigaki, K., Hikita, H., Yamaguchi, M., and Fujita, Y. (1998) The structure of dangling bonds having hydrogen at a nearby site in a-Si:H. *J. Non-Cryst. Solids*, **227–230**, 338–342.
166 Yamasaki, S. and Isoya, J. (1993) Pulsed-ESR study of light-induced metastable defects in a-Si:H. *J. Non-Cryst. Solids*, **164–166**, 169–174.
167 Branz, H. (1999) Hydrogen collision model: Quantitative description of metastability in amorphous silicon. *Phys. Rev. B*, **59**, 5498–5512.
168 Fehr, M., Schnegg, A., Rech, B., Astakhov, O., Finger, F., Bittl, R., Teutloff, C., and Lips, K. (2014) Metastable defect formation at microvoids identified as a source of light-induced degradation in a-Si:H. *Phys. Rev. Lett.*, **112**, 066403-1–5.

163 [illegible] S., Miramontes, [illegible] and [illegible] (20[illegible]) Metastable [illegible] [illegible] during melting and [illegible] hydrogenated amorphous [illegible] deposited from helium-diluted silane [illegible] [illegible]

164 [illegible] (20[illegible]) [illegible] in hydrogenated amorphous silicon. [illegible] [illegible] [illegible]

165 [illegible] (19[illegible]) [illegible] in hydrogenated amorphous silicon [illegible] [illegible] 236–2[illegible]

166 [illegible] (1985) [illegible] [illegible] [illegible] [illegible]

167 [illegible] collision model [illegible] [illegible] 58, 5[illegible]

168 [illegible] [illegible] [illegible] [illegible] [illegible]

6

Electronic and Optical Properties of Amorphous Chalcogenides

As was done in the previous chapter for hydrogenated amorphous silicon (a-Si:H), the electronic and optical properties of amorphous chalcogenides (a-Chs) will be discussed in this chapter. Let us begin with a historical overview of a-Chs.

6.1 Historical Overview of Chalcogenide Glasses

6.1.1 Applications

It is important first to provide a historical overview of the successful applications of the chalcogenide glasses (or a-Chs) that are appearing on the global market. Although science and technology (i.e., applications) are closely connected, some applications of a-Chs have proceeded without scientific knowledge in this field. It is therefore of interest to begin with the most successful applications (i.e., devices).

Electrophotography (xerography), using amorphous selenium (a-Se), was the most successful application. A Hungarian scientist first proposed the concept of the photographic process [1]. His pioneering work on electrostatic image recording formed the basis of so-called xerography, developed by Carlson and Kornei in 1938. Modern xerographic processes are the same as those proposed at that time [2]. Why a-Se was used successfully for xerography is due to the following two reasons: its high photoconductivity, and its low conductivity in dark conditions. The high contrast in conductivity is very useful for electronic charging in dark conditions and discharging in the photoilluminated state. This technology was later applied to laser printers. Owing to requirements for higher cost-effectiveness in the market, a-Se has been replaced by organic polymers [3].

The digital versatile disc (DVD) is well known as a rewritable optical data storage device and has produced a big market in the area of information

Amorphous Semiconductors: Structural, Optical, and Electronic Properties, First Edition.
Kazuo Morigaki, Sándor Kugler, and Koichi Shimakawa.

technology. This is due to the finding of an ideal material system, GeSbTe (called GST), by the Panasonic group in Japan [4]. A phase change between amorphous and crystalline states is used in DVDs, initiated by focused short laser pulses. The marked contrast in the optical properties between the two phases is so significant that this difference can be used to store information at high speeds of the order of nanoseconds. The DVD system currently has a memory capacity exceeding 50 GB/disc.

Rewritable *electrical* memory devices that use a phase-change random access memory (PRAM) have been also developed and are used in mobile phones, following the memory-switching devices originally proposed by Ovshinsky [5]. Here, a marked contrast in electrical resistivity between the two phases of GST is used for electrical data storage [6].

A very sensitive photoconductor, using an a-Se thick film, which is especially sensitive to X-rays owing to its high atomic weight, has been applied in a direct X-ray imaging device for medical use. This device incorporates a large area of thick (1 mm) a-Se evaporated onto a thin-film transistor made from an a-Si:H active matrix array [7]. The X-ray-induced carriers in the a-Se travel in a high electric field ($\sim 10^5$ V/cm) and are collected at biased electrodes with a storage capacitor. This type of digital X-ray image sensor is widely used in mammography.

Surprisingly, avalanche photomultiplication has been found in a-Se [8]. Tanioka [9] at The Japan Broadcasting Corporation (NHK) developed a "high-gain avalanche rushing amorphous photoconductor" (HARP) vidicon tube utilizing avalanche photomultiplication. The HARP vidicon color TV camera has a very high sensitivity (~1000 times that of charge-coupled-devices). Details of these commercially available devices will be given in Chapter 8.

6.1.2 Science

A major report on semiconductor switching and memory devices using a new type of semiconductor, that is, chalcogenide glasses, was presented [5]; it attracted interest from journalists (newspaper and broadcast), and initiated the vast field of disordered or amorphous semiconductors. The fundamental physical and chemical properties of these materials are in principle determined by their structures, and hence scientists wanted to know about disordered structures. In the case of tetrahedrally bonded amorphous germanium and amorphous silicon, a random network model using 440 atoms (hand-built) was first proposed by Polk [10]. Then, computer-generated structures of threefold-coordinated amorphous arsenic and twofold-coordinated a-Se were proposed by Greaves and Davis [11]. Now, the most popular technique for understanding microscopic structures may be molecular dynamics simulations [12], which have also been used for the study of photoinduced structural transformations in a-Se [13, 14]. The coordination number Z of bonded atoms plays an important role in the structural properties. The "magic numbers" $Z_c = 2.4$ [15] and 2.67 [16] are well known to dominate the optical and electronic properties of a-Chs.

We turn now to the electronic properties. A model of the electronic density of states (DOS) for noncrystalline semiconductors was first proposed by Cohen, Fritzsche, and Ovshinsky [17]. The DOS is separated by a band gap, but it extends into the gap; the parts in the gap are called the localized band tails. While the valence band in amorphous silicon (a-Si) originates from the bonding states, that in a-Chs is formed from a lone pair (LP) band. Therefore, a-Chs are often called lone-pair semiconductors [18]. The wave-vector k-selection rule does not hold in amorphous semiconductors, so Tauc [19] defined the band gap using optical absorption and *energy space* (without using a wave vector). The band gap is therefore called the Tauc gap or optical gap.

The electronic and optical properties are also dominated by bonding defects. In a-Chs, these defects are believed to be charged dangling bonds, which are electronically not neutral [20, 21]. Note that the nature of defects in a-Chs is quite different from that for a-Si (or a-Si:H).

The illumination of a-Chs, effectively with band gap light, induces various changes in the structural and electronic properties: the band gap decreases with illumination (an effect called *photodarkening*) [22], and this is accompanied by volume changes [23]. The number of defects has also been found to increase with photoillumination [24], similarly to a-Si:H [25]. Owing to the importance of both science and applications, a huge amount of work has been devoted to these topics (see the reviews by Tanaka [26], Pfeiffer, Paesler, and Agarwal [27], and Shimakawa, Kolobov, and Elliott [28]).

6.2 Basic Glass Science

Glass formation is strongly related to the glass transition temperature and the crystallization tendencies of materials. We hence briefly introduce these fundamental issues in the following.

6.2.1 Glass Formation

Glass formation can be modeled by considering the following three fundamental aspects: (i) kinetics, (ii) thermodynamics, and (iii) structure and chemical bonding. These three aspects cannot be discussed independently [29, 30]. Tammann [31] was among the first scientists who tried to understand the glass formation process in terms of these aspects [32, 33]. A brief review and discussion of the formation of chalcogenide glasses (ChGs) will be presented in this section.

Qualitatively, the most important regularities in the so-called glass-forming ability (GFA) can be understood from the following experimental findings, although there are some exceptions [32]. The glass formation area of two- and three-component chalcogenide glass alloys decreases with increasing degree of metallization of the covalent bonds, that is, with an increase in the atomic number [34]. For example, for the ternary alloy systems, a decreasing tendency is shown by the GFA in the following sequences: S > Se > Te (group VI), As > P > Sb (group V), Si > Ge > Sn (group IV) [35], and B > Ga > In (group III) [36].

A useful evaluation of an empirically proposed GFA can be performed through differential thermal analysis or differential scanning calorimetry (DSC) using

$$\mathrm{GFA} = \frac{T_\mathrm{c} - T_\mathrm{g}}{T_\mathrm{m} - T_\mathrm{c}} \tag{6.1}$$

where T_g, T_c, and T_m are the glass transition, crystallization, and melting temperatures, respectively [37].

From a theoretical point of view, Phillips [15] proposed an important correlation between his magic number of structural constraints, as already discussed in Section 6.1, and the GFA of an alloy. The optimum value of GFA is reached when the number of bond constraints is equal to the number of degrees of freedom per atom. This viewpoint has been generalized and leads to a suggestion that polymer-like structural fragments are the most important features in glass formation.

The nature of the glassy state needs to be understood together with the corresponding crystalline and liquid states. The transformations between these states, that is, phase transitions, can be characterized by the structural relaxation towards establishment of thermodynamic equilibrium, which is expected to be highly dependent on temperature. As discussed and shown in many standard textbooks, we must discuss the free energy changes in each state, as shown in Figure 6.1. Let us begin with the relaxation of a glass-forming liquid upon cooling. It is known that there is no principal structural difference between glasses (below T_g) and liquids (above T_m). It should be also noted that these structures are very similar to those of deposited films (where there is a transition from the gas phase to the solid state). Thus we know in the case of chalcogenide glasses that *glassy* and *amorphous* are not in fact distinguishable when we are discussing structural and electronic properties. The term *glassy* is usually used for a glassy state that is produced by quenching from the melted state.

6.2.2 Glass Transition Temperature

The glass transition is an important phenomenon in melt-quenched glasses [38, 39], and hence the factors that dominate T_g will be discussed in this section. Note that the following discussion applies to glasses in general (and

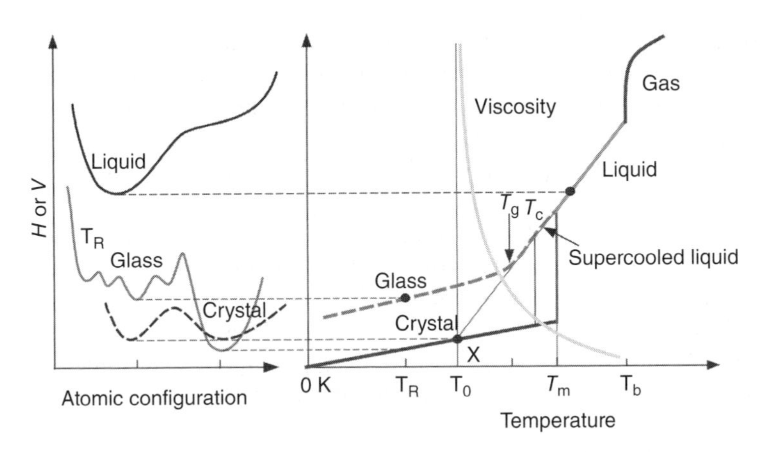

Figure 6.1 Dependence of enthalpy *H* or volume *V* on the atomic configuration and temperature of a material. T_g, T_c, and T_m are the glass transition, crystallization, and melting temperatures, respectively. *Source*: Tanaka and Shimakawa 2011 [39]. Reproduced with permission from Springer Science and Business Media.

not only to ChGs). As already stated, glasses prepared by the melt-quenching method undergo a glass transition at T_g, which is defined as the temperature of the transformation between the glassy and supercooled liquid states. A supercooled liquid is a material that remains in a liquid state below T_m. In spite of the clear meaning of the glass transition, the experimental determination of T_g is not easy. This is due to the fact that the deduced T_g depends on the experimental conditions, such as the cooling rate of the melt and the heating rate in the DSC measurement [38, 39]. In spite of this ambiguity in the determination of T_g, however, it is known that its magnitude is correlated with some physical parameters.

A well-known empirical rule is $T_g \approx 2T_m/3$ in units of kelvin [40]. Recently, it has been suggested that quantum effects may play an important role in the glass transition [41]: quantum effects lead to a significant decrease in T_g with respect to T_m, so that the ratio T_g/T_m may become much smaller than 2/3 in materials where T_g is near or below 60 K; in this case it is given by

$$\frac{T_g}{T_m} = \frac{A}{1 + B/T_g} \tag{6.2}$$

where $A \approx 2/3$ and B depends on the strength of the boson peak and some other parameters.

As T_m is primarily related to the cohesive energy, T_g is expected also to be related to the cohesive energy in covalent glasses. The glassy matrix may be destroyed at T_g, and therefore this temperature may be related to a network constraint or coordination number. While there are numerous approaches to this question [38, 42], Tanaka [43] has proposed the following empirical relation for some covalently bonded glasses: $\ln T_g \approx 1.6Z + 2.3$, where Z is the average coordination number of the atoms. The parameter Z is an important parameter (see Section 6.1) for describing the physical and chemical properties of glasses. There exist "magic numbers" $Z = 2.4$ and 2.67 in the constraint theory of glassy networks [39, 44].

For the vast majority of ChGs, T_g is known not to be simply related to Z [42, 45]. Instead of using Z, by using the overall mean bond energy $\langle E \rangle$, the empirical relation $T_g = 311(\langle E \rangle - 0.9)\,\mathrm{K}$ has been proposed for covalently bonded chalcogenide glasses [42]. The fact that T_g scales well with $\langle E \rangle$ for covalent chalcogenide glasses suggests that T_g is principally determined by the cohesive energy. The short-range structure of glasses may be an important property in relation to T_g.

The average bond energy is not a measurable quantity, and therefore other measurable parameters, such as the microhardness H, may be useful for scaling instead of $\langle E \rangle$. The microhardness H may be correlated with the cohesive energy and hence T_g versus H may show a linear relationship. In fact, Freitas *et al.* [45] found an excellent correlation between T_g and H in a covalent amorphous chalcogenide system. It should be also noted that computer simulations are also useful in understanding the nature of the glass transition [46, 47].

The viscosity η is also an important factor in glass science. η is highly dependent on temperature and it takes a value of around 1×10^{12} Pa s at T_g for most glasses, including oxides. In the region $T > T_g$, the temperature-dependent η can be expressed empirically by the Vogel–Tammann–Fulcher equation,

$$\eta(T) = A\exp\left[B/(T - T_0)\right] \tag{6.3}$$

where A and B are constants, and T_0 ($< T$) is a fictive temperature. At $T = T_g$ ($> T_0$), the supercooled liquid becomes a solid. When $\eta(T)$ is scaled by T_g/T, glasses can be classified into two groups, namely *strong* ($T_0 \sim 0$) and *fragile* glasses [48]. It is known that oxide glasses are strong and ChGs are fragile.

6.2.3 Crystallization of Glasses

When a glassy material is heated, crystallization occurs at a temperature T_c between T_g and T_m. This type of phase change is an important subject, and in fact a reversible phase change between glassy and crystalline states in chalcogenide materials has been applied in phase-change memory devices, as will be discussed in Chapter 8. The dynamics of crystallization of glasses can be fairly well understood by using the classical nucleation and growth model developed by Kolmogorov, Johnson-Mehl, and Avrami (the so-called KJMA model) [49, 50]. The KJMA model deals with the macroscopic evolution of the transformed phase under isothermal annealing conditions. Although the KJMA formalism is widely used, with some extensions, to understand the kinetics of the crystallization of glasses, it is less well recognized that the KJMA model has some difficulties. For example, fitting the KJMA results to experimental data produces unreasonable physical parameters for the crystallization process [44, 51].

Let us briefly discuss the KJMA formalism. The KJMA model assumes a random distribution of nucleation sites in a volume V and that growth of the crystalline phase ceases when two neighboring new-phase grains impinge on each other. The transformed volume V^{tr} is given by

$$dV^{\mathrm{tr}} = \left[1 - \left(\frac{V^{\mathrm{tr}}}{V}\right)\right] dV^{\mathrm{ex}} \tag{6.4}$$

where V^{ex} is called the extended volume, and is the *virtual* volume of the particles growing without any impingement. Integration of the above equation produces the familiar KJMA equation,

$$f = 1 - \exp\left(-f^{\mathrm{ex}}\right) \tag{6.5}$$

where $f = V^{\mathrm{tr}}/V$ and $f^{\mathrm{ex}} = V^{\mathrm{ex}}/V$. In the case of instantaneous nucleation of spherical grains with a number density N, the growth rate of the volume for a radius r with an interface-controlled reaction can be given as

$$\frac{dV^{\mathrm{ex}}}{dt} = 4\pi r^2 NV \frac{dr}{dt} \tag{6.6}$$

Here the effective growth rate is given by $G = dr/dt$, and then f^{ex} at time t is expressed by

$$f^{\mathrm{ex}} = \frac{4\pi}{3} Nr^3 = \frac{4\pi}{3} N(Gt)^3 \tag{6.7}$$

We then obtain the following well-known KJMA formula,

$$1 - f = \exp\left[-(kt)^n\right] \tag{6.8}$$

with $n = 3$. Here, n is called the *Avrami exponent*, and k is an effective rate constant (for the growth rate). The Avrami exponent should be an integer providing information about the dimensionality of the crystallization process: $n = 1$ for rod-like crystals, $n = 2$ for plate-like crystals, and $n = 3$ for spherical forms.

The values of n deduced from many experiments lie between 1 and 5.8 for chalcogenide phase-change materials consisting of GeSbTe (the GST system), for example, and the frequency factor ν appearing in the reaction rate $k(T)$ takes unreasonably large values ($10^{17}-10^{24}\,s^{-1}$) [51]:

$$k(T)=\nu\exp\left(-\frac{E_A}{k_B T}\right) \tag{6.9}$$

where E_A is the activation energy for the phase change and T is the temperature. Noninteger values of n indicate a violation of the KJMA formalism, which is based on the following fundamental assumptions: (i) a random distribution of the potential sites of nucleation, (ii) instantaneous nucleation, (iii) interface-controlled grain growth, and (iv) a time-independent growth rate k.

Let us modify or extend the KJMA approach in the following way [52]. We try to introduce a concept of fractal growth into the space dimension. It is generally expected that the surfaces of most aggregated materials will have a fractal dimension. The growth rate of the extended volume (Equation (6.6)) can be modified as follows:

$$\frac{dV^{ex}}{dt}=Cr^{D}NV\frac{dr}{dt} \tag{6.10}$$

where C is a constant and D is the fractal dimension of the *surface* of the grains. Finally, we obtain f:

$$1-f=\exp\left[-(kt)^{D+1}\right] \tag{6.11}$$

The Avrami exponent n can be therefore attributed to the complex shape of the crystalline grain surfaces, which has a fractal structure. In fact, $D\sim3.0$ for silica gels and $D=2.6$–2.9 for sandstones have been reported [52].

The deduced rate constant $k(T)$ is also a problem (see Equation (6.9)). The reported activation energies E_A are around 2 eV (1 – 3 eV) for GST, for example, leading to unreasonably large frequency factors ν ($10^{17}-10^{24}\,s^{-1}$) being deduced from the application of the classical KJMA model. The ν-value should be around $10^{12}\,s^{-1}$ (the phonon frequency). Why do these unreasonable values for the frequency factor appear in the KJMA formalism? In many thermally activated phenomena, such as electronic and ionic transport and structural relaxation in disordered matter, we often observe the Meyer–Neldel (MN) compensation rule [53]. In Equation (6.9), $k(T)$ needs to be modified as follows:

$$k(T)=\nu_0\exp\left(\frac{E_A}{E_{MN}}\right)\exp\left(-\frac{E_A}{k_B T}\right) \tag{6.12}$$

where ν is given by

$$\nu = \nu_0 \exp\left(\frac{E_A}{E_{MN}}\right) \tag{6.13}$$

Here E_{MN} is the MN energy, and ν_0 is a constant. Equation (6.12) is called the MN rule or compensation law [53].

The introduction of both fractal geometry and the MN rule for thermally activated processes (which we call the extended KJMA model) overcomes the difficulties with the classical KJMA model without losing its simplicity and beauty.

6.3 Electrical Properties

6.3.1 Electronic Transport

6.3.1.1 DC Transport

In this section, we discuss DC electronic transport phenomena in a-Chs. The electronic transport around room temperature is known to be dominated by band transport (by holes in the valence band) [54]. The Fermi level E_F lies nearly midgap and therefore the number of free holes p in the valence band is given by

$$p = N_v \exp\left[-\frac{E_F - E_v}{k_B T}\right] \tag{6.14}$$

where N_v is the effective density of states in the valence band and E_v is the band edge energy of the valence band. The conductivity σ is therefore given by

$$\sigma = ep\mu_p \tag{6.15}$$

where μ_p is the microscopic hole mobility ($\sim 1\,\mathrm{cm}^{2-1}\,\mathrm{s}^{-1}$) in the valence band.

Although the central issues concerning electronic transport have been widely discussed in the literature [39, 54–56], there is still an important and interesting issue that is not properly understood. This is the Meyer–Neldel rule or compensation rule, written as [53, 57]

$$\sigma = \sigma_0 \exp(-\Delta E/k_B T) \tag{6.16}$$

Note, however, that σ_0 is not a constant; it is given by

$$\sigma_0 = \sigma_{00} \exp(\Delta E/E_{MN}) \tag{6.17}$$

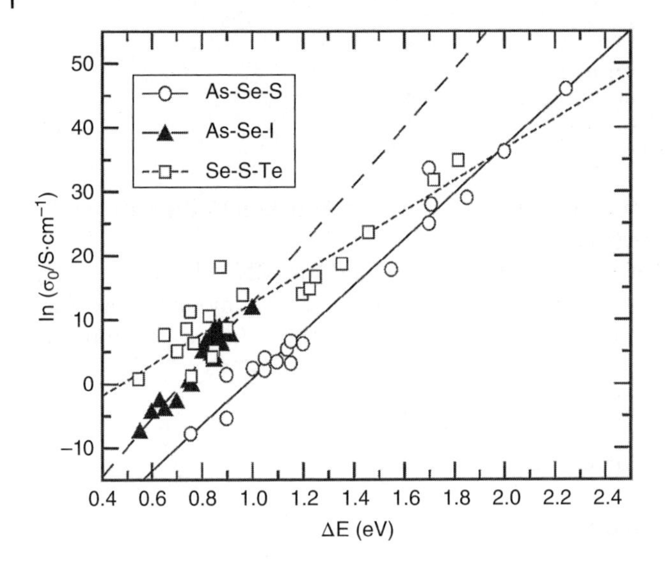

Figure 6.2 The Meyer–Neldel rule for the electronic conductivity of some glassy chalcogenides. *Source*: Shimakawa and Abdel-Wahab 1997 [59]. Reproduced with permission from AIP Publishing LLC.

where ΔE and E_{MN} are called the activation energy and the MN energy, respectively. Examples of the MN rule are shown in Figure 6.2 for some a-Chs. The prefactor itself is a function of the activation energy ΔE. The MN energy E_{MN} is around 40–50 meV.

A similar effect is also found in a-Si:H [58], which is well explained by a statistical shift in the Fermi level (i.e., a temperature variation of the Fermi level). This model is not applicable to a-Chs [59], and the reason why the MN rule is found in a-Chs is still not clear. As the MN rule is *universally* observed in a wide class of materials, Yelon *et al.* [53] have suggested that the multiexcitation entropy is important in some kinetics and thermodynamics. In our view, however, kinetics and thermodynamics are different topics, and therefore each phenomenon may have a different origin.

For example, the same MN rule as in Equation (6.17) has been found for ionic transport in crystalline solids, and this can be explained in terms of phonon absorption and emission processes with lattice distortion [60]. The MN law has been reported in both crystalline and disordered materials, although it is believed to exist only in disordered matter. A definitive solution to this unsolved problem of the MN rule is therefore extremely necessary, since this effect may depend on unknown physical principles [44].

Finally, it must be stated that there are no detectable localized (defect) states near midgap in a-Chs, which is quite different from the situation in a-Si (not a-Si:H) and amorphous carbon. In these materials, hopping (tunneling) transport by multiphonon or single-phonon processes occurs through midgap states (see, for example, [61]). The Boltzmann transport theory traditionally used for crystalline materials has been applied here also for a-Chs. However, if the carrier scattering time is very short ($<10^{-15}$ s) or the carrier mean free path approaches the interatomic separation, this theory cannot be used for analyzing experimental data without some caution [62]. In fact, the importance of this issue has been pointed out when, in particular, metallic transport and the Hall effect in disordered matter have been discussed.

6.3.1.2 Hall Effect

The most anomalous behavior in the carrier transport in amorphous semiconductors is seen in the Hall effect, which was discussed in Section 5.5: the Hall effect has the opposite sign to that expected from the thermoelectric power. This is called the *pn anomaly*: holes give a negative sign and electrons a positive sign of the Hall voltage in a-Chs [54]. As the thermoelectric power should follow the classical Boltzmann theory, the signature of this measurement predicts which type of carriers dominate the transport, either p- or n-type.

While there have been several attempts to explain the pn anomaly [54], it has remained an unsolved issue. A treatment of quantum interference effects in electron transport may be needed when the carrier mean free path is shorter than a critical value [63].

6.3.1.3 Thermoelectric Power

As already discussed in Section 5.4.4, the thermoelectric power S is related to the Peltier coefficient Π by

$$S = \Pi / T \tag{6.18}$$

The Peltier coefficient is defined as the energy carried by electrons (or holes) per unit charge, and the energy is measured relative to the Fermi level E_F. Each electron (or hole) contributes to Π in proportion to its relative contribution to the total conductivity σ. Therefore Π (and hence S) is given by

$$\Pi = -\frac{1}{e}\int (E - E_F)\frac{\sigma(E)}{\sigma}\frac{\partial f}{\partial E}dE$$

and

$$S = -\frac{k}{e}\int \frac{E - E_F}{k_B T}\frac{\sigma(E)}{\sigma}\frac{\partial f}{\partial E}dE \tag{6.19}$$

where σ is given by the energy-dependent conductivity $\sigma(E)$ as

$$\sigma = e\int \mu(E)N(E)f(1-f)dE = -\int \sigma(E)\frac{\partial f}{\partial E}dE \tag{6.20}$$

where f is the Fermi distribution function and $f(1-f) = -k_B T\, df/dE$ and $\sigma(E) = e\mu(E)N(E)k_B T$. Note that $S < 0$ for electrons at energies $E > E_F$ and $S > 0$ for holes at energies $E < E_F$.

S for holes associated with the valence band (in the case of a-Chs) is given by [54, 56, 58]

$$S = \frac{k}{e}\left(\frac{E_F - E_v}{k_B T} + 1\right) = \frac{k}{e}\left(\frac{E_s}{k_B T} + 1\right) \tag{6.21}$$

where $E_s = E_F - E_v$. It is of interest to point out that the value of E_s which appears in S should be the same as the E_σ ($=\Delta E$) which appears in the conductivity (Equation (6.14). However, E_σ is always larger than E_s by an amount ΔW in a-Chs, similarly to a-Si:H as stated in Section 5.5. ΔW is reported to be around 0.2 eV in a-Chs [64, 65]. Why such a difference ΔW appears in a-Chs and a-Si:H is a matter of debate [66–68]. If the conduction or valence band edge fluctuates energetically for some reason [68], electrical conduction occurs above a percolation threshold energy (it has to surmount "cols" using an energy ΔW). For the thermoelectric power, this energy ΔW may be cancelled out through carriers moving both up and down.

The above view is similar to the two-channel model of conduction, that is, carrier transport occurs via both tail and band states [67, 69], since the hopping energy ΔW (up and down) is not involved in the thermoelectric power, as stated above. If small polarons dominate carrier transport, the relation $E_\sigma = E_s + \Delta W$ should be found, since the hopping energy is cancelled out in the thermoelectric power. ΔW therefore corresponds to the hopping activation energy, and hence the small-polaron binding energy is $2\Delta W$ [65, 70].

6.3.1.4 AC Transport

Details of the definition of AC transport have already been given for a-Si:H (Section 5.4.2), and hence we shall discuss only briefly the AC transport observed in a-Chs. In the earlier stages of the study of AC conductivity in a-Chs, the AC and DC losses were thought to have different origins; that is, free holes in the valence band contributed to the DC transport, and two electrons, as a *bipolaron*, could hop between oppositely charged coordination defect sites (e.g., C_3^+ and C_1^-), which would induce AC loss [71–74]. To understand the bipolaron hopping mechanism in detail, some knowledge about the nature of defects is needed, which will be provided in Section 6.5.

Note again that the bipolaron hopping model is based on the pair approximation (PA), in which $\sigma_1(\omega)$ cannot give a DC conductivity, because $\sigma_1(0)$ as $\omega \to 0$ (see Section 5.4.2). A model based on the PA cannot account for the experimental data if both DC and AC transport occur by the same mechanism [71, 72, 75]. As we stated in Section 5.4.2, the proper approach for overcoming this drawback of the PA seems to be the *continuous-time random-walk* (CTRW) approximation. A simple form of the AC conductivity based on the CTRW approximation was presented by Dyre [75, 76] as

$$\sigma^*(\omega) = \sigma(0)\frac{i\omega\tau}{\ln(1+i\omega\tau)} \tag{6.22}$$

where τ is the *maximum* hopping (or tunneling) relaxation time. It was shown that the AC conductivity was directly related to the DC conductivity $\sigma(0)$. Note again that the above equation can be applied when the DC and AC transport are due to the same hopping mechanism.

As stated before, the AC conductivity of a-Chs has been discussed in terms of bipolaron hopping using the PA model. The same experimental data were analyzed by use of the Dyre equation, that is, a random walk of bipolarons was assumed instead of the PA [77]. While the AC conductivity is dominated by band transport of holes, the AC conductivity at high frequencies is also explained well by a random walk of bipolarons with a better estimation of the number of charged defects than is obtained from the PA. The AC loss mechanism in the microwave range should be the same, and an example is shown in Figure 6.3. As the details are related to defect spectroscopy, we shall discuss them in Section 6.5.

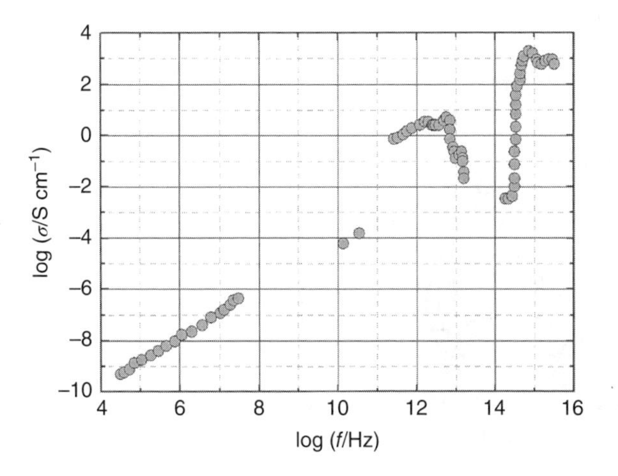

Figure 6.3 Frequency-dependent losses in the radio frequency, microwave, infrared, and ultraviolet ranges in glassy As_2Se_3 measured at 300 K. *Source*: Tanaka and Shimakawa 2011 [39]. Reproduced with permission from Springer Science and Business Media.

6.3.2 Ionic Transport

We must also discuss ionic conductors based on a-Chs, since ion-conducting chalcogenides have generated considerable interest in their application, for example in secondary ion batteries and fuel cells. Some ion-conducting a-Chs belong the class of so-called superionic conductors, whose DC conductivity exceeds 10^{-5} S cm^{-1}. While ionic conductors have been studied for a long time, the mechanism of transport in them is still not fully understood. An issue important for the application of these materials is a full understanding of their dynamics.

First, we show some typical examples of the dynamical properties of mobile ions of silver (Ag^+) in a-Chs doped with Ag [78–80]. It is known that a-As_2Se_3 doped with Ag shows ionic transport [81]. The dynamical properties of ions are usually deduced by the use of impedance spectroscopy (IS), in which the frequency-dependent (~10^{-3}–10^8 Hz) impedance is measured (see, for example, [82, 83]). A typical example of IS for glassy $Ag_{25}As_{25}S_{50}$, which is known to be a highly ionically conducting material, is shown in Figure 6.4. Although, in general, the real part Z_1 and the imaginary part Z_2 of the impedance (the Z_1–Z_2 complex plane) are used for analysis in the IS method, taking the real and imaginary parts of the resistivity ρ_1 and ρ_2, respectively, is known to be convenient for analysis in practice [80].

For relatively highly conductive materials, as seen in the figure, the ρ_1–ρ_2 complex plane consists of a high-frequency semicircle (at lower resistivity) and a low-frequency tail (at higher resistivity). It is known that the high-frequency semicircle relates to the conduction in bulk regions of the sample and the

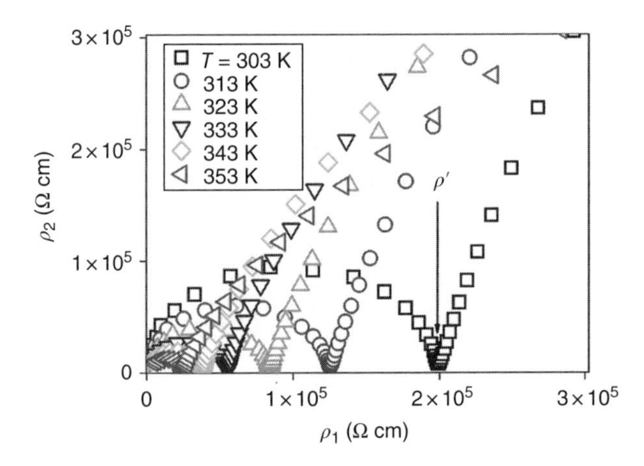

Figure 6.4 Impedance spectroscopy: ρ_1 (real part of resistivity) and ρ_2 (imaginary part of resistivity) of glassy $Ag_{25}As_{25}S_{50}$. *Source*: Patil *et al.* 2014 [80]. Reproduced with permission from AIP Publishing LLC.

low-frequency tail relates to the electrode (or interface) polarization. These properties are traditionally analyzed by using an equivalent electrical circuit (EEC). However, the EEC approach, which is macroscopic in principle, does not directly produce any physical parameters such as the number of mobile ions or the diffusion coefficient.

Recently, instead of using the EEC, a microscopic approach to analyzing the results of impedance spectroscopy has been proposed [78–80]. We briefly introduce this analysis method here. A random-walk approach to describing mobile ions was employed for both the bulk and the interface in glassy $Ag_{25}As_{25}S_{50}$, in a case that we shall use as an example. Figure 6.5 shows the real and imaginary parts of the conductivity σ_1 and σ_2, respectively, which are dependent on the angular frequency ω. The open circles and solid lines, respectively, show the experimental and calculated results. The Dyre equation (6.22) for electronic hopping transport can also be applied to ionic transport, since the basic mathematical treatment for electronic process should be the same as that for ionic process [75].

The conductivity due to DC hopping (over a potential barrier) $\sigma(0)$ in 3D space is given by [79, 80]

$$\sigma(0) = \frac{N(eR)^2}{6k_B T \tau_m} \tag{6.23}$$

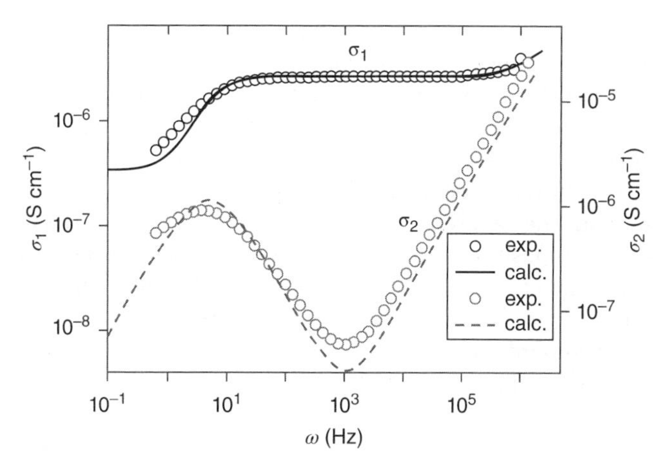

Figure 6.5 Frequency dependence of σ_1 (real part of conductivity) and σ_2 (imaginary part of conductivity) for glassy $Ag_{25}As_{25}S_{50}$. *Source*: Patil *et al.* 2014 [80]. Reproduced with permission from AIP Publishing LLC.

where N is the number of mobile ions (here Ag^+), R is the hopping distance, and τ_m is the maximum hopping time, which was discussed in Section 6.3.1. The onset frequency of the frequency-dependent σ_1 occurs at around $1/\tau_m$.

The interfacial region is connected in series with the bulk, and hence the overall complex conductivity can be estimated from the values of the bulk and interfacial conductivities. Then the overall complex conductivity $\sigma^*(\omega)$ $(= \sigma_1 + i\sigma_2)$ is given by

$$\frac{1}{\sigma^*(\omega)} = \frac{f}{\sigma_i^*(\omega)} + \frac{1-f}{\sigma_b^*(\omega)} \tag{6.24}$$

where f is the spectroscopic weight of the interfacial region, and the subscripts i and b denote the interface and bulk conductivities, respectively. It was assumed that the transport mechanism in the interfacial region was also dominated by Dyre's random-walk process [79, 80, 84]. The fitting to the experimental results for σ_1 and σ_2 produces $N_b \sim 3 \times 10^{21}\,\text{cm}^{-3}$ and $\tau_m \sim 3 \times 10^{-6}\,\text{s}$, and a diffusion coefficient $D_b \sim 1 \times 10^{-10}\,\text{cm}^{-2}\,\text{s}^{-1}$ for the bulk region, with $N_i \sim 3 \times 10^{21}\,\text{cm}^{-3}$ and $\tau_m \sim 1\,\text{s}$, and a diffusion coefficient $D_i \sim 3 \times 10^{-16}\,\text{cm}^{-2}\,\text{s}^{-1}$ for the interfacial region ($f \sim 2 \times 10^{-5}$) [80]. The new approach to IS summarized here may be very useful, and will overcome the drawbacks of the traditional EEC approach.

Next, it is of interest to discuss the so-called power-law dependence of the ionic conductivity and diffusion coefficient on the metallic content of chalcogenides and oxide glasses [85–89]. For example, in Ag–Ge–S glasses, the relations $\sigma_{DC} \propto (x_{Ag})^\alpha$ and $D_{Ag} \propto (x_{Ag})^\beta$ are found experimentally, where x_{Ag} is the Ag content in at.% (between 0.003 and 5) in the system, σ_{DC} is the DC conductivity, which is usually measured by the IS method described earlier, D_{Ag} is the tracer diffusion coefficient, and α and β are temperature-dependent constants. Note that no significant structural change has been reported in this composition range.

Figures 6.6(a) and (b) show the Ag concentration dependences of σ_{DC} and D_{Ag}, respectively, measured at 298 and 373 K, for the Ag–Ge–S system. Here, the relation $\alpha \sim \beta + 1$ is empirically obtained. The reason for this will be discussed later, together with the values of α and β. There are several models to explain the power-law composition dependence [85–89]. Among them, the simplest model that accounts for the experimental data may be the configuration entropy change model [89], which we now briefly discuss.

The Ag^+ ions must surmount a potential barrier. The Gibbs free energy G should be used for the potential barrier, as given by

$$G = H - T_0 S \tag{6.25}$$

where H is the enthalpy, S is the entropy, and T_0 is the frozen-in temperature ($= T_g$ for glasses). H may have a constant value for a moderate concentration of

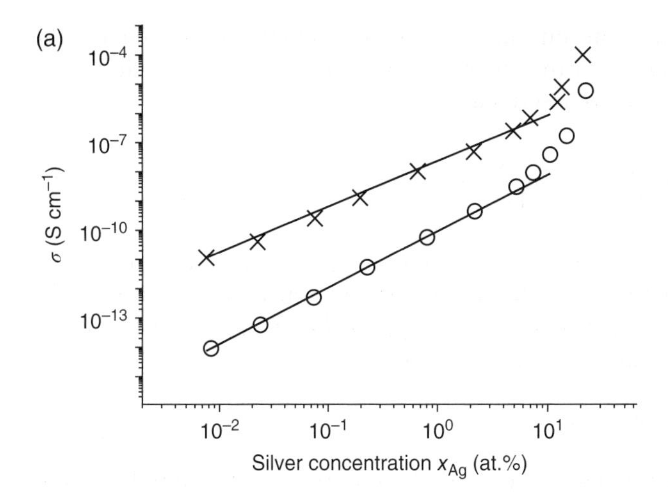

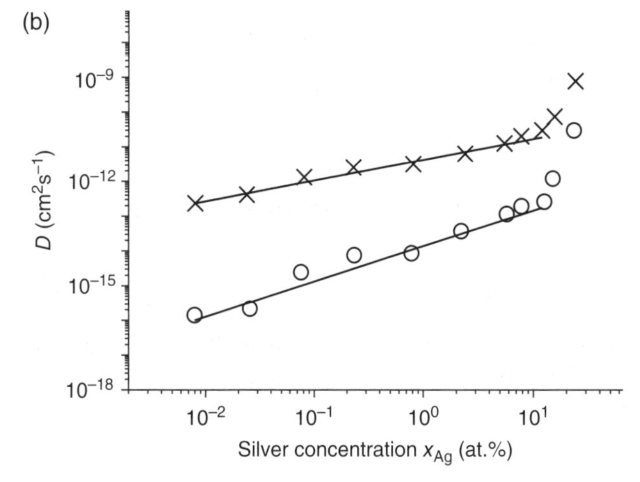

Figure 6.6 Silver concentration dependence of (a) the DC conductivity σ_{DC} and (b) the diffusion coefficient D measured at 298 and 373 K for the Ag–Ge–S system. *Source*: Shimakawa and Wagner 2013 [89]. Reproduced with permission from AIP Publishing LLC.

Ag in an ideal mixture. The network, containing a homogeneous mixture, has a change in entropy given by

$$S = k_B \ln \frac{N!}{n!(N-n)!} \tag{6.26}$$

where k_B is the Boltzmann constant, N is the total number of sites, and n is the concentration of Ag. Then the chemical potential μ_{Ag} of Ag (the Gibbs free energy for one Ag atom) is described by

$$\mu_{Ag} = \frac{\partial G}{\partial n} = \mu_{Ag}^{\ *} + k_B T_g \ln \frac{x'_{Ag}}{1 - x'_{Ag}} \tag{6.27}$$

where $\mu_{Ag}{}'$ is a constant, and $x_{Ag}{}'$ (= $x_{Ag}/100$) is equal to n/N, because x_{Ag} is in atom percent. When $x_{Ag} \ll 1$, the second term in Equation (6.27) is given by $k_B T_g \ln(x_{Ag}{}')$. Then G (the activation energy for hopping) decreases by this amount when Ag is introduced, according to

$$G = G_0 - k_B T_g \ln(x_{Ag}) \tag{6.28}$$

where G_0 is a constant and the concentration of Ag^+ is assumed to be almost the same as that of Ag.

Let us first discuss the diffusion coefficient D_{Ag}. Diffusion is thermally activated, and hence D_{Ag} can be given as

$$D_{Ag} = D_0 \exp(-G/k_B T) \tag{6.29}$$

where D_0 is a constant. By combining the above two equations, we get

$$D_{Ag} = D_0 \exp(-G_0/k_B T)(x_{Ag})^{T_g/T} \tag{6.30}$$

The above equation is just what we observe experimentally. The parameter β should be equal to T_g/T. From Equation (6.29), using the Nernst–Einstein relation, σ_{DC} is given by

$$\sigma_{DC} = \frac{e^2 N x_{Ag}}{k_B T} D_{Ag} = \sigma_0 \exp(-G_0/k_B T)(x_{Ag})^{1+T_g/T} \tag{6.31}$$

where σ_0 is a constant and the parameter α is equal to $1 + T_g/T$ ($=1+\beta$), which agrees with the experimental results.

We now understand that the power law can be simply understood by considering the compositional dependence of the Gibbs free energy change on mixing of an additive, as predicted in Equation (6.25). In fact, the activation energy follows Equation (6.28), as shown in Figure 6.7. The open circles and crosses are the experimentally observed activation energies for σ_{DC} and D_{Ag}, respectively. The solid line is a least-squares fit to the experimental data, with $k_B T_g = 0.05\,\text{eV}$ ($T_g = 580\,\text{K}$), which is close to the value obtained from thermal analysis [90].

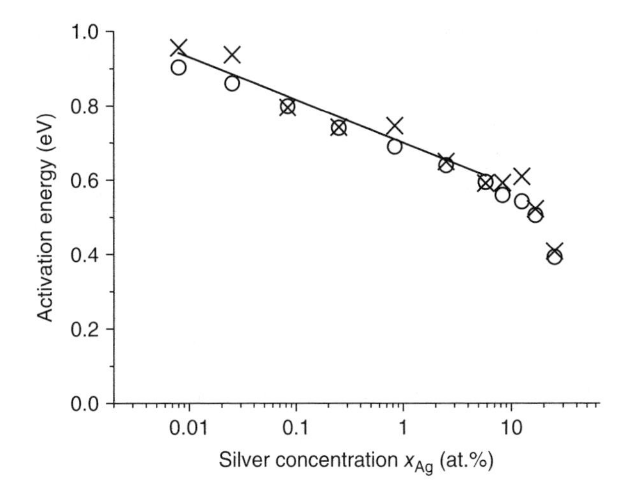

Figure 6.7 Silver concentration dependence of the activation energy for the Ag–Ge–S system. *Source*: Shimakawa and Wagner 2013 [89]. Reproduced with permission from AIP Publishing LLC.

6.4 Optical Properties

6.4.1 Fundamental Optical Absorption

As already discussed in Section 5.5, the principal optical absorption near the visible photon energy range may be produced by electronic transitions between the conduction and valence bands. This is called the fundamental optical absorption. The forms of the density of states for both the conduction and the valence bands therefore dominate the fundamental optical absorption. It is known that the valence band is formed in a-Chs from lone pair states and the conduction band from covalent antibonding states. The form of the DOS for the valence band may not always be the same as that for the conduction band. Furthermore, the DOS in disordered matter in general may need to be described by taking *fractal concepts* into account [91]. Note that the importance of fractal concepts is well recognized in the field of disordered matter [92]. Thus, it is of interest to generalize the discussion of fundamental optical absorption in terms of fractal concepts [44, 56, 93, 94].

The volume V of a sphere in a fractal space is proportional to r^D, where r is the radius and D is the fractal dimension, which is different from the Euclidean space dimension d (=1, 2, 3). The DOS is then given by

$$N(E)dE = AE^{(D-2)/2}dE, \tag{6.32}$$

where A is a constant [95]. In a normal homogeneous space, D should be the same as d; for example, $D = d = 3$ leads to $N(E)$ proportional to $E^{1/2}$, as described in Chapter 5. However, D will be smaller than the dimensions of the Euclidean space if the material contains a large amount of voids or contains inhomogeneities.

Let us discuss the fundamental optical absorption in a fractal space of dimension D. As described in Section 5.5, the optical absorption coefficient $\alpha(\omega)$ for interband electronic transitions is given as follows [54]:

$$\alpha(\omega) = B\int \frac{N_c(E)N_v(E - \hbar\omega)dE}{\hbar\omega} \tag{6.33}$$

where B is a constant and $N_c(E)$ and $N_v(E)$ are the DOS for the conduction band and valence band, respectively. Let us express each DOS as

$$N_c(E) = \text{const}(E - E_c)^{\alpha} \tag{6.34}$$

$$N_v(E) = \text{const}(E - E_v)^{\beta} \tag{6.35}$$

where $\alpha = (D_c - 2)/2$ and $\beta = (D_v - 2)/2$. Here D_c and D_v are the fractal dimensionalities of the conduction and valence bands, respectively. Then Equation (6.33) produces the following equations [93]:

$$\alpha(\omega)\hbar\omega = B'(\hbar\omega - E_0)^{\alpha+\beta+1}, \tag{6.36}$$

where B' is another constant, which yields

$$\left[\alpha(\omega)\hbar\omega\right]^n = B'^{1/n}(\hbar\omega - E_0)s \tag{6.37}$$

where $1/n = \alpha + \beta + 1$. If the forms of both $N_c(E)$ and $N_v(E)$ are parabolic ($\alpha = \beta = 1/2$ for 3D space), then Equation (6.37) becomes

$$\left[\alpha(\omega)\hbar\omega\right]^{1/2} = B(\hbar\omega - E_0) \tag{6.38}$$

This is the well-known Tauc relation, and E_0 is often called the Tauc gap or optical gap (see Section 5.5).

As shown in Figure 6.8, the Tauc relation for the fundamental optical absorption can in general be applied to a-Chs, although there are some exceptions: for example, n may deviate from 1/2 and take values between 0.3 and 1. In glassy chalcogenides, the fundamental optical absorption follows the Tauc law ($n = 0.5$). However, as shown in Figure 6.9, the optical absorption of a-As_2S_3 films follows Equation (6.46) with $n = 0.7$ (Figure 6.9(a)) and 0.59 (Figure 6.9(b)) for as-deposited and well-annealed films, respectively. A value of n greater than 0.5 means that the value of D_c and/or D_v is smaller than 3 (the value for Euclidean space), suggesting that the as-deposited films may have a more fractal nature structurally.

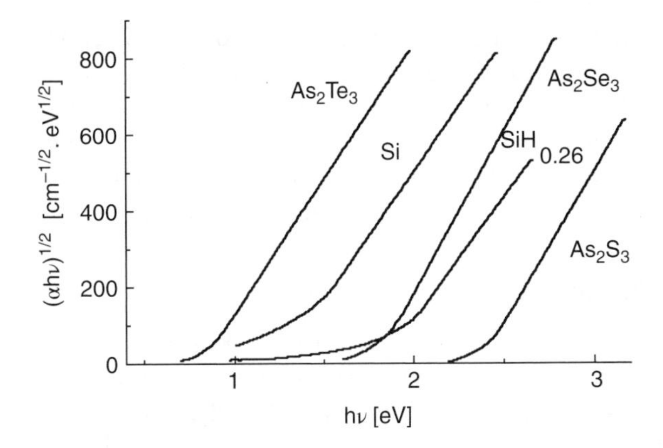

Figure 6.8 Tauc plots for various amorphous semiconductors. *Source*: Singh and Shimakawa 2003 [56]. Reproduced with permission of Taylor & Francis Ltd.

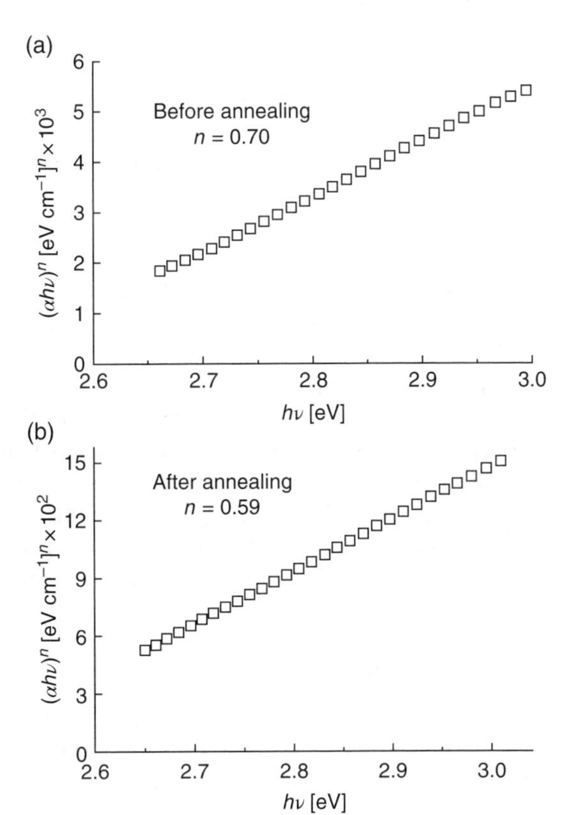

Figure 6.9 Optical absorption spectra of obliquely deposited a-As_2Se_3, fitted to $(\alpha h\nu)^n$ versus $(h\nu - E_0)$. (a) $n = 0.7$ before annealing; (b) $n = 0.59$ after annealing at 170 °C for 2 h.

It should be mentioned that $n \sim 1$ for a-Se has been reported [54]. When we use the fractal concept for this behavior, the value of $(D_c + D_v)$ should be around 4. The reason for this is not very clear. However, this peculiar effect in a-Se may be related to its chainlike structure: the antibonding states (conduction band) have a 1D nature, as long as no interaction between the chains is assumed. It is therefore expected that $\alpha = -1/2$ and $\beta = 1/2$, producing $n = 1$ [93].

6.4.2 Urbach and Weak Absorption Tails

As stated for a-Si:H (Section 5.5.2), an exponential optical absorption tail below the energy of the fundamental optical absorption edge, which is often called the Urbach tail [96], is also found in a-Chs. Many disordered materials, including organic materials, exhibit an Urbach tail that can be expressed as

$$\alpha = \alpha_0 \exp\{(\hbar\omega - E_0)/E_U\} \tag{6.39}$$

where E_0 is the Urbach focus and E_U, representing the exponential steepness, is called the Urbach energy. α is found to be between 10^0 and $10^3\,\mathrm{cm}^{-1}$ and its origin has long been an unclear issue in this field [97]. E_U is around 50–100 meV for a-Chs [97]. It is known that E_U increases with some kinds of disorder [56, 98].

It has been shown that E_U increases with the average atomic coordination number Z, for example in the Ge–As–Se and Si–P–S systems [99], suggesting that the topological strains in covalent networks become higher with an increase in the coordination number, producing an increase in the disordered nature. The relation $E_U \approx 1.3 k_B T_g$ has also been theoretically predicted [100, 101]. As stated in Section 6.2.1, T_g is empirically proportional to Z, and E_U being proportional to Z may be a natural consequence. As E_U seems to be strongly dependent on the disordered nature of materials and the Urbach tail is widely observed in different material systems, it is of interest to present the reported E_U values for other material systems, as shown in Figure 6.10. There may be no E_U values smaller than around 50 meV; that is, there seems to exist a minimal Urbach energy of $E_U \approx 50$ meV in disordered materials. This energy can be related to the intrinsic density fluctuations which are inherent in the medium-range structure, with a scale of 1 – 2 nm. Other disordered structures such as those with compositional disorder and heterogeneity, depending on the material system, tend to show a further increase in the Urbach energy [97].

Below $\alpha < 10^0\,\mathrm{m}^{-1}$ at lower photon energies, we find another exponential tail:

$$\alpha \sim A \exp(\hbar\omega / E_W) \tag{6.40}$$

This is called the weak absorption tail, and the tail factor E_W is reported to be ~ 200 meV for a-As_2S_3, for example [97]. This weak absorption can be attributed to the existence of impurities even if the sample can be regarded as

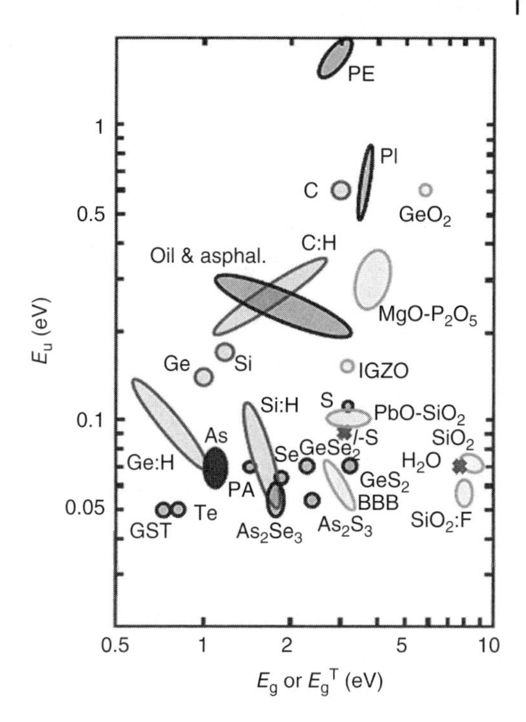

Figure 6.10 Urbach energy versus band gap for various materials. *Source*: Tanaka 2014 [97]. Reproduced with permission from Elsevier.

intrinsic. Figure 6.11 shows the weak absorption tail for a-As_2S_3 doped with Fe [102]. The open circles show measurements for Tanaka's purest As_2S_3 with the same E_W [39]. We hence suggest that the weak absorption is attributed to impurity effects. It is not clear why the absorption takes the form of an exponential function like the Urbach tail.

6.4.3 Photoluminescence

The study of photoluminescence (PL) in chalcogenide glasses has continued for several decades after the pioneering work on this topic [103, 104]. Continuous-wave (CW) light sources were used for these early PL studies on *glassy* chalcogenides, and the first comprehensive review of PL was provided by Street [105]. To obtain a clearer insight into PL dynamics, time-resolved PL using pulsed excitation [106] can be used. Frequency-resolved spectroscopy (FRS) may give direct information about the lifetime and its distribution. In particular, *quadrature frequency-resolved spectroscopy* (QFRS) is known to be more suitable for studying amorphous semiconductors that have a broad PL lifetime distribution [107].

The model first proposed for CW PL was based on the model of charged dangling bonds, which will be discussed in Section 6.5. Here, a photoinduced electron (or hole) is trapped by a charged defect D^+ (or D^-) and then a hole (or

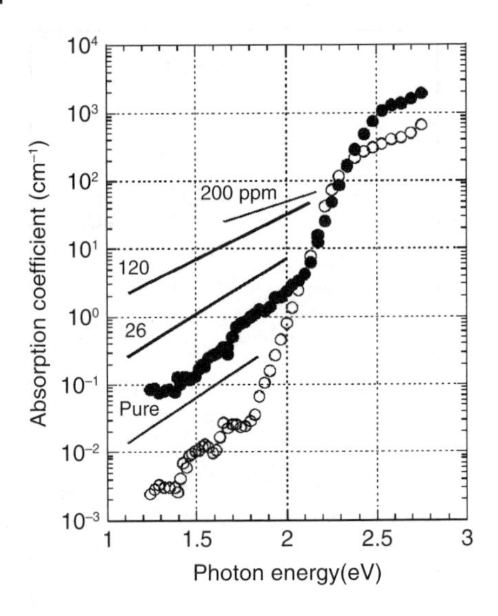

Figure 6.11 Purity-dependent weak absorption tail for glassy As_2S_3. *Source:* Tanaka and Shimakawa 2011 [39]. Reproduced with permission from Springer Science and Business Media.

electron) recombines with the trapped electron (or hole) with emission of a photon. A large Stokes shift, that is, an energy difference between the absorption and emission energies, should be observed if structural relaxation occurs upon charge trapping at defects D^+ or D^- [105]. Subsequent results from time-resolved PL (PL decay) are not consistent with the above picture, however. As will be discussed below, a short-time decay component with a polarization memory ($<10^{-6}$ s) exists, which originates from the anisotropy of the optical transition dipole moment of the PL center, suggesting that an electron–hole pair created at the center in such an anisotropic structure recombines radiatively without transferring to another site [106].

A possible mechanism for the origin of short-time PL with polarization memory is based on a self-trapped exciton (STE), which can be induced by direct excitation of electrons into conduction band tail states. A schematic illustration of the STE model is shown in Figure 6.12. STE formation in a-As_2Se_3 and As_2S_3 after optical excitation has been proposed [108–110], in which an electron is self-trapped by an As–Se or As–S bond. Optical excitation (I) to an excited state can be followed by one of two nonradiative decay channels, either directly back to the ground state (III) or to the metastable STE state (IV). Radiative recombination of this kind of STE occurs on a short timescale [38, 106]. The time decay behavior is described by a stretched exponential function, and the origin of this behavior has been discussed in the context of a model of localized excitons [111].

We summarize here two characteristic features of PL in chalcogenides: (i) the large Stokes shift, and (ii) the PL lifetime distribution. In the case of (i), when chalcogenides are excited by light with $\hbar\omega \approx E_0$ (the Tauc gap), the

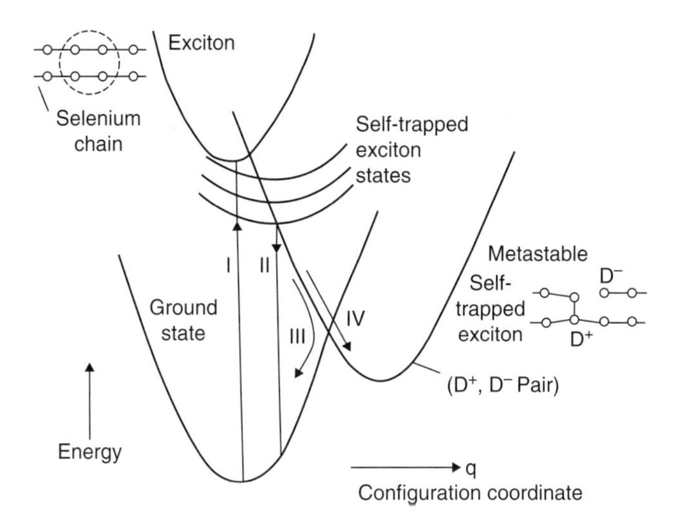

Figure 6.12 A model for the self-trapping of excitons in chalcogenide glasses. *Source*: Shimakawa *et al.* 1995 [28]. Reproduced with permission of Taylor & Francis.

luminescence appears at a peak energy $E_{PL} \approx E_0/2$ with a broad spectrum. This is often called the half-band-gap rule (HBR). Figure 6.13 shows such behavior observed in a-Chs. For comparison with other systems, experimental data from oxide glasses (SiO_2 and GeO_2) have also been inserted. We can hence understand that this rule is commonly obeyed by both chalcogenides and oxides, and may be the most puzzling effect in relation to PL. It should be noted that the PL behavior observed in the amorphous and crystalline states is almost the same [105, 112] (HBR is not a unique feature of a-Chs!).

If we simply assume that there are PL centers near midgap, HBR is just a result of this excitation–recombination process. Then, a question may arise. What is this process? The most commonly accepted answer to this question is given by the *charged defect* concept, as stated already [105]. In this case, nearly half of the band gap energy is expected to be consumed by phonon emission (owing to electron–lattice interaction). Similarly to the charged dangling bond model of PL in the case of STE recombination, although the short-time PL is easy to explain, the origin of the HBR is still not clear.

Let us now go into (ii) the PL lifetime distribution. There are three main peaks in the lifetime distribution ($\sim 10^{-6}$, 10^{-3}, and 10^{2} s) for a-Chs, which are similar to those for a-Si:H. Figure 6.14 shows results from QFRS in glassy As_2S_3 [113]. The lifetimes, $\sim 10^{-8}$ and $\sim 10^{-4}$ s, are thought to originate from exciton-like singlet (S) and triplet (T) recombination, respectively, while time-resolved PL studies [106] on the same material show three decay components, $\sim 10^{-8}$,

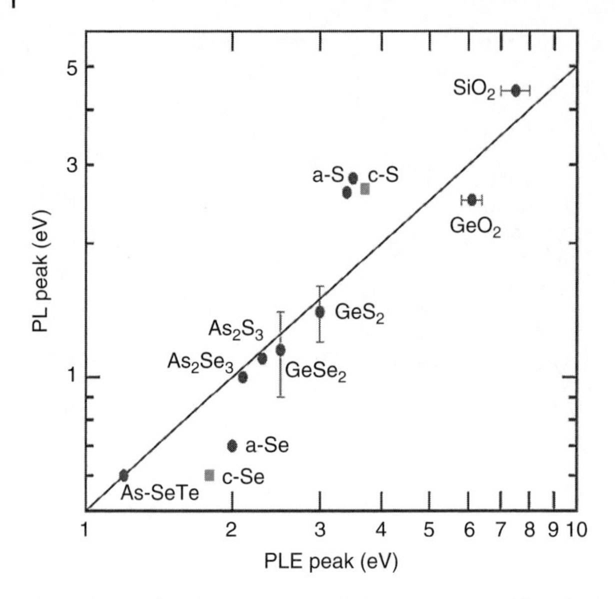

Figure 6.13 Photoluminescence peak energy versus photoluminescence excitation (PLE) peak energy for chalcogenide and oxide glasses and crystalline chalcogenides. *Source*: Tanaka and Shimakawa 2011 [39]. Reproduced with permission from Springer Science and Business Media.

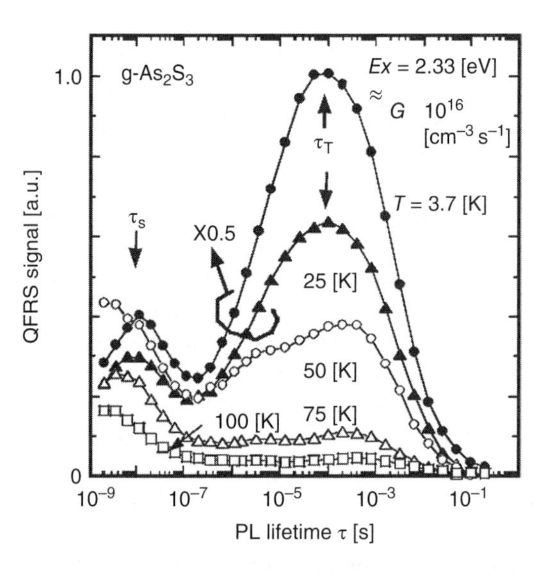

Figure 6.14 Frequency-resolved photoluminescence in glassy As_2S_3. *Source*: Aoki *et al.* 2005 [113]. Reproduced with permission.

~10^{-6}, and ~10^{-4} s. The 10^{-6} s component is missing in the QFRS results, which may be due to the different excitation intensity. Note that a high light excitation (intense pulses of 10 ns width) is required for time-resolved PL measurements, while in the QFRS technique there is no need for high-intensity excitation, which ensures a linear response [114].

Finally, we should discuss recent photonic applications of a-Chs. Chalcogenide glass optical fiber amplifiers and other devices can be made. So-called broadband excitation, that is, the excitation of rare earth ions by photoexciting a host chalcogenide glass, has been demonstrated [115]. The rare earth ions are not directly excited by photons, and the mechanisms of energy transfer from the host material are still not clear [116]. Recently, *upconversion* of PL (an anti-Stokes shift) has been demonstrated, also using rare earth ions in a-Chs [117]. Two-step photoexcitation, through an intermediate energy state, and subsequent photoemission produce PL of a higher energy than the excitation. Studies of QFRS are important for understanding the dynamics of this process, and are being conducted intensively [118].

6.4.4 Photoconduction

The methods used to investigate photoconductivity are classified into *primary* and *secondary* photoconductivity measurements. So-called time-of-flight (TOF) spectroscopy belongs to the class of primary photoconductivity measurements; here, photocarriers transit from an illuminated electrode to a counterelectrode (see also Section 5.5.5). The more common steady-state or time-dependent (usually photoconductivity decay) spectroscopy techniques are called secondary photoconductivity measurements, in which the whole sample is illuminated. TOF experiments are therefore completely different from secondary photoconduction experiments and involve photocarriers far from thermal equilibrium. This technique is used for deducing the carrier *drift* mobility. We hence discuss the two types of photoconductivity measurement separately.

6.4.4.1 Primary Photoconduction Measurements

A sample with a sandwich configuration is used in TOF measurements, as shown in Figure 6.15. Both electrons and holes are produced in a thin sheet at an optically transparent electrode using a short-pulse laser. We therefore need light of energy much higher than the band gap (large α), which produces a thin sheet of electron–hole pairs. One type of carriers (electrons or holes) drifts across the relatively thick sample (thickness d) under an external applied voltage V.

While the packet of carriers (electrons or holes) is moving through the sample at a speed v_d, a constant current (a plateau region with a time duration t) is induced in the external circuit (an appropriate resistance R connected in

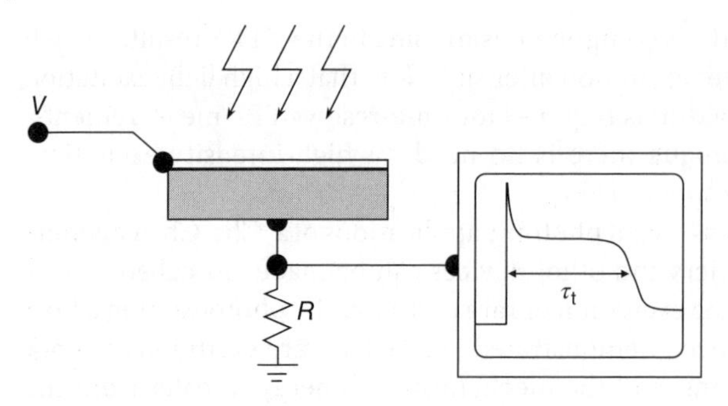

Figure 6.15 Measurement system for the transient photoresponse (time of flight) of photoconductive insulating films. τ_t is called the transit time.

series with the sample). When the carrier packet reaches the counterelectrode at time τ_t, the induced current immediately decays to zero. The drift mobility μ_d is then estimated from

$$\mu_d = \frac{v_d}{V/d} = \frac{d/\tau_t}{V/d} = \frac{d^2}{V\tau_t} \tag{6.41}$$

where τ_t is called the transit time. When the charge packet has a Gaussian form, that is, the carrier packet is broadened by a Gaussian profile from the mean position by $l_m = (2Dt)^{1/2}$ at time t, where D is the diffusion coefficient of carriers passing through the medium, the shape of the current has a relatively sharp edge (a sharp, approximately rectangular shoulder) [119].

Hole transport at room temperature in a-Se shows a sharp edge in the photocurrent [119]. However, in the majority of circumstances, including low-temperature measurements on the same sample, TOF studies reveal characteristics markedly different from those mentioned above [119]: no plateau region in the photocurrent is observed, that is, there is no clear shoulder in the i_p versus t curve, indicating a spread in the arrival time [120, 121]. We call this non-Gaussian or dispersive transport. A plot of log i_p versus log t, fortunately, reveals a break (shoulder) in the curve, and $i_p(t)$ is given empirically as

$$i_p(t) = At^{\beta_1 - 1} \qquad (t < \tau_t) \tag{6.42}$$

and

$$i_p(t) = Bt^{-(1+\beta_2)} \quad (t > \tau_t) \tag{6.43}$$

where β_1 (<1.0) and β_2 (<1.0) are called the dispersion parameters, and τ_t the transit time (a shoulder in the log i_p versus log t curve).

Scher and Montroll [122] first gave the theoretical background to dispersive transport in terms of the CTRW approximation. The most popular model to explain dispersive transport as described in Equations (6.42) and (6.43) is the multitrapping (MT) model [123, 124], in which photoexcited carriers, under thermal equilibrium, repeatedly experience trapping and detrapping between band and tail states. In this model, $\beta_1 = \beta_2$ $(= k_B T/kT_c)$, where $k_B T_c$ is a characteristic energy of the tail states (the width of the tail on an energy scale), fits the experimental data well for a-Si:H [125].

It is known, however, that the MT model cannot be applied to a-Chs [56]: unlike the case for a-Si:H, the values of β_1 and β_2 are not generally the same (e.g., $\beta_1 = 0.68$ and $\beta_2 = 0.77$ for a-As_2Se_3 [120]), and these parameters are almost temperature-independent (not given by $k_B T/kT_c$). An example of TOF data for a-As_2Se_3 is shown in Figure 6.16. The drift mobility of holes at room temperature ($\sim 10^{-5}\,cm^2\,V^{-1}\,s^{-1}$) in a-Chs is very much smaller than that ($\sim 1\,cm^2\,V^{-1}\,s^{-1}$) of electrons in a-Si:H. The low value of the drift mobility in a-Chs may be due to deep trapping events. In fact, the large activation energy of the drift mobility, for example 0.55 eV for a-As_2Se_3 and 0.25 eV for a-Se, suggests that the carrier drift in a-Chs is dominated by deep localized states. Note that the activation energy of the drift mobility of a-Si:H is reported to be ~0.1 eV. The number of deep states is expected to be higher than that in a-Si:H. As a-Se is physically and technologically a very important material, we will discuss the relation between the drift mobility and the electronic density of states in a-Se in Section 6.5.

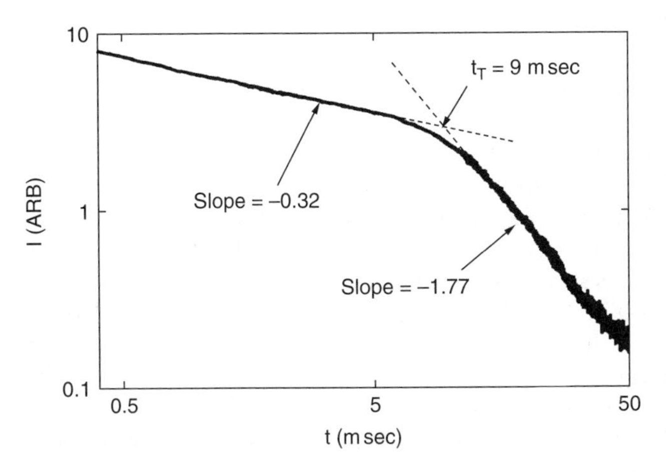

Figure 6.16 Time-of-flight hole current in a-As_2Se_3 on a logarithmic scale. t_T here corresponds to τ_t in Figure 6.15. *Source*: Singh and Shimakawa 2003 [56]. Reproduced with permission of Taylor & Francis.

6.4.4.2 Secondary Photoconduction Measurements

The (secondary) photoconductivity σ_p is defined as

$$\sigma_p = eG\tau\mu \tag{6.44}$$

where G is the free-carrier generation rate (in cm^{-3} s^{-1}), τ is the recombination time (in s), and μ is the free-carrier mobility (or microscopic mobility) (in cm^{-2} V^{-1} s^{-1}). When transport occurs in the band tail states, μ represents the hopping mobility, which is very much smaller than the free-carrier mobility. The product of G and τ gives the number of photogenerated carriers under steady-state illumination. When a thin film of thickness d is used, G may be given by

$$G = \eta N_0 (1 - R)[1 - \exp(-\alpha d)]/d \tag{6.45}$$

where η is the quantum efficiency of generation of photocarriers, N_0 is the number of incident photons per unit area (in cm^{-2} s^{-1}), R is the reflectivity, and α is the optical absorption coefficient (in cm^{-1}). Note that multiple reflection is not taken into consideration here.

If the condition $\alpha d \ll 1$ is satisfied, photons can be absorbed uniformly throughout the sample, and σ_p is then expressed by

$$\sigma_p = e\eta\alpha N_0 (1 - R)\mu\tau \tag{6.46}$$

In general, τ is proportional to $G^{(\gamma-1)}$ and thus σ_p varies as G^γ.For $\gamma < 1.0$, the dependence of σ_p on G is *sublinear*, and for $\gamma > 1.0$ it is *superlinear*. The value of γ provides background information about the recombination processes which dominate the photoconduction in materials [126], and hence it will be discussed later.

The following two parameters can be used for characterizing photoconductors. The first is the *photosensitivity*, which is defined by the product of μ and τ, called the $\mu\tau$ product and given by $\mu\tau = \sigma_p/eG$ (see Equation (6.44)). Another definition of the photosensitivity is given macroscopically by $(\sigma_p - \sigma_d)/\sigma_d$, where σ_d is the dark conductivity. The second factor for characterizing photoconductors is the *response time*: when the illumination is turned on or off, the photocurrent reaches its steady-state value or zero in an exponential manner, according to $1 - \exp(-t/\tau)$ or $\exp(-t/\tau)$, respectively. The time delay is defined as the response time. If there are no localized states, the recombination time τ is the same as the response time, otherwise the response time is dominated by trapping/detrapping processes at localized states [126].

Usually, for a-Chs, the rise or decay of the photocurrent is not given by a simple exponential curve of the kind mentioned above, suggesting that additional processes of trap filling and emptying may be involved, and hence the response time becomes longer than the recombination time [126]. Empirically, in most cases, the response is given by a stretched exponential, $\exp[-(t/\tau)^{\beta}]$, where β (<1.0) is called the dispersion parameter, instead of a simple exponential time response. This issue will be also addressed later on.

Before proceeding with the discussion, let us define the *quasi-Fermi levels* (or steady-state Fermi levels) E_{Fp} for holes and E_{Fn} for electrons, under steady-state illumination. These are defined by

$$E_{\mathrm{Fp}} - E_{\mathrm{v}} = k_{\mathrm{B}}T\ln\left(N_{\mathrm{v}}/p\right) \tag{6.47}$$

where E_{v} is the valence band edge, N_{v} is the effective density of states for the valence band, and p is the number of free holes, and

$$E_{\mathrm{Fn}} - E_{\mathrm{c}} = k_{\mathrm{B}}T\ln\left(N_{\mathrm{c}}/n\right) \tag{6.48}$$

where E_{c} is the conduction band edge, N_{c} is the effective density of states for the conduction band, and n is the number of free electrons. It is known that the locations of the quasi-Fermi levels play a key role in recombination processes [56].

Figure 6.17 shows a typical example of the temperature variation of σ_{p} as a function of light intensity for the well-known photoconductive material a-Sb_2Te_3. The photoconductivity has a maximum at a specific temperature T_{m}. For $T < T_{\mathrm{m}}$, σ_{p} obeys the relation $\sigma_{\mathrm{p}} \propto G^{1/2}$ at relatively high G and decreases with decreasing temperature as [121]

$$\sigma_{\mathrm{p}} \propto \exp\left(-W/k_{\mathrm{B}}T\right) \tag{6.49}$$

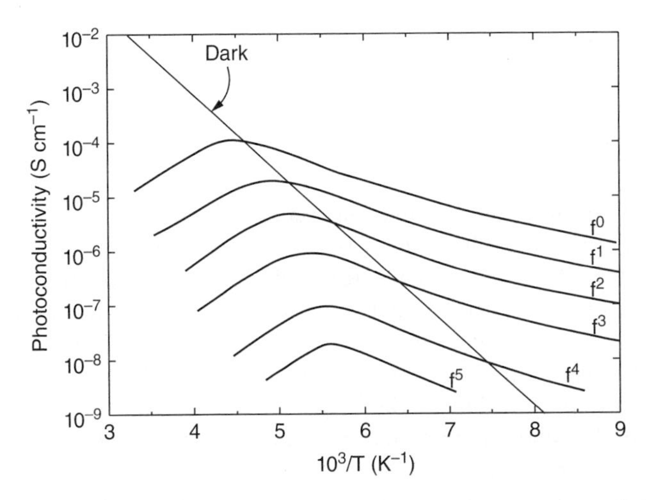

Figure 6.17 Temperature variation of photoconductivity of an a-Sb_2Te_3 film. The highest light intensity f^0 corresponds to 2.5×10^7 photons/cm^2 s. *Source*: Singh and Shimakawa 2003 [56]. Reproduced with permission of Taylor & Francis.

where W is about 0.1 eV for a-Sb_2Te_3. The activation energy W appears to be independent of G. Eventually, at very low temperature, σ_p becomes temperature independent for amorphous semiconductors (a-Chs, a-Si:H, and others) [127], as shown in Figure 6.17. Hopping of photocarriers through a particular energy level in the band tail, called the *transport energy level*, has been suggested to dominate the low-temperature behavior [128, 129].

Traditionally, the photoconductivity in the medium-temperature range has been interpreted as follows. The quasi-Fermi level mentioned above moves towards the respective band tail states at higher G. Then bimolecular tail-to-tail recombination becomes dominant. Let us denote the numbers of localized holes and electrons in their respective band tails by Δp_t and Δn_t. Then, the relation $\Delta p_t \Delta n_t = \Delta p_t^2 \propto G$ leads to $\Delta p_t \propto G^{1/2}$. Note that holes, which have a larger mobility than electrons, are the dominant carrier in a-Chs. In thermal equilibrium, Δp_t and the number of excess free holes Δp are given by

$$\Delta p_t = N_{tv} \exp\left[-\left(E_{Fp} - E_{tv}\right)/k_B T\right] \tag{6.50}$$

where N_{tv} is the effective density of tail states and E_{tv} its energy location, and

$$\Delta p = N_v \exp\left[-\left(E_{Fp} - E_v\right)/k_B T\right] \tag{6.51}$$

where N_v is the effective density of states for the valence band and E_v is the band edge energy of the valence band. Using above two equations, Δp can be written as

$$\Delta p = \Delta p_t N_v \exp\left[-\left(E_{tv} - E_v\right)/k_B T\right] \tag{6.52}$$

and then

$$\sigma_p = G^{1/2} \exp\left[-\left(E_{tv} - E_v\right)/ k_B T\right] \tag{6.53}$$

a-Se is the most important photosensitive material, and hence we show the temperature-dependent photoconductivity of a-Se in Figure 6.18, in which W is estimated to be ~0.12 eV. It should be noted that in a-Se the significant nonphotoconductive optical absorption, that is, optical absorption associated with near-band-gap illumination, does not create photocarriers. This may be due to geminate pair creation (and geminate recombination), which does not contribute to phototransport, although the origin of this is still not clear. Exciton formation under photoillumination may be a candidate for such nonphotoconductive optical absorption.

One further important remark should be made, about the temperature dependence of σ_p observed in other a-Chs. For a-Sb_2Te_3, as discussed above, W is around 0.1 eV. Some other materials, for example a-As_2Se_3 and a-$GeSe_2$, have

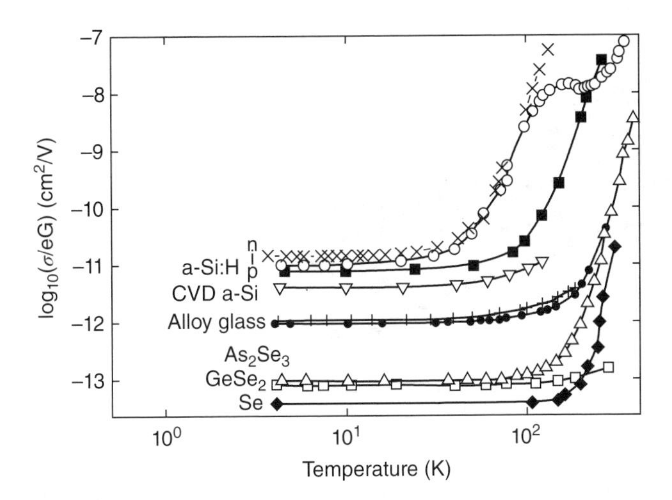

Figure 6.18 Temperature dependence of normalized photoconductivity for typical amorphous semiconductors. *Source*: Fritzshe 1989 [127]. Reproduced with permission from Elsevier.

$W \approx 0.52$ and 0.48 eV, respectively [130], suggesting that deep centers rather than shallow states (band tails) are involved in the recombination processes. In this case, the energy E_{tv} discussed above should be replaced by E_t for deep localized (defect) states. The nature of defects in a-Chs will be discussed in Section 6.5.

Finally, let us discuss the dynamical response of the photoconductivity after illumination has ceased. It has been found that the photocurrent $I_p(t)$ in most a-Chs decays as

$$I_p(t) = I_0 \exp\left(-Ct^{\beta}\right) \tag{6.54}$$

where β (<1.0) is the dispersion parameter [130, 131]. As we have already discussed for a-Si:H (Section 5.7), this form of decay is called a stretched exponential function and is observed in many phenomena in disordered matter [130, 132]. An example for a-As_2Se_3, with $\beta = 0.16$, is shown in Figure 6.19. The long-term photocurrent decay in a-Chs can be explained in the following manner [130]. The majority of excess electrons (Δn) and holes (Δp) are trapped by charged defects (D^+ and D^-), which will be discussed in Section 6.5.

When deeply trapped holes described by Δp_t decay by recombination given by the following dispersive reaction kinetics,

$$\frac{d\Delta p_t}{dt} = -B(T)t^{-(1-\beta)}\Delta p_t \tag{6.55}$$

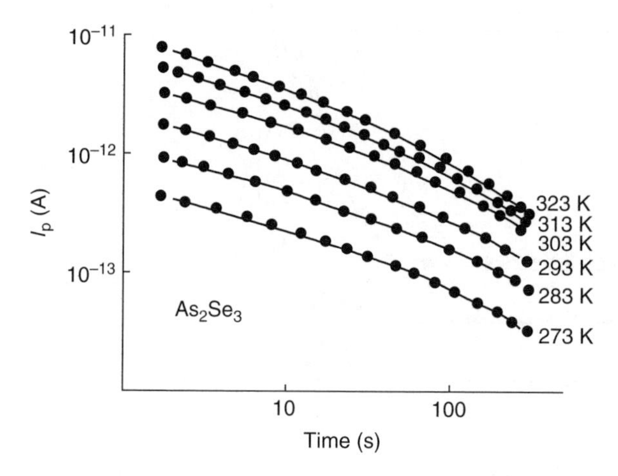

Figure 6.19 Decay of residual photocurrent in a-As_2Se_3 after steady illumination is stopped.

where $B(T)$ is a temperature-dependent constant, Δp_t is given by

$$\Delta p_t \propto \exp\left[-B(T)t^{\beta}/\beta\right] \tag{6.56}$$

As we know from Equation (6.52), Δp is proportional to Δp_t, and $I_p(t)$ is then proportional to Δp_t given above, which is of the same form as in Equation (6.54). The question arises: Why should Δp_t follow Equation (6.56)? The answer to this question may be stated as follows. Geminate-like (monomolecular) recombination, that is, intrapair recombination, rather than interpair (bimolecular) recombination, dominates the long-term decay of this *residual* photoconductivity. Intrapair recombination between randomly distributed sites has a distribution $P(R)$ that can be given as

$$P(R) = 4\pi NR^2 \exp\left(-4\pi NR^3/3\right) \tag{6.57}$$

where N is the trapping-site density and R is the separation of sites. It is not clear why the reaction rate is dispersive in nature. A random distribution of trapped holes and electrons may be the origin of the stretched exponential function [132]. We shall not discuss its origin in more detail here.

Finally, we discuss the transient photocurrent (fast decay) in the short-time range (10^{-8}–10^{-2} s). Figure 6.20 shows the transient photocurrent as a function of temperature in a-As_2Se_3 [123]. We observe a power-law decay following $i_p(t) \propto t^{\alpha-1}$, where the parameter α (<1.0) is temperature-dependent. Multiple trapping of free photocarriers at localized states is assumed as follows [123, 124]. The photoexcited carriers, for example holes in the valence band, drop down to the valence band tail states in a time range of the order of 10^{-12} s. With repeating

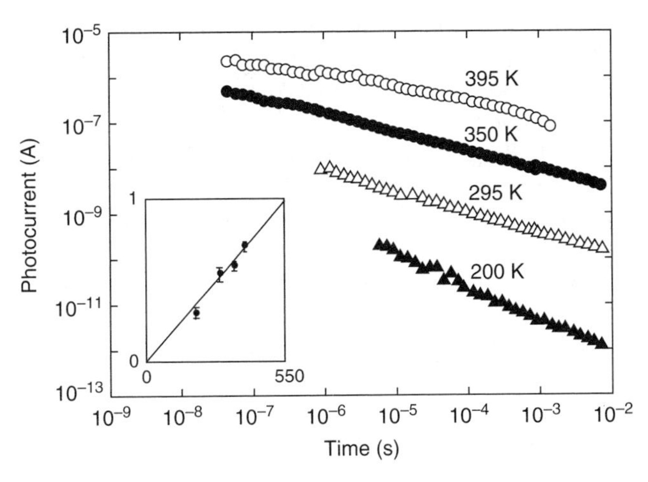

Figure 6.20 Photocurrent decay after pulsed excitation. The temperature-dependent β is shown in the inset. *Source*: Orenstein and Kastner 1981 [123]. Reproduced with permission from the American Physical Society.

trapping and detrapping, holes move down to deeper tail states as time proceeds. For an exponential distribution of the DOS in the tail states, that is,

$$g(E) = \frac{N_0}{\varepsilon_0} \exp(-\varepsilon/\varepsilon_0) \tag{6.58}$$

where N_0 is the total density of tail states and ε_0 is the characteristic energy of the tail, the drift mobility μ_d, under the assumption of no recombination events in this time range, is given as

$$\mu_d \propto (v_0 t)^{\alpha-1} \tag{6.59}$$

where ν_0 is of the order of the phonon frequency ($10^{12}\,\mathrm{s}^{-1}$) and $\alpha = k_B T/\varepsilon_0$. The value of ε_0 for a-As_2Se_3 is reported to be 50 meV [123]. Note here that the energy ε is measured from the band edge.

6.5 The Nature of Defects, and Defect Spectroscopy

As already stated in Section 5.2, when an atom has a distinct bonding state different from its full coordination, it is called a coordination defect. In a-Chs, for example a-Se, such defects should have onefold and threefold coordination. These coordination defects can be characterized by a *negative* electronic correlation energy and are frequently called *negative-U* defects, which will be discussed below.

Strictly speaking, there is no clear experimental evidence for the presence of negative-U defects in a-Chs [39]. In a-Si:H, for example, one of the principal defects is the so-called dangling bond, with threefold coordination, which has a neutral spin state, and can be monitored by the electron spin resonance (ESR) technique. However, no such ESR signal has been observed in a-Chs except in a special case [54]. ESR signals only appear during and after illumination. We call these signals light-induced ESR (LESR). Based on these experimental findings, a *charged-dangling bond* (CDB) model with negative U was proposed [20].

If significant atomic relaxations around defects are involved, the local distortion energies of the electronic ground state energy of the defect configurations should be taken into consideration [133]. Electron–phonon coupling induces the network to relax to a new equilibrium state lower than the state without such coupling.

Spin pairing of electrons at a dangling bond, that is, a two-electron (or empty) state at a dangling bond, can be favored by lattice distortions. The effective correlation energy U_{eff} for this condition can be given as [133]

$$U_{\text{eff}} = U_{\text{c}} - \lambda^2 / c \tag{6.60}$$

where U_{c} is the normal correlation energy (electron–electron interaction), λ is the electron–phonon coupling constant, and c is related to the phonon frequency by $\omega = (c/M)^{1/2}$ in the Einstein approximation, where M is the atomic (or molecular) mass. U_{eff} is negative if $U_{\text{c}} < \lambda^2/c$. Thus we call the defect a negative-U defect.

The above idea was applied to defects in a-Chs [20] and later applied to a chemical-bond argument for a-Se [21]. Such defects are charged owing to pairing of electrons (negative) or holes (positive) and hence they are not ESR-active centers. In the following, the chemical-bond argument in the CDB model is shown in Figure 6.21. The structure of a-Se is twofold coordinated using the outer two p-orbital electrons, forming mostly 1D chains. A dangling bond due to a chain end or bond breaking may contain one unpaired electron if the electron–phonon coupling is not strong, and it is written as C_1^0, where C denotes the chalcogen (S, Se, or Te), the subscript 1 represents the onefold coordination, and the superscript 0 the neutral electronic state. Note that C_2^0 would be a normal bond in this description. When electron pairing is possible at a C_1^0 center, the transfer of an electron from one C_1^0 to another produces a C_1^- (negatively charged) and a C_3^+ (positively charged), which is a threefold-coordinated defect. This reaction is exothermic with a negative U and is written as

$$2C_2^0 = C_1^- + C_3^+ \tag{6.61}$$

A configuration coordinate diagram as illustrated in Figure 6.21(b) may help to understand the energy states in this configuration. Thus C_1^0 is unstable, and C_1^- and C_3^+ are thermodynamically stable; the latter two are often called a

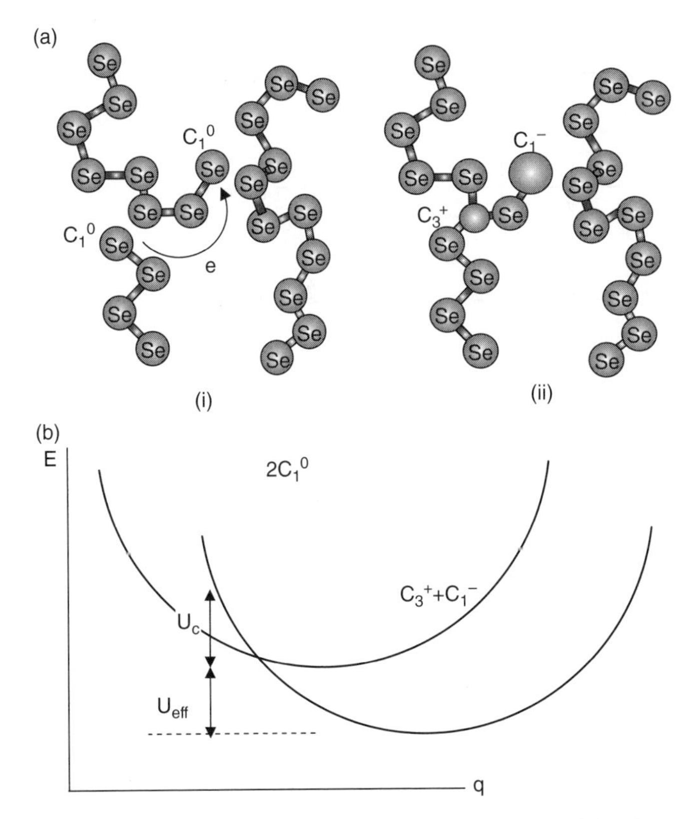

Figure 6.21 (a) Formation of charged defects in a-Se; (b) configuration coordinate diagram for the formation of a D^+ (C_3^+)–D^- (C_1^-) pair. The overall energy is lowered by the effective correlation energy U_{eff}. *Source*: Singh and Shimakawa 2003 [56]. Reproduced with permission of Taylor & Francis.

valence-alternation pair (VAP) [21]. A molecular orbital model, which is useful for understanding which atomic orbitals are used in chemical bonds, is shown in Figure 6.22. We can see that for C_3^+ (Se_3^+) only three bonding p-orbitals are used, while for C_1^- (Se_1^-) one p-orbital and two lone pairs are used. As C_3^+ and C_1^- are produced by two dangling bonds, these correspond to the notation D^+ and D^- (positive and negative dangling bonds). These states are expected to produce deep localized states.

The D^+ state is located near the conduction band and the D^- near the valence band. The thermal energy levels associated with the D^+ and D^- states are shown schematically in Figure 6.23. The Fermi level is fixed at an energy midway between the D^+ and D^- levels. When an electron is trapped in a D^+, this becomes a D^0, and for detrapping of an electron from this D^0, an energy W_2 is needed. The same applies to D^-: W_1 is the energy needed for detrapping from D^-.

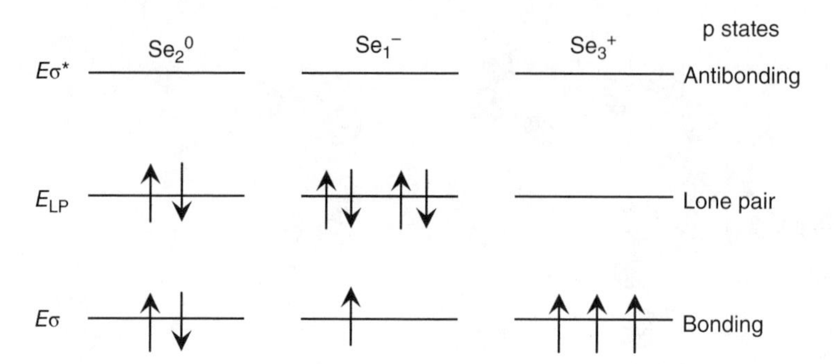

Figure 6.22 A molecular orbital diagram for the electronic structure of different bonding states in Se.

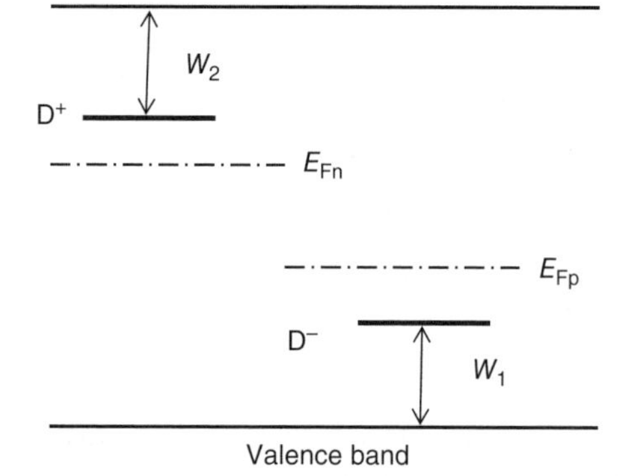

Figure 6.23 Thermal energy levels and configuration coordinate diagram for D^+ and D^-. E_{Fn} and E_{Fp} are the quasi-Fermi levels for photoexcited electrons and holes, respectively.

Let us discuss the number of VAP centers from a thermodynamic point of view. The *law of mass action* for the reaction in Equation (6.61), that is, $2C_2^{\,0} + E_{VAP} = C_1^{\,-} + C_3^{\,+}$, may be useful for this purpose, where E_{VAP} is the creation energy of VAPs. Then, at the frozen-in temperature, that is, at T_g, the concentrations of $C_1^{\,-}$ and $C_3^{\,+}$ are given by

$$\left[C_1^{\,-}\right]\left[C_3^{\,+}\right] = \left[C_2^{\,0}\right]^2 \exp\left(-E_{VAP}/k_B T_g\right) = N_d^{\,2} \tag{6.62}$$

and then

$$\left[C_1^{\,-}\right] = \left[C_3^{\,+}\right] = N_0 \exp\left(-E_{VAP}/2k_B T_g\right) = N_d \tag{6.63}$$

where N_0 is the total number of atoms $[C_2^{\,0}]$, N_d is the number of $C_1^{\,-}$ or $C_3^{\,+}$, and E_{VAP} is the creation energy of a VAP. Since the two VAP centers, $C_1^{\,-}$ and $C_3^{\,+}$, have equal numbers and opposite charges, they can pair up owing to their mutual Coulomb interaction, producing an intimate VAP (IVAP). IVAPs may not act as trapping or recombination centers, unlike VAPs, since IVAPs may act as neutral centers and hence the trapping or recombination cross section for these may be smaller than for charged centers.

The absence of an ESR signal in dark conditions can be explained in the following way. Note that the C_1^0 (ESR-active) center is produced by the reverse reaction, $C_1^- + C_3^+ + U_{eff} = 2C_1^0$ [74]. The concentration of C_1^0 is therefore predicted by the law of mass action as

$$\left[C_1^0\right] = \sqrt{\left[C_1^-\right]\left[C_3^+\right]}\exp\left(-U_{eff}/2k_BT\right) \tag{6.64}$$

In a-As_2Se_3, for example, for $U_{eff} = 0.7$ eV and $[C_1^-] = [C_3^+] = 2\times10^{17}\,cm^{-3}$ [77], $[C_1^0]$ at room temperature is estimated to be $1.5\times10^{11}\,cm^{-3}$. This is actually impossible to detect by an ESR signal. In liquid chalcogenide glasses, ESR signals are observable [134]. Roughly, assuming that the basic defect structure is retained in the liquid state, for example at 800 K, $[C_1^0] = 3\times10^{15}\,cm^{-3}$ can be estimated, which approaches a feasible value for ESR measurement ($\sim10^{16}\,cm^{-3}$).

We know from Section 5.4.5 that a-Si:H shows remarkable effects of doping with impurities (such as P and B). We shall now discuss what happen in a-Chs. Experimentally, unlike a-Si:H, the effect of doping impurities is not significant. This may be due to the relatively high density of charged defects in a-Chs. The approach using the law of mass action is useful [135]. Let us consider, for example, n-type doping, since a-Chs are naturally p-type semiconductors (in the absence of any doping). Charge neutrality requires the following relation:

$$\left[C_1^-\right] + n = \left[C_3^+\right] + \left[A^+\right] + p \tag{6.65}$$

where A^+ is the ionized dopant (if an appropriate dopant is used), and n and p are the numbers of free electrons and holes, respectively. Substituting Equation (6.64) into Equation (6.61) produces

$$\left[C_3^+\right] = -\frac{1}{2}\left[A^+\right] + \sqrt{N_d^2 + \frac{1}{4}\left[A^+\right]^2} \tag{6.66}$$

and

$$\left[C_1^-\right] = \frac{1}{2}\left[A^+\right] + \sqrt{N_d^2 + \frac{1}{4}\left[A^+\right]^2} \tag{6.67}$$

Note here that n and p are negligible compared with N_d. It is predicted from the above equations that n-type doping induces negatively charged defects and p-type doping increases the number of positively charged defects, similarly to the doping effects observed in a-Si:H.

Ge–S(Se) systems doped with Bi show n-type behavior, although the doping efficiency is not good as in a-Si:H. [136, 137]. For n-type doping, as predicted from Equations (6.66) and (6.67), the relation $[C_1^-] \approx [Bi^+] \gg [C_3^+]$ is expected, in which the Fermi level shifts upward. This self-compensation effect makes it difficult to produce heavily doped materials.

Finally, let us discuss how to deduce *experimentally* the values of W_1 and W_2 and the numbers of these centers. This is called *defect spectroscopy*. Unlike the case for a-Si:H, there have not been many attempts to develop defect spectroscopy for a-Chs. In the following, we briefly introduce several well-known results.

6.5.1 Electron Spin Resonance

As we already know, ESR signals are not generally detected under dark conditions in a-Chs. ESR signals are only detected under band gap illumination and/or after illumination. This is called light-induced ESR [138, 139]. An exception is the Ge–S glass system, which exhibits ESR signals without photoillumination [140–144]. Figure 6.24 shows LESR signals observed in a-Se, a-As_2Se_3, and a-As_2S_3 at 4.2 K [138]. The density of spins with a g-value of 2.0023 saturates at $\sim 10^{16}\,\text{cm}^{-3}$ for a-Se and at $\sim 10^{17}\,\text{cm}^{-3}$ for a-As_2Se_3 and a-As_2S_3, at low illumination intensity ($\sim 1\,\text{mW}\,\text{cm}^{-2}$). The centers responsible for this are believed to be due to electrons and holes occupying charged defects, that is, $D^+ + e \rightarrow D^0$ and $D^- + h \rightarrow D^0$. The saturation at 10^{16}–$10^{17}\,\text{cm}^{-3}$ suggests that the density of charged defects is in this range. These LESR signals are bleached by thermal annealing at around 200 K or by midgap illumination.

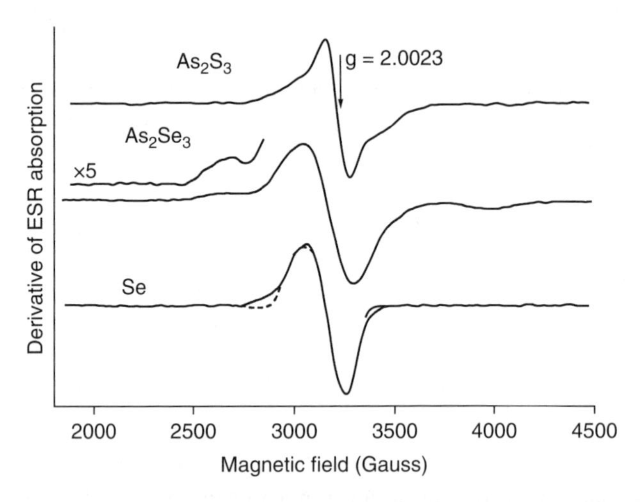

Figure 6.24 Optically induced ESR signals from a-Chs at 4.2 K. The dashed curve superimposed on the Se curve is a computer simulation. *Source*: Bishop *et al.* 1977 [139]. Reproduced with permission from the American Physical Society.

At high illumination intensity ($\sim$100 mW cm^{-2}), the spin density does not saturate but instead increases to 10^{20} cm^{-3}, which may be due to new creation of defects by prolonged intense illumination [145]. A detailed discussion of this issue will be given in Section 6.6.

Optically detected magnetic resonance (ODMR) measurements, which basically belong to ESR technology, have also been done on a-Chs [109, 146–150]. Unlike the case for a-Si:H, this technique cannot provide information about defects in a-Chs, and hence we shall not discuss this issue.

6.5.2 Optical Absorption

The weak optical absorption ($\alpha < 10$ cm^{-1}) in the midgap energy region can be attributed to transitions between deep defect states and the conduction band (or valence band). Illumination at low temperature increases the weak absorption. Similarly to the LESR signals, these induced centers disappear after thermal annealing at around 200 K or after infrared (IR) illumination [138, 139]. It is therefore expected that the midgap weak absorption is due to the existence of the same centers as those responsible for the LESR signals. If the origin of these centers is the same as for the LESR, prolonged intense band gap illumination should also produce an increase in the midgap absorption. This is actually observed, and these photoinduced effects will be discussed in Section 6.6.

6.5.3 Primary Photoconductivity

TOF data can be used to deduce the DOS. A theoretical analysis of multitrapping transport has been done by use of a Laplace transform formalism. Both hole and electron TOF measurements and subsequent analysis of the resulting data have been systematically performed on a-Se [151, 152]. The DOS near the valence band is a featureless, monotonically decreasing distribution with respect to energy up to +0.4 eV from E_V. The DOS near the conduction band obtained from electron TOF data, as shown in Figure 6.25, shows two broad peaks ($\sim 1 \times 10^{17}$ cm^{-3} eV^{-1} at $\sim$0.3 eV and $\sim 1 \times 10^{14}$ cm^{-3} eV^{-1} at 0.5 eV below the conduction band). Note that so-called stabilized a-Se (containing 0.3 at.% of As) was used for electron TOF measurements. It is known, however, that the localized states are not much affected by doping with a small amount of As. It was also found that the deep states lying below 0.65 eV had an integral concentration in the range of 10^{11}–10^{14} cm^{-3}, which was also confirmed by electrophotographic spectroscopy as discussed later on.

6.5.4 Secondary Photoconductivity

The constant photocurrent method (CPM), mentioned in Section 5.5.4, is useful for obtaining the weak optical absorption coefficient, from which some information can be obtained about deep gap states. There have been several

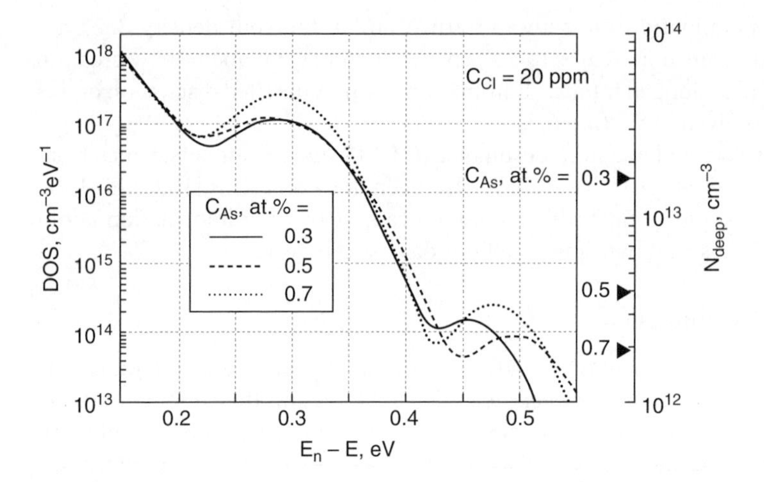

Figure 6.25 Influence of As addition on the DOS distribution. The concentration of Cl was 20 ppm. *Source*: Koughia *et al.* 2005 [151]. Reproduced with permission from AIP Publishing LLC.

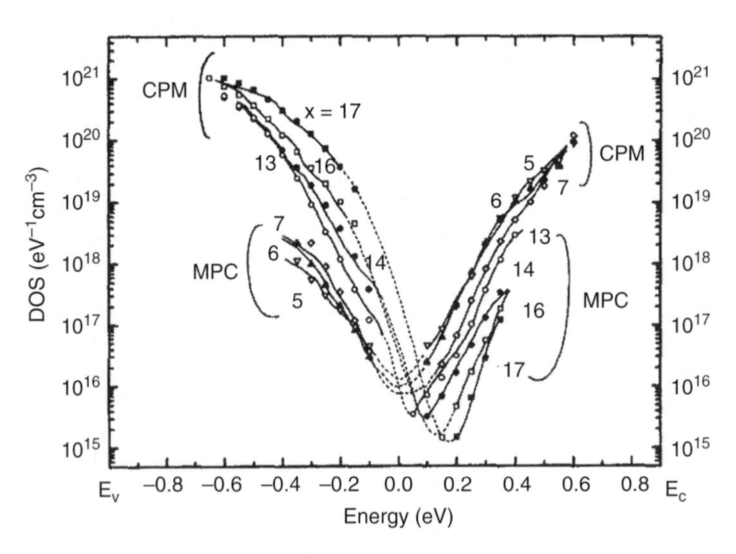

Figure 6.26 Electronic DOS in a-Ge–Se–Bi films. *Source*: Kounavis 2001 [153]. Reproduced with permission from the American Physical Society.

studies using the CPM on a-Chs [153–157]. The DOS in the band gap was deduced from the results and no DOS peak was found for the a-Ge–Se–Bi system [154], as shown in Figure 6.26. As will be shown below, the DOSs deduced from different techniques are not always the same.

6.5.5 Electrophotography

Electrophotography is often called xerography, and this technologically important method has also been applied to deduce the DOS. The DOS for hole traps has been obtained for a-Se [158]. The DOS for a-Si:H has also been obtained using electrophotography [56, 159–161]. After repeated charging and discharging in a xerographic sample configuration (with a positive corona charge for monitoring hole traps; see the original papers cited above), hole trapping occurs uniformly throughout the sample. These trapped holes produce an intense surface potential, which then decays with time. By analyzing the decay behavior, the energy-dependent trap profile, that is, the DOS, can be deduced.

Figure 6.27 shows the hole trap distribution in a-Se deduced from xerography [158]. The energy is referred to the valence band edge. The DOS peak appears at 0.87 eV above the valence band. The remarkably small integral density of traps near midgap ($\sim 10^{14}\,\text{cm}^{-3}$) may not be consistent with the values obtained in the other studies of a-Chs cited earlier. The reason for this is not clear. To detect the surface potential correctly, complete hole blocking is needed to avoid leakage of carriers.

6.5.6 Electronic Transport

Not much work on the measurement of electronic transport by defect spectroscopy has been done for a-Chs, since, for example, good Schottky barriers cannot be produced on a-Chs, partly because of the relatively high density of defects. Thus, even in principle, field effect and capacitance measurements are not easy to perform. If a high density of gap states exists, space charge effects of currents and space-charge-limited currents are also difficult to measure, since the Fermi level is not moved by current injection [162].

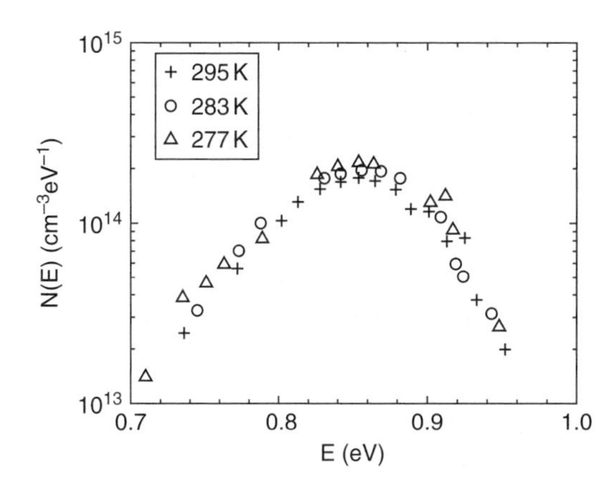

Figure 6.27 Hole trap distribution *N(E)* deduced from isothermal decay of the xerographic residual potential. The energy is referred to the valence band edge. *Source*: Abkowitz and Markov 1984 [158]. Reproduced with permission of Taylor & Francis.

This technique of impedance spectroscopy, in which the real and imaginary parts of the conductivity in an AC electric field are monitored, is well known in the field of ionic transport in materials [78–80]. There are excellent reviews of the so-called AC conduction (or AC loss) in a-Chs [71, 72]. Correlated barrier hopping (CBH) of bipolarons, that is, two electrons hopping between charged defects (D^+ and D^-), has been proposed by Elliott [72], and this model has been extended to the CBH of single polarons (electron hopping between D^0 and D^+ or D^-) [74].

The density of charged defects (10^{18}–$10^{19}\,cm^{-3}$) estimated from the CBH model is found to be larger than that (10^{16}–$10^{17}\,cm^{-3}$) expected from other studies. The reason for this is clear: the CBH model is based on the pair approximation, which is basically an extension of Debye relaxation in which carriers are confined into special (nearest-neighbor) pairs. The CTRW approximation may be an alternative approach that can be applied to the AC loss of materials [75, 76]. The CBH approach has been applied to the AC loss in a-Chs, and fitting to the experimental data produces a reasonable density of charged defects (10^{17}–$10^{18}\,cm^{-3}$) in a-Chs [77].

6.6 Light-Induced Effects in Chalcogenides

Significant light-induced changes have been also found in amorphous chalcogenides [26–28]. There are two types of changes, *reversible* and *irreversible.* Irreversible changes, in which the material cannot be returned to the original dark state by thermal annealing below T_g, can be induced. Here, however, we will discuss only reversible changes, in which the material can be returned to its original state by thermal annealing or another photoexcitation (IR irradiation). Unlike the case for a-Si:H, significant changes in structure have been found experimentally, and hence these will be discussed separately from the changes in electronic properties.

6.6.1 Electron Spin Resonance

As already discussed in Section 6.5, ESR signals in a-Chs (except for some systems) are observed only during and after band gap illumination (LESR) [139]. The LESR centers saturate at around 10^{16}–$10^{17}\,cm^{-3}$ at low irradiation intensity, and at high excitation intensities, the spin density exceeds $10^{20}\,cm^{-3}$ [145, 163]. Detailed studies of LESR in As_xS_{1-x} [163] show two types of LESR center associated with four different metastable centers. The two centers (one of electron and the other of hole origin) which anneal at lower temperatures are labeled type-I LESR centers, and the other two centers (again one of electron and the other of hole origin), which are thermally stable, are labeled type-II LESR centers, as shown in Figure 6.28, together with the annealing behavior of the midgap optical absorption, which will be discussed later.

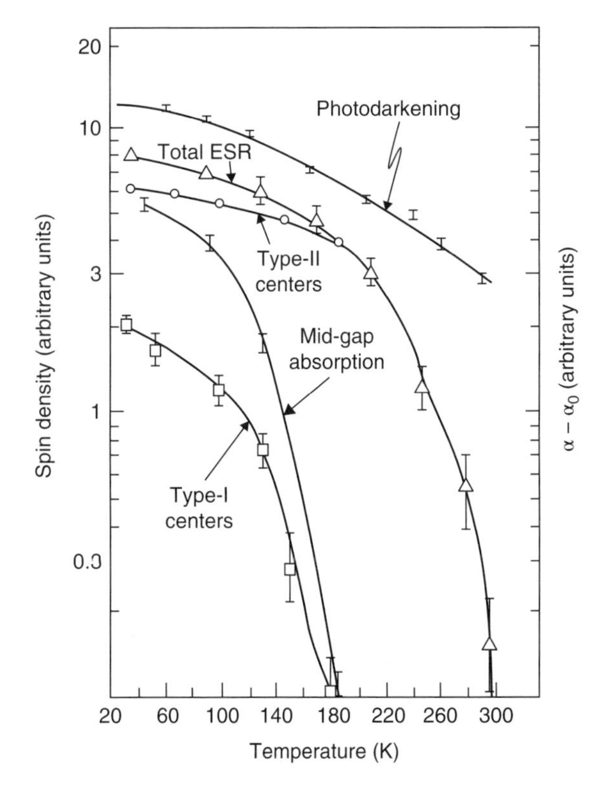

Figure 6.28 Comparison of thermal decay of the LESR, midgap optical absorption, and photodarkening for a-As_2S_3. *Source*: Hautala *et al.* 1988 [163]. Reproduced with permission from the American Physical Society.

The bond-breaking mechanisms and microscopic details of both type-I LESR and type-II LESR centers have been discussed on the basis of the above observations [164]. While there are several possible configurations for bond breaking due to optical excitation, the following two configurations may be candidates for the two types of LESR centers, as shown in Figures 6.29 and 6.30. The type-I centers are identified with electrons or holes trapped at $As_2^+(2S)$ or $S_1^-(As)$, respectively, just after bond breaking, as shown in Figure 6.29, and hence these are denoted by $As_2^0(2S)$ or $S_1^0(As)$. These are called intimate pairs (IPs). These IP centers are stabilized by successive bond switching and then form randomly distributed pairs (RPs). These centers should be the same as those observed as "LESR" centers under weak illumination [139], in which $As_2^+(2S)$ or $S_1^-(As)$ are previously existing charged defects (under either IP or RP conditions). The instability of the type-I LESR centers, which anneal out at around 180 K, may be due to IPs.

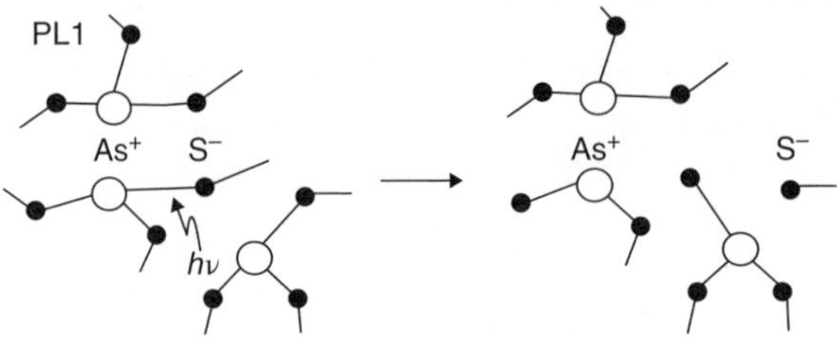

PL1

As+ S−

hν

As+

S−

Figure 6.29 Photoinduced defects in a-As_2S_3.

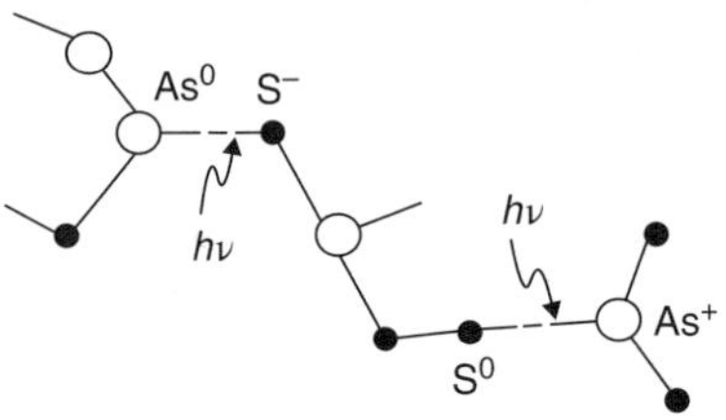

As0 S−

hν

hν

As+

S0

Figure 6.30 Photoinduced LESR centers in As_2S_3.

It is known that the As_xS_{1-x} system contains homopolar bonds, that is, As–As and S–S bonds, and the As_2^0(As,S)–S_1^- pair and the S_1^0(S)–As_2^+(2S) pair are produced by bond breaking [164], as shown in Figure 6.30. These are type-II LESR centers. These close pairs of defect configurations will be relatively stable and are annealed out at around room temperature. Note that the neutral centers As_2^0 and S_1^0 are not produced just by occupation (trapping) by electrons and holes, unlike type-I LESR centers.

We have now understood the difference in thermal stability between type-I and type-II LESR centers. There are further differences between their behavior. The concentration of the type-I defects (both S- and As-centered) is independent of the composition of As_xS_{1-x}, whereas As_{II} centers dominate in materials with an As-rich composition, and S_{II} centers dominate in S-rich materials. Note also that the ESR signatures of S_I and S_{II} are very similar (nonbonding p-orbitals), while the signatures of As_I and As_{II} are different, that is, the As_{II} is almost purely p-like and As_I has a delocalized s-orbital contribution [163].

6.6.2 Optical Absorption

Prolonged illumination (2.41 eV, 100 mW cm^{-2}) of a-As_2S_3 at low temperature [145] also increases the midgap optical absorption. The midgap optical absorption coefficient at 1.6 eV, for example, is about six times larger than for illumination with 10 mW cm^{-2}. As shown in Figure 6.28, thermal annealing decreases this additional midgap optical absorption, and this absorption is annealed out at around 200 K, showing that the thermal behavior of the midgap absorption is similar to that of the type-I LESR centers. This suggests that the type-I LESR centers lie near the midgap position and hence contribute to midgap absorption; that is, the centers that contribute to type-I LESR and midgap optical absorption are the same.

6.6.3 Photoluminescence

The intensity of PL is reduced by above-band-gap illumination; this is called PL fatigue [104, 165]. The rate of fatigue depends on the illumination intensity and the excitation energy. It increases with increasing irradiation intensity. The time evolution of the fatigue follows a stretched exponential function, which will be discussed later in this section. The fatigue is stable at low temperatures and is recovered after thermal annealing or IR irradiation. It was found for a-As_2S_3 that fatigue in PL intensity induced by short-time irradiation (with a small total dose) was recovered after annealing at 150 K, while fatigue induced by long-time irradiation was not recovered at 150 K and needed room-temperature annealing to anneal out completely, suggesting that there are two distinct induction mechanisms.

In addition to the above effects in a-As_2S_3, a new low-energy PL peak (~0.85 eV) was induced by prolonged irradiation with band gap light, accompanying the fatigue of the main PL peak (~0.11 eV), as shown in Figure 6.31 [166]. The high- and low-energy peaks are called PL1 and PL2, respectively, similarly to LESR-I and LESR-II. Note that PL2 was excited for detection with lower-energy light (1.58 eV). Interestingly, there is a linear relation between the decrease in PL1 and the increase in PL2; that is, fatigue itself creates new PL centers.

The PL fatigue may have two reasons: one may be a decrease in PL centers themselves, and the other an increase in nonradiative centers. We thus discuss the nonradiative channel, before proceeding further with PL fatigue. As already shown in Figure 6.12 (Section 6.4), nonradiative recombination occurs through a metastable STE, which can be regarded as an intimate D^+–D^- pair proposed originally by Street [108] (see also Sections 6.4 and 6.5). There are similar annealing properties in PL1 fatigue, type-I LESR, and midgap optical absorption, suggesting that the same process contributes to all these phenomena. This close relation suggests that PL1 fatigue occurs by induction of nonradiative centers, which are responsible for LESR and subgap absorption. It should be noted that after prolonged irradiation PL1 fatigue is not recovered by annealing at around 150 K [145]. The centers induced by prolonged illumination are more stable than those induced by short-time illumination.

An alternative model for fatigue of PL1 is a reduction in the number of PL centers [167]; for example, the number of As–S bonds which are converted to STEs acting as radiative recombination centers may decrease because of the breaking of As–S bonds accompanied by strong lattice relaxation. This bond breaking induces metastable STEs, which act as nonradiative recombination centers (see Figure 6.12).

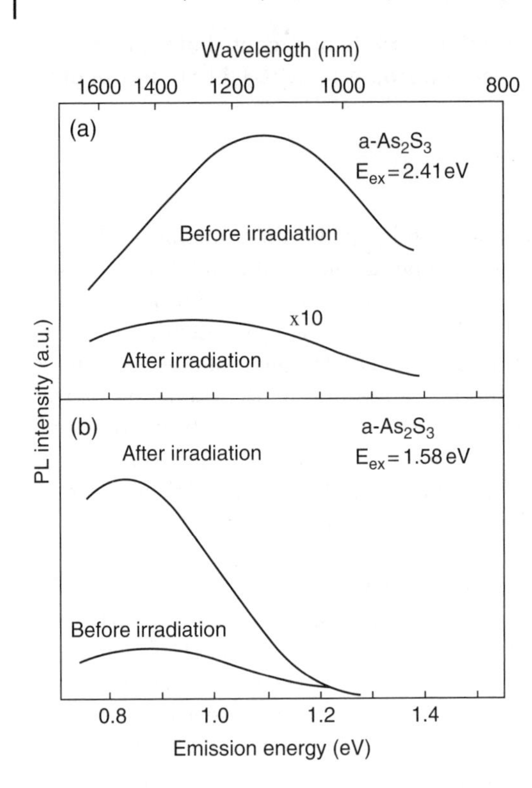

Figure 6.31 PL spectra of a-As_2S_3: (a) with excitation energy 2.41 eV before and after illumination (PL1); (b) with excitation energy 1.58 eV before and after illumination. *Source*: Shimakawa *et al.* 1993 [171]. Reproduced with permission from Elsevier.

Let us now summarize the photoinduced changes in PL. The fatigue of PL1 induced by short-time irradiation is due to the induction of a nonradiative channel associated with the D^0 state, resulting from the occupation of preexisting native charged defects by electrons and/or holes. This D^0 state is involved in type-I LESR and midgap absorption. The fatigue following prolonged illumination is caused by the formation of new metastable STEs (nonradiative), which destroys radiative STE centers. The thermal stability of metastable STEs is greater than that of nonradiative D^0s, which are induced just by occupation of D^+ and D^- by electrons and holes, respectively. Note also that the D^0 states formed by the occupation of a metastable STE by an electron and/or a hole, that is, intimate pairs of D^+ and D^-, act as type-I LESR and midgap absorption centers. Therefore prolonged illumination increases both type-I LESR and midgap absorption.

The recovery of the fatigue of PL1 after short-time irradiation and the disappearance of type-I LESR and midgap absorption, which occur at around 200 K, may be the result of the thermal release of electrons or holes from native charged defects and from metastable STEs. The recovery of PL1 after prolonged illumination can be attributed to the annealing out of the metastable STEs themselves.

6.6.4 Photoconductivity

A decrease in photoconductivity after photoirradiation was first reported in a-Si:H [168], as stated in Section 5.7. This effect is often called photodegradation, or the Staebler–Wronski effect [28]. A similar decrease in photoconductivity during and after prolonged illumination has been observed in a-Chs [130, 169, 170], and many common features have been observed in the time-dependent decrease in photoconductivity between a-Si:H and a-Chs [171]. These changes in a-Chs are removed by thermal annealing near T_g. Note, however, that in a-Se the photoconductivity does not show such a change at room temperature, although it decreases at low temperatures [130].

Figure 6.32 shows an example of the decrease in photocurrent in a-As_2S_3 during band gap irradiation at several temperatures. The photodegradation reaches a maximum at around 300 K [171]. The solid curves represent results calculated using a stretched exponential function (see Equation (6.78) below). The kinetics of the origin of the stretched exponential function will be discussed later in Section 6.6.6.

Newly created defects acting as recombination centers may be responsible for the decrease in photoconductivity. A microscopic model for the creation of light-induced metastable defects (LIMDs), for example in As_2S_3, was shown in Figure 6.30. First we should mention that IPs are not expected to act as efficient recombination centers for photocarriers, since they may be "felt" as electrically neutral and hence the capture cross section for free carriers may be small. The RPs shown in Figure 6.30 may instead be responsible for the decrease in photoconductivity. Such RPs result from defect-conserving

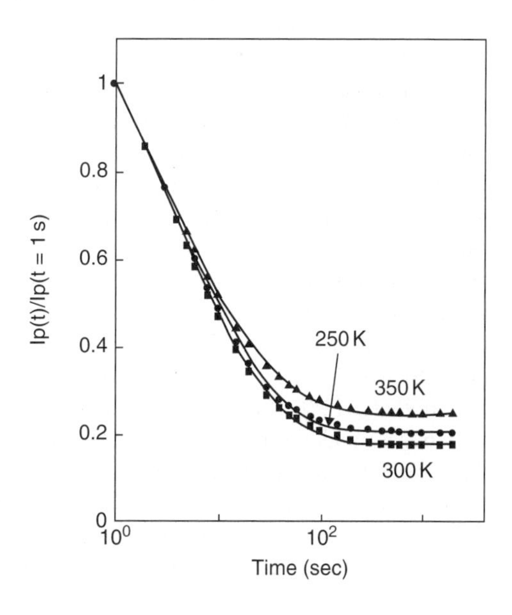

Figure 6.32 Time-dependent change in photocurrent in a-As_2S_3 measured at different temperatures. *Source*: Ganjoo *et al.* 2002 [182]. Reproduced with permission from Elsevier.

bond-switching reactions at optically induced IPs, for example $P_2^+–C_1^-$, shown in Figure 6.29, where P and C refer to the pnictogen and chalcogen, respectively. RPs are annealed at higher temperature than IPs, as stated already, and hence thermal annealing at around T_g is needed for complete recovery to the original photoconductivity (before illumination). It should be mentioned that no photocurrent decrease is observed in narrow-band-gap materials [172].

6.6.5 Electronic Transport

After illumination, the DC dark conductivity σ_{DC} of a-Si:H is known to become smaller than it is before illumination, as discussed in Section 5.7. However, in a-Chs, σ_{DC} takes the same value before and after illumination, with some exceptions [172, 173], indicating that the position of E_F in the band gap is unchanged even when new defects are induced by irradiation. E_F can be pinned by charged defects [54]. If the energy levels of the newly created defects are the same as those of preexisting charged defects, E_F cannot be shifted by irradiation.

As discussed already in Section 6.4, a study of the AC conductivity is useful for defect spectroscopy. If new defects are induced, an increase in the AC conductivity should be observed. Photoinduced changes in the AC conductivity are discussed in the following. The so-called AC conductivity is the real part of the conductivity under an AC electric field, and hence it is often called the AC loss. On the other hand, the imaginary part of the conductivity reflects *capacitance* when hopping transport is dominant. The usual technique for measuring this is impedance spectroscopy, as already mentioned.

Figure 6.33 shows the frequency-dependent AC loss σ_{AC} in a-As_2S_3 measured at 300 K (before and after illumination at 300 K) and at 90 K (before and after illumination at 90 K) [174]. σ_{AC} is proportional to ω^s (with $s = 1.0$). Although the change induced by low-temperature (90 K) illumination is annealed out at around 200 K, the change caused by room-temperature illumination is removed by annealing near T_g. Similarly, an increase in the capacitance by about 15% in a-As–Se after prolonged irradiation at room temperature, which is annealed out around 420 K (near T_g), has been reported [175]. The above results suggest that two kinds of center are induced, one at low temperatures and the other at high temperatures.

Illumination induces IP and RP centers, as shown in Figure 6.30, and the $As_2^+–S_1^-$ center is an interconversion pair (i.e., two electrons hop between the two defects) [28]. The large increase in σ_{AC} caused by illumination at 90 K, which is annealed out at 200 K, may be due to photoinduced $As_2^+–S_1^-$ centers. The change induced by illumination at 300 K, which is annealed out at around T_g, can be attributed to RP centers (separated As_2^+ and S_1^-). These centers could act as recombination centers, and hence the induced AC loss is related to the decrease in photoconductivity. Figures 6.29 and 6.30 may help one to

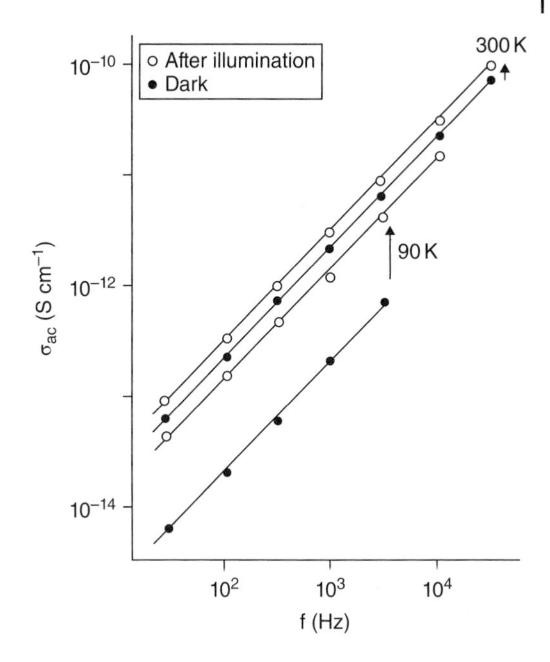

Figure 6.33 Real part of AC conductivity of a-As_2S_3 measured at 90 and 300 K, before and after illumination. *Source*: Shimakawa and Elliott 1988 [174]. Reproduced with permission from the American Physical Society.

understand the overall features of photoinduced effects in a-Chs, so that readers will be able to understand which defect configuration dominates in various effects.

6.6.6 Defect Creation Kinetics

The photocurrent I_p under conditions of thermal equilibrium can be expressed as

$$I_p = \frac{C}{N_0 + N_d} = \frac{I_s}{1 + N_d/N_0} \tag{6.68}$$

where C is a constant, N_0 and N_d are the numbers of preexisting and photoinduced RP centers, respectively, and I_s is a constant current reached after prolonged illumination. Fitting to experimental data, for example the data shown in Figure 6.32, produces a time-dependent N_d which can be expressed empirically as [176]

$$N_d = N_s\left[1 - \exp\left\{-(t/\tau)^{\beta}\right\}\right] \tag{6.69}$$

where N_s is the saturated number, τ is the effective time constant, and β is the dispersion parameter (<1.0). From the curve of N_d versus Gt, where G is the

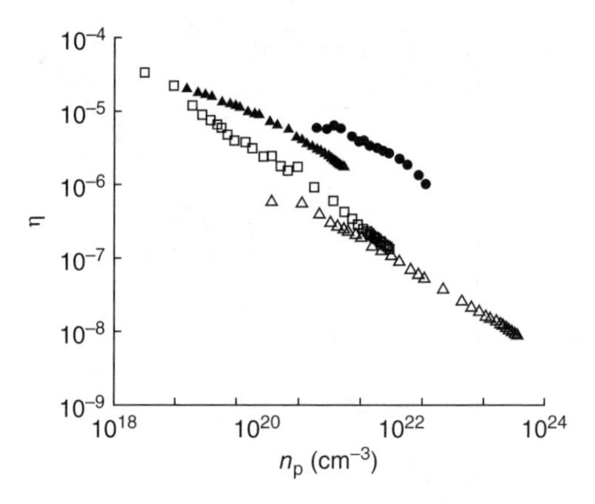

Figure 6.34 Quantum efficiency of defect creation in a-As_2Se_3 (solid symbols) and in a-Si:H (open symbols) as a function of number of photons. Open squares and solid triangles are for subgap illumination and open circles and solid circles are for band gap illumination. *Source*: Shimakawa *et al*. 2004 [176]. Reproduced with permission of Taylor & Francis.

illumination intensity and t the illumination time, the quantum efficiency η for defect creation has been deduced for a-As_2Se_3 [176], as shown in Figure 6.34, together with that for a-Si:H. The values of η for a-As_2Se_3 are larger than those for a-Si:H. This may be due to the larger structural flexibility of a-As_2Se_3 than of a-Si:H.

The type of solution expressed by Equation (6.69) can be given by the following rate equation:

$$\frac{dN_\mathrm{d}}{dt} = k_\mathrm{p}\left(N_\mathrm{T} - N_\mathrm{d}\right) - k_\mathrm{r} N_\mathrm{d} \tag{6.70}$$

where N_T is the total participating density, k_p is the promotion rate (forward reaction), and k_r is the recovery rate (back reaction). When both k_p and k_r are of a time-dispersive type, that is,

$$k_\mathrm{p} = At^{\beta-1} \quad \text{and} \quad k_\mathrm{r} = Bt^{\beta-1} \tag{6.71}$$

where A and B are constants, Equation (6.69) will be the solution of Equation (6.70).

It is not clear, however, why k_p and k_b are given by time-dispersive rates. The following considerations, about how and why a time-dispersive reaction is involved in the mechanism of photoinduced defect creation (PDC), may be useful [177]. Let us begin with the Poisson waiting-time probability distribution for a random event, given by [178]

$$\Psi(t) = \nu e^{-\nu t} \tag{6.72}$$

where ν is the average frequency (in s^{-1}). The average waiting time for a random event is given by

$$\langle t \rangle = \int_0^\infty t\psi(t)\,dt = \frac{1}{\nu} \tag{6.73}$$

This corresponds to a single relaxation process with a reaction rate ν. When the reaction requires the surmounting of a potential barrier U, ν can be written as

$$\nu = \nu_0 \exp\left(-\frac{U}{k_B T}\right) \tag{6.74}$$

where ν_0 is a characteristic frequency.

When the height U of the potential barrier for producing RPs is distributed in an exponential manner as

$$P(U) = \frac{1}{U_0}\exp\left(-\frac{U - U_{\min}}{U_0}\right), \quad \left(U - U_{\min} \geq 0\right) \tag{6.75}$$

where $U_{\min}$ is the minimum potential barrier height that needs to be surmounted, the distribution function of ν, $P(\nu)$, is then given by [178]

$$\begin{aligned} P(\nu) &= \frac{P(U)}{\left|\frac{d\nu}{dU}\right|} = \frac{k_B T}{\nu} P(U) = \frac{1}{\nu}\frac{k_B T}{U_0}\exp\left(-\frac{U - U_{\min}}{U_0}\right) \\ &= \frac{1}{\nu}\frac{k_B T}{U_0}\exp\left(\frac{U_{\min}}{U_0}\right)\exp\left(-\frac{U}{U_0}\right) = \frac{1}{\nu}\exp\left(\frac{U_{\min}}{U_0}\right)\beta\exp\left(\beta\ln\left(\frac{\nu}{\nu_0}\right)\right) \\ &= \frac{1}{\nu}\exp\left(\frac{U_{\min}}{U_0}\right)\beta\left(\frac{\nu}{\nu_0}\right)^{\beta} = \beta\exp\left(\frac{U_{\min}}{U_0}\right)\nu_0^{-\beta}\nu^{\beta-1} = C\nu^{\beta-1} \end{aligned} \tag{6.76}$$

where $\beta = k_B T / U_0$ and $C = \beta\exp(U_{\min}/U_0)\nu_0^{-\beta}$.

Under the above conditions, the Poisson waiting-time distribution can be modified as follows:

$$\Psi(t) = \int_0^\infty P(\nu)\nu e^{-t}\,d\nu = C\int_0^\infty \nu^{\beta} e^{-\nu t}\,d\nu = C' t^{-\beta-1} \tag{6.77}$$

The *average* waiting time is then given by

$$\langle t \rangle = \int_0^t t\Psi(t)\,dt = C' t^{1-\beta} \tag{6.78}$$

Here, the upper limit of the integral is t and not infinity. Note that $\langle t \rangle$ becomes infinity if we take this limit as infinity. This is a characteristic future of time-dispersive reactions. To avoid this difficulty, $\langle t \rangle$ is conventionally defined in the time interval between 0 and t. The average reaction rate is then given as

$$\langle v \rangle = \frac{1}{\langle t \rangle} = C'' t^{\beta - 1} \tag{6.79}$$

which takes the same form as k_p and k_r. It is now clear that the exponential distribution of the potential barrier produces a time-dispersive reaction for LIMD creation. Equation (6.79) is of the same form as Equation (6.71). The dispersion parameter β is related to the extent of potential fluctuations.

6.6.7 Structure-Related Properties

One of the unique features of a-Chs is their ability to undergo *reversible* structural changes on photoirradiation. Photoinduced metastability involving changes in the structure is discussed in the following sections.

6.6.7.1 Photodarkening

In the early 1970s, reversible photoinduced changes were observed, for example a decrease in the optical band gap (~2%) of a-$As_2Se(S)_3$ after band gap illumination [179]. As the optical absorption edge shifts to lower energies, the illuminated area becomes darker in color and hence this effect is called photodarkening (PD). Subsequent annealing near T_g leads to recovery of the initial physical parameters.

First, we shall show that there is a strong correlation between the temperature T_{ill} at which the photoillumination is done and the magnitude of the PD in various a-Chs, as shown in Figure 6.35 [180]. Note here that T_{ill} has been

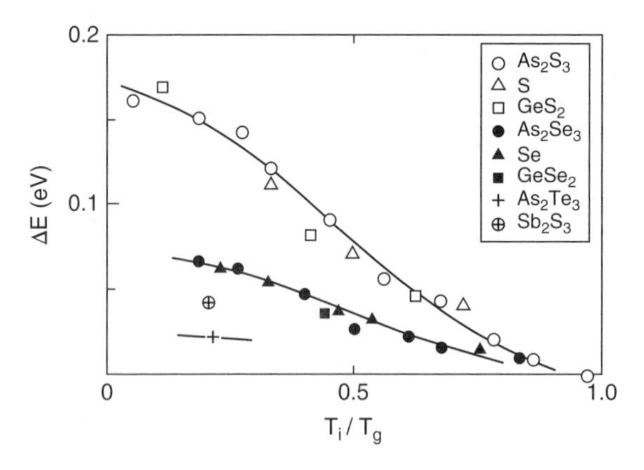

Figure 6.35 Temperature dependence of PD in various a-Chs normalized to the glass transition temperature T_g. *Source*: Tanaka 1983 [180]. Reproduced with permission from Elsevier.

normalized to T_g (i.e., T_{ill}/T_g). The results indicate that PD is promoted at lower temperatures, and hence it can be understood to be a photon-assisted effect. When scaled by T_{ill}/T_g, ΔE_0 decreases in the order S, Se, Te. With increasing Te content, the metallic character is increased, suggesting that a dual bonding nature (mixing covalent and van der Waals bonds) is essential for the occurrence of PD in a-Chs [39]. With increasing T_{ill}, back reaction from the photodarkened states occurs and hence the PD reaches a saturated value ΔE_0 under thermal equilibrium. It has been suggested that a change in the interaction of chalcogen lone pair electrons is required for PD to occur, since band gap photoillumination excites lone pair electrons in chalcogenides.

Next, we discuss the dynamics of the occurrence of PD. Detailed *in situ* measurements of the time evolution and decay of PD have been made on a-As_2Se_3, a-As_2S_3, and a-Se [181, 182]. Two laser beams, one providing strong irradiation with Ar laser light and the other providing weak probing light for measurement of the transmittance T, were used in the *in situ* measurements. The changes $\Delta\alpha$ in the optical absorption coefficient were defined as $\Delta\alpha = (-1/d)\ln(T/T_0)$, where T/T_0 is the transmitted signal relative to that measured before illumination, and d is the sample thickness. Figure 6.36 shows the time evolution and decay of the changes $\Delta\alpha$ during various cycles of strong irradiation at 50 and 300 K for a-As_2Se_3, for example [181, 182]. Illumination at 50 K clearly induces a larger $\Delta\alpha$ than it does at 300 K. When the Ar laser irradiation is switched off, a decrease (decay) in $\Delta\alpha$ is observed which quickly

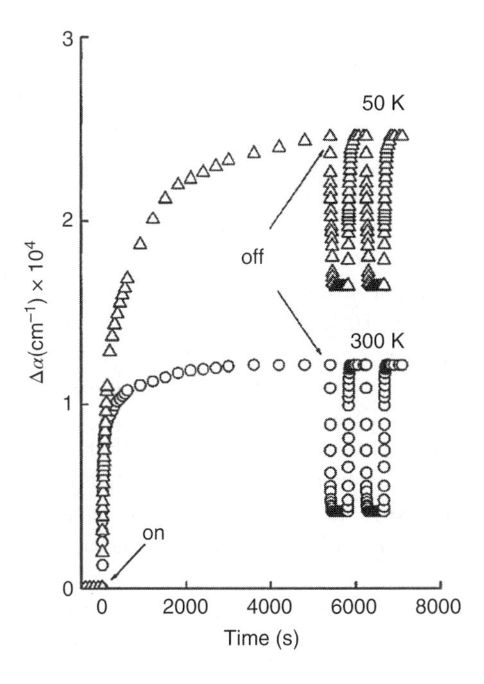

Figure 6.36 Time evolution and decay of the changes $\Delta\alpha$ during various cycles of strong irradiation at 50 and 300 K for a-As_2Se_3. *Source*: Ganjoo *et al.* 2002 [182]. Reproduced with permission from Elsevier.

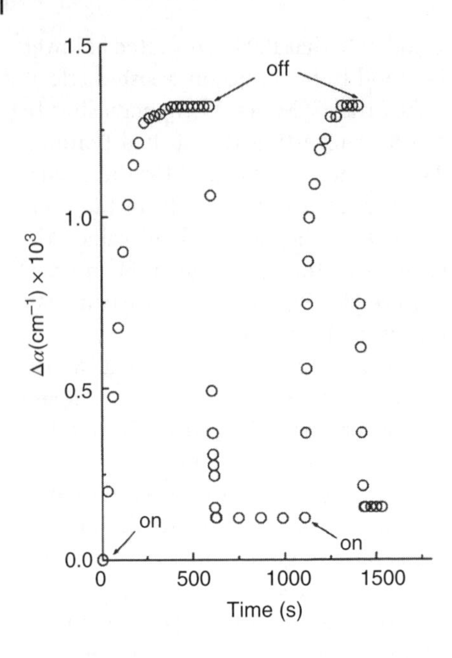

Figure 6.37 Time evolution and decay of transient portion of $\Delta\alpha$ induced at 300 K in a-Se. *Source*: Ganjoo *et al.* 2002 [182]. Reproduced with permission from Elsevier.

reaches a constant value; this value is usually referred to as the *metastable* photodarkening. The magnitude of the decay of $\Delta\alpha$ should be therefore called the *transient change* in the PD. The total change during illumination consists of a transient and a metastable portion. It is of interest that the transient portions induced at 50 and 300 K are almost the same. Interestingly, for a-Se at 300 K, while the metastable PD portion is very small, the transient portion is significant, as shown in Figure 6.37 [182]. This suggests that the transient portion of the PD is temperature independent, unlike the metastable portion of the PD.

The short-time dynamics of PD in the *nanosecond* time domain has also been studied in these materials [183], as shown in Figure 6.38. Here, the photoexcitation was done with a Nd:YAG laser with a 7 ns pulse width. The change in transmittance $\Delta T/T$ (<10%), where ΔT is the induced change, corresponds to $\Delta\alpha < 10^3\,cm^{-1}$, which may be consistent with the initial change in $\Delta\alpha$ induced by CW excitation (see Figure 6.36), and this should be considered as the transient portion. We now know that transient PD occurs on a nanosecond timescale.

Finally, we should mention that significant PD has also been found after vacuum ultraviolet irradiation [184], suggesting that extra excitation of core electrons also participates in the occurrence of PD: excitation of core electrons produces holes in the valence band; that is, it indirectly excites lone pair electrons. The following experimental findings are of interest: PD does not occur in (i) a-$As_2Se(S)_3$ alloyed with Cu [185, 186], (ii) ionic transport materials such as Ag–As–S and Na–Ge–S [187], and (iii) ultrathin (~10 nm) As_2S_3 films [188].

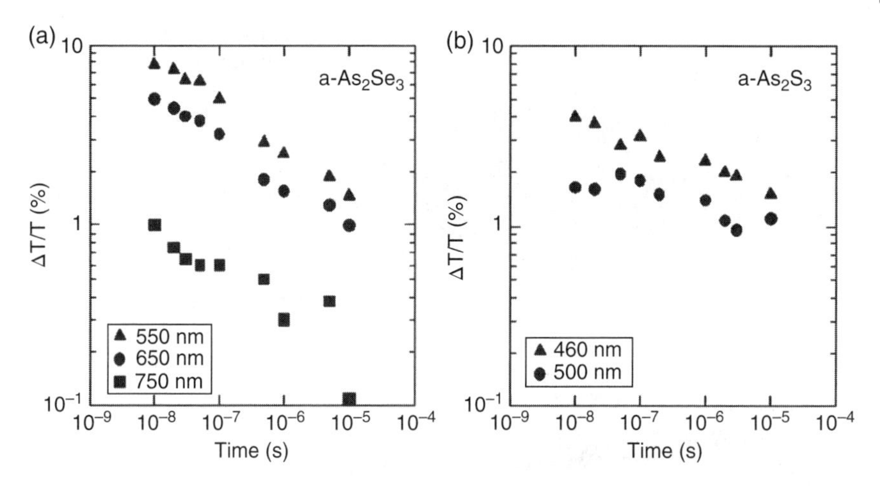

Figure 6.38 Short-time dynamics of PD in nanosecond time domain for (a) a-As_2Se_3 and (b) a-As_2S_3. *Source*: Sakaguchi and Tamura 2008 [183]. Reproduced with permission from Elsevier.

6.6.7.2 Photoinduced Volume Expansion

Photoinduced volume expansion (PVE), accompanied by PD, is usually observed in most a-Chs, though with some exceptions [28]. This PVE, observed as a significant change in the thickness of thin films, has also been shown to have transient and metastable components [181, 189].

Figure 6.39 shows the transient relative change in thickness $\Delta d/d$ as a function of time for a-As_2Se_3, using a technique based on spectral analysis of optical interference fringes [189]. The data, taken at room temperature, show that $\Delta d/d$ increases rapidly on switching on of the illumination, reaches a maximum, and then decreases slowly with time. The decrease is similar to that of the transient PD reported for a-As_2S_3 and to the degradation of the photocurrent with illumination for a-Chs [28, 130]. When the illumination is stopped, there is a slight decrease in $\Delta d/d$ followed by a slow decay, which again parallels the photocurrent decay [130], which was discussed in Section 6.4. The final state is the metastable PVE. The fact that the decrease in $\Delta d/d$ has a similar form to that for the photocurrent suggests that the number of photocarriers dominates both the PVE and the PD, as will be discussed later in the following section.

An alternative new technique which can detect a very small change in PVE has been developed, in which the surface height is measured in real time *in situ*. Here, Twyman–Green interferometry, that is, a fringe-phase-shifting method and image analysis, is employed [190]. Using this technique, a surface height map of the sample was obtained every 1/4 s with ±1 nm accuracy. Figure 6.39(a) shows an example of a surface height map of a-As_2Se_3. Note that

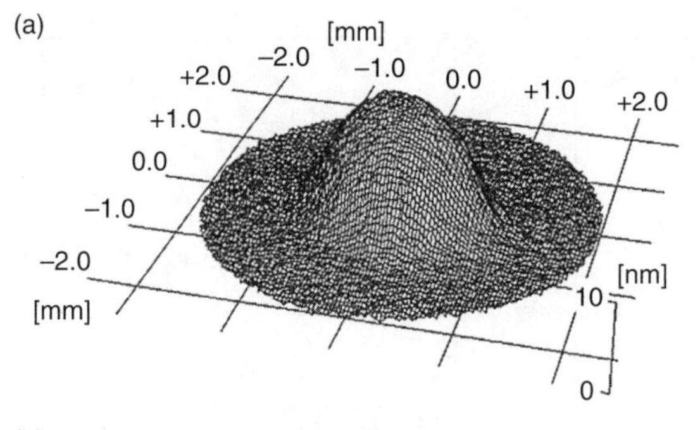

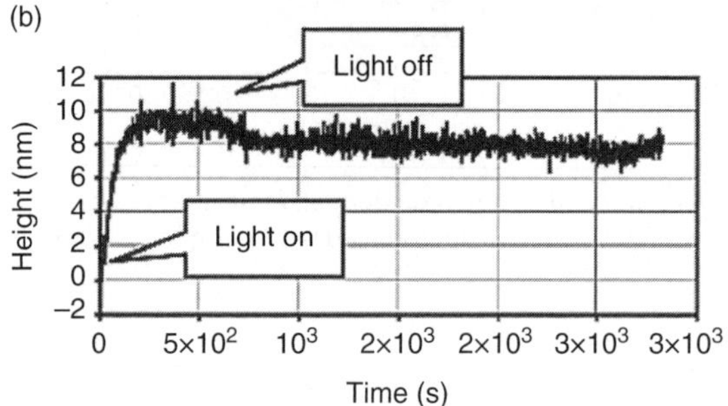

Figure 6.39 (a) Surface height map after band gap illumination of a-As_2Se_3. The height scale is enlarged by about 10^5 times relative to the horizontal scale. (b) Time evolution of induced height change. *Source*: Ikeda and Shimakawa 2004 [190]. Reproduced with permission from Elsevier.

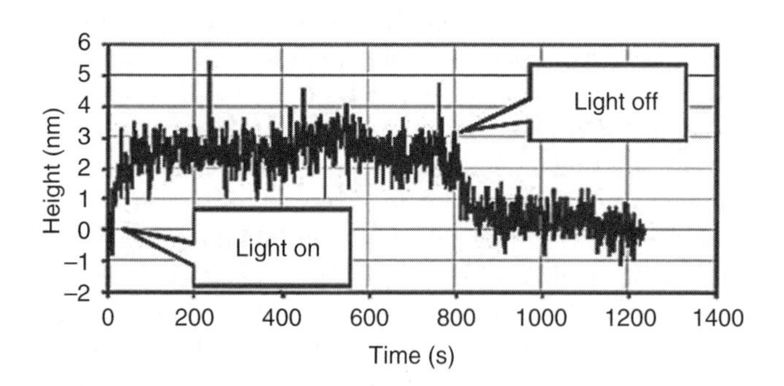

Figure 6.40 Time evolution of surface height change of a-Se measured at room temperature. *Source*: Ikeda and Shimakawa 2004 [190]. Reproduced with permission from Elsevier.

the height scale is enlarged to about 10^5 times the horizontal scale. The maximum height (the top of the "hat") is around 8 nm (metastable) at room temperature, which corresponds to $\Delta d/d \approx 2\%$; the time evolution of the height change is shown in Figure 6.39(b). After the illumination was stopped, the surface height started to decrease and then reached a constant value which corresponds to the metastable PVE state, as we have already observed. Figure 6.40 shows the time evolution of the surface height change measured at room

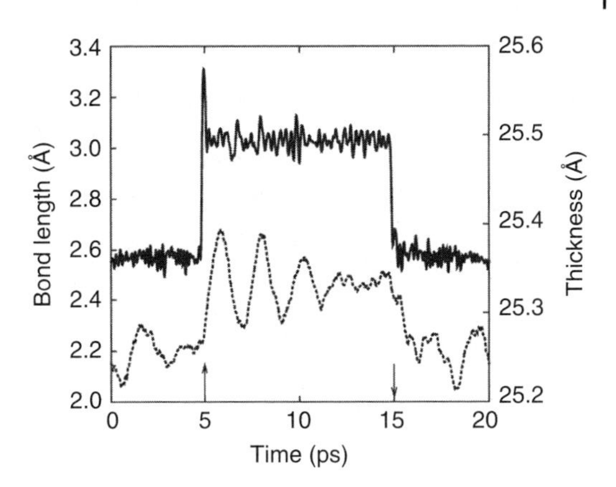

Figure 6.41 Simulated time evolution of bond length (and thickness) in a-Se using tight-binding molecular dynamics. *Source*: Hegedüs *et al.* 2005 [14]. Reproduced with permission from the American Physical Society.

temperature for a-Se. The height increases rapidly by 2.5 nm ($\Delta d/d \approx 0.5\%$). After the illumination is turned off, the induced change disappears, that is, no metastable PVE, as well as no PD, is induced at room temperature in a-Se.

Atomic-scale computer simulations of PVE have also been performed for a-Se. Figure 6.41 shows the simulated time evolution of the bond length (and thickness), obtained using tight-binding molecular dynamics, in a-Se [14]. A sample containing 162 atoms, produced by a "cook and quench" procedure, was used. The thickness change is defined in one direction in a particular cell. As the detailed simulation procedures have been described elsewhere [14], we give only a brief summary of the simulation process for a-Se here.

Photoexcitation was modeled by transferring an electron from the highest occupied molecular orbital to the lowest unoccupied molecular orbital (LUMO). It was assumed in the simulation that the electron and hole separated in space on a femtosecond timescale. Elongation of a bond and bond breaking occurred when an additional electron was put into the LUMO, producing PVE. It was confirmed in the simulation that the majority of bond breaking occurred between twofold- and threefold-coordinated Se sites. As shown in Figure 6.41, before the excitation at a time of 5 ps, the bond length was approximately 0.255 nm, and it increased by 10–20% during illumination, accompanied by damped oscillations. After deexcitation of the electrons (when the excitation was stopped) at 15 ps, the bond length were restored to their original values. This is a reversible effect and can be regarded as a transient PVE, as stated already. The relative thickness change $\Delta d/d \approx 0.5\%$ deduced from the simulation is consistent with the value experimentally obtained for a-Se, as mentioned above.

6.6.7.3 Is There Any Direct Relation among PD, PVE, and PDC?

As already mentioned, PD, PVE, and PDC are the principal photon-induced changes in a-Chs. All of these have metastable portions that are recovered after thermal annealing near T_g. Now, we need to answer the following fundamental

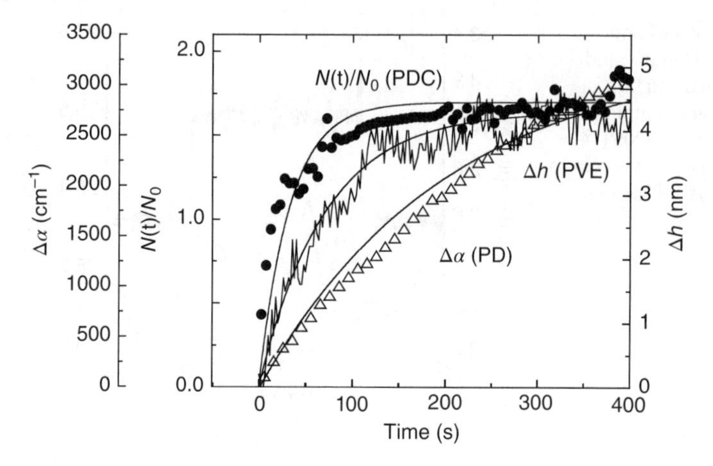

Figure 6.42 Time evolution of the changes in $\Delta\alpha$ (PD), $N(t)/N_0$ (PDC), and Δh (PVE) for a-As_2Se_3 of thickness $d = 200$ nm. *Source*: Nakagawa *et al.* 2010 [191]. Reproduced with permission from John Wiley & Sons.

question: Is there a direct relation among PD, PVE, and PDC? To answer this question, *in situ* simultaneous measurements of all three phenomena have been performed with a-As_2Se_3 [191]. We may obtain a clear insight into the correlation among these photoinduced effects through a discussion of time-dependent dynamics.

As the details of the *in situ* simultaneous measurement system have been described elsewhere [191], we shall only review the results on a-As_2Se_3 obtained at room temperature. Figure 6.42 shows the time evolution of the changes in $\Delta\alpha$ (PD), $N(t)/N_0$ (PDC), and Δh (PVE) for a thickness $d = 200$ nm. All of these physical parameters have been discussed in previous sections. The optical penetration depth, α^{-1}, was 600 nm and almost uniform photoexcitation throughout the film could be assured. All of the changes are given empirically by

$$y = A\left[1 - \exp\left\{-\left(t/\tau\right)^{\beta}\right\}\right] \tag{6.80}$$

where A is a constant equal to the total amount of change, τ is the effective response time, β is the dispersion parameter (<1.0), and y is the value of $\Delta\alpha$, $N(t)/N_0$, or Δh at time t.

The results can be summarized as follows: $\tau = 18$ s with $\beta = 0.55$ for PDC, $\tau = 70$ s with $\beta = 1.0$ for PVE, and $\tau = 200$ s with $\beta = 1.0$ for PD were deduced. We can understand why the effective response times are different; the reactions proceed in time in the order PDC, PVE, PD. The dynamics of PDC is expressed by a stretched exponential function, while that of PVE is expressed by just an exponential function. Note, however, that the dynamics of PD for relatively

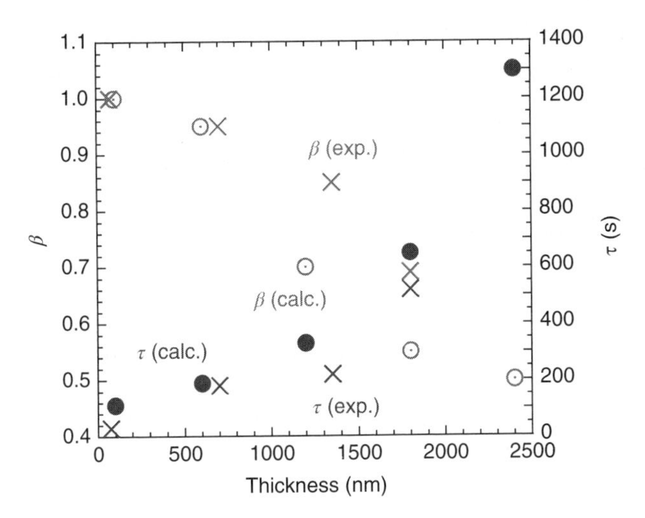

Figure 6.43 Thickness dependence of the parameters τ and β for a-As_2Se_3. *Source:* Shimakawa *et al.* 2009 [192]. Reproduced with permission from AIP Publishing LLC.

thick films is dispersive and given by a stretched exponential function (see Section 6.6.7.1), suggesting that the dynamics of PD depends on the thickness *d*.

Figure 6.43 shows the thickness dependence of the parameters τ and β for a-As_2Se_3 [192]. The values of τ and β increase and decrease, respectively, with the thickness of the film. A sequential process of growth of the PD through the thickness of the film, starting from the surface, may be the origin of the stretched exponential function: that is, the fact that the PD follows a stretched exponential function is not an intrinsic effect but depends on the geometry (i.e., the thickness). It thus turns out that PDC is the only phenomenon that has dynamics represented by a stretched exponential function (dispersive in nature), as we have already discussed.

Before closing this chapter, we should discuss the models which account for the experimental results on PD. There are three principal models to account for the occurrence of PD: (i) a change in the atomic positions [26], (ii) bond breaking and/or alternation [193, 194], and (iii) a macroscopic charged-layer model [195].

The third of these proposals, the charged-layer model, may be the best one for explaining all of the phenomena in a unified way [191]. In materials composed of clustered layers, such as a-Chs, the van der Waals interaction between clusters can be weakened by electronic repulsion if the clusters are charged negatively by electrons; that is, photoinduced holes diffuse away to the unilluminated region and the electrons remain. This induces expansion and slip motion between layers; the PVE results from expansion and the PD from slip motion, as supported by a DOS calculation [196]. As already stated, some material systems do not show PD. If no clustered layers are formed or a rigid network is formed by introducing additives such as Cu, expansion and slip motion does not occur easily, leading to the occurrence of no PD. In ultrathin As_2S_3, no holes escape from layers (there is almost uniform excitation) and hence the layers cannot be charged, inducing no PVE and no PD.

On the other hand, PDC corresponds to bond breaking, which occurs in intralayer regions, that is, bond breaking occurs between As and Se (or S). However, intralayer bond breaking may not be important for PD and PVE, since the extent to which such events occur is only of the order of parts per million (the number of induced recombination events is not very large). It was previously considered that there was a certain correlation among these photoinduced effects in a-Chs. However, it can now be concluded that there is no direct relation among PD, PVE, and PDC.

References

1 Selényi, P. (1935) Methoden, Ergebnisse und Aussichten des elektrostatischen Aufzeichnungsverfahrens. (Elektrographie), *Z. Tech. Phys.*, **12**, 607–614.
2 Pai, D.M. and Springett, B.E. (1993) Physics of electrophotography. *Rev. Mod. Phys.*, **65**, 163–211.
3 Weiss, D.S. and Abkowitz, M. (2006) Organic photoconductors, in *Springer Handbook of Electronic and Photonic Materials* (eds S.O. Kasap and P. Capper), Springer, New York, Chapter 39.
4 Yamada, N., Ohno, E., Nishiuchi, K., Akahira, N., and Takao, M. (1991) Rapid phase-transitions of GeTe–Sb_2Te_3 pseudobinary amorphous thin films for an optical disk memory. *J. Appl. Phys.*, **69**, 2849–2856.
5 Ovshinsky, S.R. (1968) Reversible electrical switching phenomena in disordered structures. *Phys. Rev. Lett.*, **21**, 1450–1453.
6 Wuttig, M. and Yamada, N. (2007) Phase-change materials for rewritable data storage. *Nature Mater.*, **6**, 824–832.
7 Rawlands, J.A. and Kasap, S.O. (1997) Amorphous semiconductors usher in digital X-ray imaging. *Phys. Today*, **50**, 24–31.
8 Juška, G., Arlauskas, K., and Montrimas, E. (1987) Features of carriers at very high electric fields in a-Se and a-Si:H. *J. Non-Cryst. Solids*, **97–98**, 559–561.
9 Tanioka, K. (2007) The ultra sensitive TV pickup tube from conception to recent development. *J. Mater. Sci.*, **18**, S321–S325.
10 Polk, D.E. (1971) Structural model for amorphous silicon and germanium. *J. Non-Cryst. Solids*, **5**, 365–376.
11 Greaves, G.N. and Davis, E.A. (1974) A continuous random network model with three-fold coordination. *Philos. Mag.*, **29**, 1201–1206.

12 Greaves, G.N. and Sen, S. (2007) Inorganic glasses, glass-forming liquids and amorphizing solids. *Adv. Phys.*, **56**, 1–166.
13 Drabold, D.A., Zhang, X., and Li, J. (2003) First-principles molecular dynamics and photostructural response in amorphous silicon and chalcogenide glasses, in *Photo-induced Metastability in Amorphous Semiconductors* (ed. A.V. Kolobov), Wiley-VCH, Weinheim, pp. 260–278.
14 Hegedüs, J., Kohary, K., Pettifor, D.G., Shimakawa, K., and Kugler, S. (2005) Photoinduced volume changes in amorphous selenium. *Phys. Rev. Lett.*, **95**, 206803-01–04.
15 Phillips, J.C. (1979) Topology of covalent non-crystalline solids I: Short –range order in chalcogenide alloys. *J. Non-Cryst. Solids*, **34**, 153–181.
16 Tanaka, K. (1989) Structural phase transitions in chalcogenide glasses. *Phys. Rev. B*, **39**, 1270–1279.
17 Cohen, M.H., Fritzshe, H., and Ovshinsky, S.R. (1969) Simple band model for amorphous semiconducting alloys. *Phys. Rev. Lett.*, **22**, 1065–1068.
18 Kastner, M.A. (1972) Bonding bands, lone-pair bands, and impurity states in chalcogenide semiconductors. *Phys. Rev. Lett.*, **28**, 355 357.
19 Tauc, J. (1968) Optical properties and electronic structure of amorphous Ge and Si. *Mater. Res. Bull.*, **3**, 37–46.
20 Street, R.A. and Mott, N.F. (1975) States in the gap in glassy semiconductors. *Phys. Rev. Lett.*, **35**, 1293–1296.
21 Kastner, M., Adler, D., and Fritzshe, H. (1976) Valence-alternation model for localized gap states in lone-pair semiconductors. *Phys. Rev. Lett.*, **37**, 1504–1507.
22 DeNeufville, J.P., Moss, S.C., and Ovshinsky, S.R. (1973/1974) Photostructural transformations in amorphous As_2Se_3 and As_2S_3 films. *J. Non-Cryst. Solids*, **13**, 191–223.
23 Hamanaka, H., Tanaka, K., Matsuda, A., and Iizima, S. (1976) Reversible photo-induced volume changes in evaporated As_2S_3 and $As_4Se_5Ge_1$ films. *Solid State Commun.*, **19**, 499–501.
24 Shimakawa, K. (1986) Persistent photocurrent in amorphous chalcogenides. *Phys. Rev. B*, **34**, 8703–8708.
25 Staebler, D.L. and Wronski, C.R. (1977) Amorphous silicon solar cell. *Appl. Phys. Lett.*, **28**, 671–673.
26 Tanaka, K. (1990) Photoinduced structural changes in chalcogenide glasses. *Rev. Solid State Sci.*, **4**, 641–659.
27 Pfeiffer, G., Paesler, M.A., and Agarwal S.C. (1991) Reversible photodarkening of amorphous arsenic chalcogens. *J. Non-Cryst. Solids*, **130**, 111–143.
28 Shimakawa, K., Kolobov, A.V., and Elliott, S.R. (1995) Photoinduced effects and metastability in amorphous semiconductors and insulators. *Adv. Phys.*, **44**, 475–588.
29 Rawson, H. (1967) *Inorganic Glass-Forming Systems*, Academic Press, London.
30 Uhlman, D.R. (1977) Glass-formation. *J. Non-Cryst. Solids*, **25**, 43–85.

31 Tammann, G. (1935) *Vitreous State*, ONTI, Moscow.
32 Minaev, V.S. and Timoshenkov, S.P. (2004) Glass-formation in chalcogenide systems and periodic system, in *Semiconducting Chalcogenide Glass I* (eds R. Fairman and B. Ushkov), Elsevier Academic, Amsterdam, pp. 1–50.
33 Minaev, V.S. (2004) Concept of polymeric polymorphous-crystalloid structure of glass and chalcogenide systems: Structure and relaxation of liquid and glass, in *Semiconducting Chalcogenide Glass I* (eds R. Fairman and B. Ushkov), Elsevier Academic, Amsterdam, pp. 139–175.
34 Goryunova, N.A. and Kolomiets, B.T. (1960) Glassy semiconductors IX. Glass-formation in compound chalcogenides based on arsenic sulphide and selenide. *Solid State Phys.*, **2**, 280–283.
35 Hilton, A.R., Jones, C.E., and Brau, M. (1966) Non-oxide IVA–VA–VIA chalcogenide glasses. *Phys. Chem. Glasses*, **7**, 105–126.
36 Borisova, Z.U. (1972) *Chemistry of Glassy Semiconductors*, LGU, Leningrad, p. 248.
37 Hruby, A. (1972) Evaluation of glass-forming tendency by means of DTA. *Czech. J. Phys. B*, **22**, 1187–1193.
38 Elliott, S.R. (1990) *Physics of Amorphous Materials*, 2nd edn, Longman Scientific & Technical, Harlow.
39 Tanaka, K. and Shimakawa, K. (2011) *Amorphous Chalcogenide Semiconductors and Related Materials*, Springer, New York.
40 Kauzmann, W. (1948) The nature of the glassy state and the behaviour of liquids at low temperatures. *Chem. Rev.*, **43**, 219–256.
41 Novikov, V.N. and Sokolov, A.P. (2013) Role of quantum effects in the glass transition temperature. *Phys. Rev. Lett.*, **110**, 065701-1–5.
42 Tichy, L. and Ticha, H. (1995) Covalent bond approach to the glass-transition temperature of chalcogenide glasses. *J. Non-Cryst. Solids*, **189**, 141–146.
43 Tanaka, K. (1985) Glass transition of covalent glasses. *Solid State Commun.*, **54**, 867–869.
44 Kugler, S. and Shimakawa, K. (2015) *Amorphous Semiconductors*, Cambridge University Press, Cambridge.
45 Freitas, R.J., Shimakawa, K., and Kugler, S. (2013) Some remarks on glass-transition temperature in chalcogenide glasses: A correlation with the microhardness. *Chalcogenide Lett.*, **10**, 39–43.
46 Langer, J.S. (2006) Excitation chains at the glass transition. *Phys. Rev. Lett.*, **97**, 115704-1–4.
47 Wilson, M. and Salmon, P.S. (2009) Network topology and the fragility of tetrahedral glass-forming liquids. *Phys. Rev. Lett.*, **103**, 157801-1–4.
48 Angell, C.A. (1995) Formation of glasses from liquids and biopolymers. *Science* **267**, 1924–1935.
49 Pineda, E.P. and Crespo, D. (1999) Microstructure development in Kolmogorov, Johnson-Mehl, and Avrami nucleation and growth kinetics. *Phys. Rev. B*, **60**, 3104–3112.

50 Tonchev, D. and Kasap, S.O. (2006) Thermal properties and thermal anaysis, in *The Springer Handbook of Electronic and Photonic Materials* (ed. S.O. Kasap and P. Capper), Springer, Heidelberg, pp. 385–408.
51 Senkader, S. and Wright, C.D. (2004) Models for phase-change of $Ge_2Sb_2Te_5$ in optical and electrical memory devices. *J. Appl. Phys.*, **95**, 504–511.
52 Shimakawa, K. (2013) Dynamics of crystallization with fractal geometry: Extended KJMA approach in glasses. *Phys. Status Solidi B*, **249**, 2024–2027.
53 Yelon, A., Movaghar, B., and Crandall, R.C. (2006) Multi-excitation entropy: Its role in thermodynamics and kinetics. *Rep. Prog. Phys.*, **69**, 1145–1194.
54 Mott, N.F. and Davis, E.A. (1979) *Electronic Processes in Non-Crystalline Materials*, 2nd edn, Clarendon Press, Oxford.
55 Morigaki, K. (1999) *Physics of Amorphous Semiconductors*, World Scientific, London.
56 Singh, J. and Shimakawa, K. (2003) *Advances in Amorphous Semiconductors*, Taylor & Francis, London.
57 Yelon, A. (2013) The fallacy of Meyer–Neldel temperature as a measure of disorder. *Monatsh. Chem.*, **144**, 91–95.
58 Overhof, H. and Thomas, P. (1989) *Electronic Transport in Hydrogenated Amorphous Semiconductors*, Springer, Berlin.
59 Shimakawa, K. and Abdel-Wahab, F. (1997) The Meyer–Neldel rule in chalcogenide glasses. *Appl. Phys. Lett.*, **70**, 652–654.
60 Shimakawa, K. and Aniya, M. (2013) Dynamics of atomic diffusion in condensed matter: Origin of the Meyer–Neldel compensation law. *Monatsh. Chem.*, **144**, 67–71.
61 Shimakawa, K. and Miyake, K. (1988) Multiphonon tunnelling conduction of localized π electrons in amorphous carbon films. *Phys. Rev. Lett.*, **61**, 994–996.
62 Mott, N.F. (1993) *Conduction in Non-Crystalline Materials*, 2nd edn, Oxford University Press, Oxford.
63 Okamoto, H., Hattori, K., and Hamakawa, Y. (1993) Hall effect near the mobility edge. *J. Non-Cryst. Solids*, **164–166**, 445–448.
64 Seager, C.H., Emin, D., and Quinn, R.K. (1973) Electrical transport of structural properties of bulk As-Te-I, As-Te-Ge, and As-Te chalcogenide glasses. *Phys. Rev. B*, **8**, 4746–4760.
65 Seager, C.H. and Quinn, R.K. (1975) DC electronic transport in binary arsenic chalcogenide glasses. *J. Non-Cryst. Solids*, **17**, 386–400.
66 Emin, D. (1975) Phonon-assisted transition rates I. Optical-phonon-assisted hopping in solids. *Adv. Phys.*, **24**, 305–348.
67 Nagel, P. (1979) Electronic transport in amorphous semiconductors, in *Amorphous Semiconductors* (ed. M.H. Brodsky), Springer, Berlin, pp. 113–158.
68 Overhof, H. and Beyer, W. (1983) Electronic transport in hydrogenated amorphous silicon. *Philos. Mag. B*, **47**, 377–392.
69 Nagels, P., Callaerts, R., and Denayer, M. (1972) D.C. conductivity, Hall effect and thermopower of amorphous $As_2Se_{3-}xAs_2Te_3$, in *Proceedings of the 11th*

International Conference on Physics of Semiconductors (ed. M. Miasek), PWN-Polish Scientific Publishers, Warsaw, pp. 549–554.

70 Emin, D. (2008) Generalized adiabatic polaron hopping: Meyer–Neldel compensation and Poole–Frenkel behaviour. *Phys. Rev. Lett.*, **100**, 166602-1–4.

71 Long, A.R. (1982) Frequency-dependent loss in amorphous semiconductors. *Adv. Phys.*, **31**, 553–637.

72 Elliott, S.R. (1987) AC conduction in amorphous chalcogenide and pnictide semiconductors. *Adv. Phys.*, **36**, 135–218.

73 Elliott, S.R. (1977) A theory of a.c. conduction in chalcogenide glasses. *Philos. Mag.*, **36**, 1291–1304.

74 Shimakawa, K. (1982) On the temperature dependence of a.c. conduction in chalcogenide glasses. *Philos. Mag. B*, **46**, 123–135.

75 Dyre, J.C. (1988) The random free-energy barrier model for ac conduction in disordered solids. *J. Appl. Phys.*, **64**, 2456–2468.

76 Dyre, J.C. (1993) Universal low-temperature ac conductivity of macroscopically disordered nonmetals. *Phys. Rev. B*, **48**, 12511–12526.

77 Ganjoo, A. and Shimakawa, K. (1994) Estimation of density of charged defects in amorphous chalcogenides from a.c. conductivity: Random-walk approach for bipolaron based on correlated barrier hopping. *Philos. Mag. Lett.*, **70**, 287–291.

78 Stehlik, S., Shimakawa, K., Wagner, T., and Frumar, M. (2012) Diffusion of Ag ions under random potential barriers in silver-containing chalcogenide glasses. *J. Phys. D: Appl. Phys.*, **45**, 205304.

79 Patil, D.S., Shimakawa, K., Zima, V., Macak, J., and Wagner, T. (2013) Evaluation of impedance spectra of ionic-transport materials by a random-walk approach considering electrode and bulk response. *J. Appl. Phys.*, **113**, 147305–4.

80 Patil, D.S., Shimakawa, K., Zima, V., and Wagner, T. (2014) Quantitative impedance analysis of solid ionic conductors: Effects of electrode polarization. *J. Appl. Phys.*, **115**, 143707-6.

81 Shimakawa, K. and Nitta, S. (1978) Influence of silver additive on electronic and ionic natures in amorphous As_2Se_3. *Phys. Rev. B*, **18**, 4348–4354.

82 Raistrick, I.D., Donald, D.R., and Macdonald, J.R. (2005) Theory, in *Impedance Spectroscopy* (eds E. Barsoukov and J.R. Macdonald), John Wiley & Sons, Inc., Hoboken, NJ, pp. 27–75.

83 Banos, N., Steele, B.C.H., and Butler, E.P. (2005) Applications of impedance spectroscopy, in *Impedance Spectroscopy* (eds E. Barsoukov and J.R. Macdonald) John Wiley & Sons, Inc., Hoboken, NJ, pp. 205–258.

84 Serghei, A., Tress, M., Sangoro, J.R., and Kremer, F. (2009) Electrode polarization and charge transport at solid interfaces. *Phys. Rev. B*, **80**, 184301-1–5.

85 Hunt, A. (1994) Statistical and percolation effects on ionic conduction in amorphous systems. *J. Non-Cryst. Solids*, **175**, 59–70.

86 Bunde, A. Ingram, M.D., and Maas, P. (1994) The dynamic structure model for ion transport in glasses. *J. Non-Cryst. Solids*, **172–174**, 1222–1236.

87 Bychkov, E. and Price, D.L. (2000) Neutron diffraction studies of $Ag_2S–As_2S_3$ glasses in the percolation and modifier-controlled domains. *Solid State Ionics*, **136–137**, 1041–1048.

88 Bychkov, E. (2009) Superionic and ionic conducting chalcogenide glasses: Transport regimes and structural features. *Solid State Ionics*, **180**, 510–516.

89 Shimakawa, K. and Wagner, T. (2013) Origin of power-law composition dependence in ionic transport glasses. *J. Appl. Phys.*, **113**, 143701–5.

90 Borisova, Z.U. (1981) *Glassy Semiconductors*, Plenum, New York.

91 Mandelbrot, B.B. (1982) *The Fractal Geometry of Nature*, Freeman, New York.

92 Zallen, R. (1983) *The Physics of Amorphous Solids*, John Wiley & Sons, New York.

93 Meherun-Nessa, Shimakawa, K., Ganjoo, A., and Singh, J. (2000) Fundamental optical absorption on fractals: A case example for amorphous chalcogenides. *J. Optoelectron. Adv. Mater.*, **2**, 133–138.

94 Shimakawa, K., Singh, J., and O'Leary, S.K. (2006) Optical properties of disordered condensed matter, in *Optical Properties of Condensed Matter and Applications* (ed. J. Singh), John Wiley & Sons, Ltd, Chichester, pp. 47–62.

95 He, X.-F. (1990) Fractional dimensionality and fractional derivative spectra of interband optical transitions. *Phys. Rev. B*, **42**, 11751–11756.

96 Urbach, F. (1953) The long-wavelength edge of photographic sensitivity and of the electronic absorption. *Phys. Rev.*, **92**, 1324.

97 Tanaka, K. (2014) Minimal Urbach energy in non-crystalline materials. *J. Non-Cryst. Solids*, **389**, 35–37.

98 Abe, S. and Toyozawa, Y. (1981) Interband absorption spectra of disordered semiconductors in the coherent potential approximation. *J. Phys. Soc. Japan*, **50**, 2185–2194.

99 Davis, E.A. and Greaves, G.N. (1976) Amorphous arsenic – properties and structural model, in *Proceedings of the 6th International Conference on Amorphous Liquid Semiconductors* (ed. B.T. Kolomiet), Nauka, Leningrad, pp. 212–220.

100 Chan, C.T., Louie, S.G., and Phillips, J.C. (1987) Potential fluctuations and density of gap states in amorphous semiconductors. *Phys. Rev. B*, **35**, 2744–2749.

101 Bacalis, N., Economou, E.N., and Cohen, M.H. (1988) Simple derivation of exponential tails in the density of states. *Phys. Rev. B*, **37**, 2714–2717.

102 Tauc, J. (1975) Highly transparent glasses, in *Optical Properties of Highly Transparent Solids* (ed. S.S. Mitra and B. Bendow), Plenum, New York, pp. 245–260.

103 Kolomiets, B.T. and Lyubin, V.M. (1973) Photoelectric phenomena in amorphous chalcogenide semiconductors. *Phys. Stat. Sol. (a)*, **17**, 11–46.

104 Cernogola, J., Mollot, F., and Benoit À La Guillaume, C. (1973) Radiative recombination in amorphous As_2Se_3. *Phys. Stat. Sol. (a)*, **15**, 401–407.

105 Street, R.A. (1976) Luminescence in amorphous semiconductors. *Adv. Phys.*, **25**, 397–453.

106 Murayama, K. (1983) Time-resolved photoluminescence in chalcogenide glasses. *J. Non-Cryst. Solids*, **59–60**, 983–990.

107 Aoki, T. (2006) Photoluminescence, in *Optical Properties of Condensed Matter and Applications* (ed. J. Singh), John Wiley & Sons, Ltd, Chichester, pp. 75–106.

108 Street, R.A. (1977) Non-radiative recombination in chalcogenide glasses. *Solid State Commun.*, **24**, 363–365.

109 Ristein, J., Taylor, P.C., Ohlsen, W.D., and Weiser, G. (1990) Radiative recombination center in As_2Se_3 as studied by optically detected magnetic resonance. *Phys. Rev. B*, **42**, 11845–11856.

110 Ristein, J. (1991) Unification of geminate and distant pair recombination statistics: Low temperature photoelectronic properties in a-Si:H. *J. Non-Cryst. Solids*, **137–138**, 563–566.

111 Shimakawa, K. (1985) Exciton recombination in amorphous chalcogenides. *Phys. Rev. B*, **31**, 4012–4014.

112 Weinstein, B.A. (1984) Anomalous pressure response of luminescence in c-As_2S_3 and a-As_2SeS_2: Consequences for defect structure in chalcogenides. *Philos. Mag. B*, **50**, 709–729.

113 Aoki, T., Saito, D., Ikeda, K., Kobayashi, S., and Shimakawa, K. (2005) Radiative recombination processes in chalcogenide glasses deduced by lifetime measurements over 11 decades. *J. Opt. Adv. Mater.*, **7**, 1749–1757.

114 Aoki, T. (2012) Photoluminescence spectroscopy, in *Characterization of Materials* (ed. E.N. Kaufmann), John Wiley & Sons, Inc., Hoboken, NJ, pp. 1–12.

115 Bishop, S.G., Turnbull, D.A., and Aitken, B.G. (2000) Excitation of rare earth emission in chalcogenide glasses by broadband Urbach edge absorption. *J. Non-Cryst. Solids*, **266–269**, 876–883.

116 Zhugayevych, A. and Lubchenko, V. (2010) An intrinsic formation mechanism for midgap electronic states in semiconductor glasses. *J. Chem. Phys.*, **132**, 044508-1–6.

117 dos Santos, P.V., Gouveria, E.A., de Araujo, M.T., Gouveria-Neto, A.S., Ribeiro, S.J.L., and Benedicto, S.H.S. (2000) IR–visible upconversion and thermal effects in Pr^{3+}/Yb^{3+}-codoped Ga_2O_3: La_2S_3 chalcogenide glasses. *J. Phys.: Condens. Matter*, **12**, 10003–10010.

118 Strizik, L., Hrabovsky, J., Wagner, T., and Aoki, T (2015) Dynamics of upconversion photoluminescence in Ge–Ga–S: Er^{3+}: Application of quadrature frequency resolved spectroscopy. *Philos. Mag. Lett.*, **95**, 466–473.

119 Marshall, J.M. (1983) Carrier diffusion in amorphous semiconductors. *Rep. Prog. Phys.*, **46**, 1235–1282.

120 Pfister, G. (1976) Dispersive low-temperature transport in a-selenium. *Phys. Rev. Lett.*, **36**, 271–273.

121 Pfister, G. and Scher, H. (1978) Dispersive (non-Gaussian) transient transport in disordered solids. *Adv. Phys.*, **27**, 747–798.

122 Scher, H. and Montroll, E.W. (1975) Anomalous transit-time dispersion in amorphous solids. *Phys. Rev. B*, **12**, 2455–2477.

123 Orenstein, J. and Kastner, M. (1981) Photocurrent transient spectroscopy: Measurements of the density of localized states in a-As_2Se_3. *Phys. Rev. Lett.*, **46**, 1421–1424.

124 Tiedje, T. and Rose, A. (1980) A physical interpretation of dispersive transport in disordered semiconductors. *Solid State Commun.*, **37**, 49–52.

125 Tiedje, T. (1984) Information about band-tail states from time-of-flight experiments, in *Semiconductors and Semimetals*, vol. **21**, Part C (ed. J.I. Pankove), Academic Press, Orlando, FL, pp. 207–238.

126 Bube, R.H. (1960) *Photoconductivity of Solids*, John Wiley &Sons, Inc., New York.

127 Fritzshe, H. (1989) Low temperature electronic transport in non-crystalline semiconductors. *J. Non-Cryst. Solids*, **114**, 1–6.

128 Shklovskii, B.I., Fritzshe, H., and Baranovskii, S.D. (1989) Electronic transport and recombination in amorphous semiconductors at low temperatures. *Phys. Rev. Lett.*, **62**, 2989–2992.

129 Shklovskii, B.I., Levin, E.I., Fritzshe, H., and Baranovskii, S.D. (1990) Hopping photoconductivity in amorphous semiconductors: Dependence on temperature and electric field and frequency, in *Advances in Disordered Semiconductors* (ed. H. Fritzshe), World Scientific, Singapore, pp. 161–191.

130 Shimakawa, K. (1985) Residual photocurrent decay in amorphous chalcogenides. *J. Non-Cryst. Solids*, **77–78**, 1253–1256.

131 Shimakawa, K. (1986) Persistent photocurrent in amorphous chalcogenides. *Phys. Rev. B*, **34**, 8703–8708.

132 Freitas, R.J., Shimakawa, K., and Wagner, T. (2014) The dynamics of photoinduced defect creation in amorphous chalcogenides: The origin of the stretched exponential function. *J. Appl. Phys.*, **115**, 013704–4.

133 Anderson, P.W. (1975) Model for the electronic structure of amorphous semiconductors. *Phys. Rev. Lett.*, **34**, 953–955.

134 Kawazoe, H., Yanagita, H., Watanabe, Y., and Yamane, Y. (1988) Imperfections in amorphous chalcogenides. I. Electrically neutral defects in liquid sulphur and arsenic. *Phys. Rev. B*, **38**, 5661–5667.

135 Fritzshe, H. and Kastner, M.A. (1978) The effect of charged additives on the carrier concentrations in lone-pair semiconductors. *Philos. Mag.*, **37**, 285–292.

136 Tohge, N., Yamamoto, Y., Minami, T., and Tanaka, M. (1979) Preparation of n-type semiconducting $Ge_{20}Bi_{10}Se_{70}$ glass. *Appl. Phys. Lett.*, **34**, 640–641.

137 Tohge, N., Minami, T., Yamamoto, Y., and Tanaka, M. (1980) Electrical and optical properties of n-type semiconducting chalcogenide glasses in the system Ge–Bi–Se. *J. Appl. Phys.*, **51**, 1048–1053.
138 Bishop, S.G., Strom, U., and Taylor, P.C. (1975) Optically induced localized paramagnetic states in chalcogenide glasses. *Phys. Rev. Lett.*, **36**, 1346–1350.
139 Bishop, S.G., Strom, U., and Taylor, P.C. (1977) Optically induced metastable paramagnetic states in amorphous semiconductors. *Phys. Rev. B*, **15**, 2278–2294.
140 Arai, K. and Namikawa, H. (1973) ESR in Ge–S glasses. *Solid State Commun.*, **13**, 1167–1170.
141 Cerny, V. and Frumar, M. (1979) ESR study and model of paramagnetic defects in Ge–S glasses. *J. Non-Cryst. Solids*, **33**, 23–39.
142 Kordas, G., Weeks, R.A., and Kinser, D.L. (1985) Paramagnetic conduction electrons in GeS_x-glasses. *J. Non-Cryst. Solids*, **71**, 157–161.
143 Kumeda, M., Kawachi, G., and Shimizu, T. (1985) Photoconductivity and photoluminescence and their relation to light-induced ESR in $(Ge_{0.42}S_{0.58})_{1-x}(Sb_{0.4}S_{0.6})_x$ glasses. *Philos. Mag. B*, **51**, 591–602.
144 Watanabe, Y., Kawazoe, H., and Yamane, M. (1988) Imperfections in amorphous chalcogenide. II. Detection and structure determination of neutral defects in As–S, Ge–S, and Ge–As–S glasses. *Phys. Rev. B*, **38**, 5668–5676.
145 Biegelsen, D.K. and Street, R.A. (1980) Photoinduced defects in chalcogenide glasses. *Phys. Rev. Lett.*, **44**, 803–806.
146 Suzuki, H., Murayama, K., and Ninomiya, T. (1979) Optical detection of ESR in amorphous As_2S_3. *J. Phys. Soc. Japan*, **46**, 693–694.
147 Depina, S.P. and Cavenett, B.C. (1981) Observation of bound exciton and distant pair resonance in amorphous phosphorus. *Solid State Commun.*, **40**, 813–818.
148 Depina, S.P. and Cavenett, B.C. (1982) Exciton and pair recombination at intimate valence-alternation in As_2S_3. *Phys. Rev. Lett.*, **48**, 556–559.
149 Tada, T., Suzuki, H., Murayama, K., and Ninomiya, T. (1984) Optically detected ESR in luminescence centers in a-As_2S_3, in *AIP Conference Proceedings*, No. 120 (eds P.C. Taylor and S.G. Bishop), AIP, New York, pp. 326–329.
150 Robins, L.H. and Kastner, M.A. (1987) Anomalous magnetic properties of triplet excited states in crystalline and amorphous arsenic triselenide. *Phys. Rev. B*, **35**, 2867–2885.
151 Koughia, K., Shakoor, Z., Kasap, S.O., and Marshall, J.M. (2005) Density of localized electronic states in a-Se from electron time-of-flight measurements. *J. Appl. Phys.*, **97**, 033706–11.
152 Koughia, K. and Kasap, S.O. (2006) Density of states of a-Se near the valence band. *J. Non-Cryst. Solids*, **352**, 1539–1542.
153 Kounavis, P. (2001) Analysis of the modulated photocurrent experiment. *Phys. Rev. B*, **64**, 045204-1–21.

154 Kounavis, P. and Mylitneau, E. (1996) The defect states in sputtered Ge–Se–Bi films. *J. Non-Cryst. Solids*, **201**, 119–127.

155 Kounavis, P. and Mylitneau, E. (1999) Photostructural changes in amorphous semiconductors studied by the modulated photocurrent method. *J. Phys.: Condens. Matter*, **11**, 9105–9114.

156 Tanaka, K. and Nakayama, S. (1999) Band-tail characteristics in amorphous semiconductors studied by the constant-photocurrent method. *Japan. J. Appl. Phys.*, **38**, 3986–3992.

157 Tanaka, K. (2000) Sub-gap excitation in As_2S_3 glass. *J. Non-Cryst. Solids*, **266–269**, 889–893.

158 Abkowitz, M. and Markov, J.M. (1984) Evidence of equilibrium native defect populations in amorphous chalcogenides from analysis of xerographic spectra. *Philos. Mag. B*, **49**, L31–L36.

159 Imagawa, O., Akiyama, T., and Shimakawa, K. (1984) Localized density of states in amorphous silicon determined by electrophotography. *Appl. Phys. Lett.*, **45**, 438–439.

160 Imagawa, O., Iwanishi, M., Yokoyama, S., and Shimakawa, K. (1985) Electrophotographic spectroscopy of gap states in hydrogenated amorphous silicon. *J. Non-Cryst. Solids*, **77&78**, 359–362.

161 Imagawa, O., Iwanishi, M., Yokoyama, S., and Shimakawa, K. (1986) The localized density of states in amorphous silicon determined by electrophotography. *J. Appl. Phys.*, **60**, 3176–3181.

162 Shimakawa, K. and Katsuma, Y. (1986) Extended step-by-step analysis in space-charge-limited current: Application to hydrogenated amorphous silicon. *J. Appl. Phys.*, **60**, 1417–1421.

163 Hautala, J., Ohlsen, W.D., and Taylor, P.C. (1988) Optically-induced electron-spin resonance in As_xS_{1-x}. *Phys. Rev. B*, **38**, 11048–11060.

164 Elliott, S.R. and Shimakawa, K. (1990) Model for bond-breaking mechanisms in amorphous arsenic chalcogenides leading to light-induced electron-spin-resonance. *Phys. Rev. B*, **42**, 9766–9770.

165 Morigaki, K., Hirabayashi, I., Nakayama, M., Nitta, S., and Shimakawa, K. (1980) Fatigue effect in luminescence of glow discharge amorphous silicon at low temperatures. *Solid State Commun.*, **33**, 851–856.

166 Tada, T. and Ninomiya, T. (1991) Optically induced metastable defects in a-As_2S_3. *J. Non-Cryst. Solids*, **137–138**, 997–1000.

167 Tada, T. and Ninomiya, T (1989) Photoluminescence from optically induced metastable states in a-As_2S_3. *J. Non-Cryst. Solids*, **114**, 88–90.

168 Staebler, D.L. and Wronski, C.R. (1977) Reversible conductivity changes in discharge-produced amorphous Si. *Appl. Phys. Lett.*, **31**, 292–294.

169 Shimakawa, K., Inami, S., and Elliott, S.R. (1990) Reversible photoinduced change of photoconductivity in amorphous chalcogenide films. *Phys. Rev. B*, **42**, 11857–11861.

170 Shimakawa, K., Inami, S., Kato, T., and Elliott, S.R. (1992) Origin of photoinduced metastable defects in amorphous chalcogenides. *Phys. Rev. B*, **46**, 10062–10069.

171 Shimakawa, K., Kondo, A., Hayashi, K., Akahori, S., Kato, T., and Elliott, S.R. (1993) Photoinduced metastable defects in amorphous semiconductors: Communality between hydrogenated amorphous silicon and chalcogenides. *J. Non-Cryst. Solids*, **164–166**, 387–390.

172 Hayashi, K., Hikida, Y., Shimakawa, K., and Elliott, S.R. (1997) Absence of photodegradation in amorphous chalcogenide films with a narrow optical bandgap. *Philos. Mag. Lett.*, **76**, 233–236.

173 Aoki, T., Shimada, H., Hirao, N., Yoshida, N., Shimakawa, K., and Elliott, S.R. (1999) Reversible photoinduced changes of electronic transport in narrow-gap amorphous Sb_2Se_3. *Phys. Rev. B*, **59**, 1579–1581.

174 Shimakawa, K. and Elliott, S.R. (1988) Reversible photoinduced change of ac conduction in amorphous As_2S_3 films. *Phys. Rev. B*, **38**, 12479–12482.

175 Kolomiets, B.T. and Lyubin, V.M. (1973) Photoelectronic phenomena in amorphous chalcogenide semiconductors. *Phys. Stat. Sol.*, **17**, 11–46.

176 Shimakawa, K., Mehern-Nessa, Ishida, H., and Ganjoo, A. (2004) Quantum efficiency of light-induced defect creation in hydrogenated amorphous silicon and amorphous As_2Se_3. *Philos. Mag. B*, **84**, 81–89.

177 Morigaki, K., Hikita, H., and Ogihara, C. (2015) *Light-Induced Defects in Semiconductors*, Pan Stanford, Singapore.

178 Papaulis, A. (1965) *Probability, Random Variables and Stochastic Processes*, McGraw-Hill Kogakusha, Tokyo.

179 DeNeufville, J.P., Moss, S.C., and Ovshinsky, S.R. (1974) Photostructural transformations in amorphous As_2Se_3 and As_2S_3 films. *J. Non-Cryst. Solids*, **13**, 191–223.

180 Tanaka, K. (1983) Mechanisms of photodarkening in amorphous chalcogenides. *J. Non-Cryst. Solids*, **59–60**, 925–928.

181 Ganjoo, A., Shimakawa, K., Kamiya, H., Davis, E.A., and Singh, J. (2000) Percolative growth of photodarkening in amorphous As_2S_3 films. *Phys. Rev. B*, **62**, R14601–R14604.

182 Ganjoo, A., Shimakawa, K., Kitano, K., and Davis, E.A. (2002) Transient photodarkening in amorphous chalcogenides. *J. Non-Cryst. Solids*, **299–302**, 917–923.

183 Sakaguchi, Y. and Tamura, K. (2008) Nanosecond dynamics of photodarkening in amorphous chalcogenide. *J. Non-Cryst. Solids*, **354**, 2679–2682.

184 Hayashi, K. and Shimakawa, K. (1996) Photoinduced effects in amorphous chalcogenides films by vacuum ultra-violet light. *J. Non-Cryst. Solids*, **198**, 696–699.

185 Liu, J.Z. and Taylor, P.C. (1987) Absence of photodarkening in glassy As_2S_3 and As_2Se_3 alloyed with copper. *Phys. Rev. Lett.*, **59**, 1938–1941.

186 Aniya, M. and Shimojo. F. (2006) Atomic structure and bonding properties in amorphous $Cux(As_2S_3)_{1-}x$ by ab initio molecular-dynamics simulations. *J. Non-Cryst. Solids*, **352**, 1510–1513.

187 Tanaka, K., Nemoto, N., and Nasu, H. (2003) Photoinduced phenomena in Na_2S–GeS_2 glasses. *Japan. J. Appl. Phys.*, **42**, 6748–6752.

188 Hayashi, K. and Mitsuishi, N. (2002) Thickness effect of the photodarkening in amorphous chalcogenide films. *J. Non-Cryst. Solids*, **299**, 949–952.

189 Ganjoo, A., Ikeda, Y., and Shimakawa, K. (1999) *In situ* photoexpansion measurements of amorphous As_2S_3: Role of photocarriers. *Appl. Phys. Lett.*, **74**, 2119–2121.

190 Ikeda, Y. and Shimakawa, K. (2004) Real-time *in-situ* measurements of photoinduced volume changes in chalcogenide glasses. *J. Non-Cryst. Solids*, **338–340**, 539–542.

191 Nakagawa, N., Shimakawa, K., Itoh, T., and Ikeda, Y. (2010) Dynamics of principal photoinduced effects in amorphous chalcogenides: *In-situ* simultaneous measurements of photodarkening, volume changes, and defect creation. *Phys. Stat. Sol. (c)*, **7**, 857–860.

192 Shimakawa, K., Nakagawa, N., and Itoh, T. (2009) The origin of stretched exponential function in dynamic response of photodarkening in amorphous chalcogenides. *Appl. Phys. Lett.*, **95**, 051908–3.

193 Elliott, S.R. (1986) A unified model for reversible photostructural effects in chalcogenide glasses. *J. Non-Cryst. Solids*, **81**, 71–98.

194 Kolobov, A.V., Oyanagi, H., Tanaka, K., and Tanaka, K. (1997) Structural study of amorphous selenium by *in situ* EXAFS: Observation of photoinduced bond alternation. *Phys. Rev. B*, **55**, 726–734.

195 Shimakawa, K., Yoshida, N., Ganjoo, A., Kuzukawa, Y., and Singh, J. (1998) A model for the photostructural changes in amorphous chalcogenides. *Philos. Mag. Lett.*, **77**, 153–158.

196 Watanabe, T., Kawazoe, H., and Yamane, Y. (1988) Imperfections in amorphous chalcogenides. III. Interacting lone-pair model for localized gap states based on a tight-binding energy-band calculation of As_2S_3. *Phys. Rev. B*, **38**, 5677–5683.

7

Other Amorphous Material Systems

In addition to hydrogenated amorphous silicon (a-Si:H) and amorphous chalcogenides (a-Chs), two other systems, that is, amorphous carbon-related materials and amorphous oxides, have reached commercial markets. We hence briefly introduce these materials in Sections 7.1 and 7.2. Some chalcogenides, for example Ag-containing materials, show high ionic conductivity (called superionic conduction), and hence we must also discuss the ionic properties of these materials, since the nature of ionic transport in these systems is still not very clear (see also Section 6.3.2).

7.1 Amorphous Carbon and Related Materials

Amorphous carbon and related materials are known to be basically formed from a mixture of sp^2 (graphitic) and sp^3 (diamond) bonding configurations [1]. Diamond is a good insulator, while graphite is usually a good conductor (a semimetal). In a diamond crystal, electrons in sp^3 hybrid orbitals form four σ bonds similar to those in group IV semiconductors (see Chapter 4). The graphitic configuration breaks the $(8-N)$ rule because the atoms have only three next neighbors in this structure, forming sp^2 hybrids. The sp^3 hybrid has already been treated in Chapter 4.

Hydrogenated amorphous silicon carbide (a-SiC:H) has a wide optical band gap [2] and hence is an important window material for photovoltaics. Fundamental changes in the electrooptical characteristics of a-SiC:H occur with doping, for example p-type doping with boron and n-type doping with phosphorus [2]. The mechanical hardness of a-SiC:H is highly dependent on the fraction of sp^3 hybridization. Hydrogen is known to assist sp^3 hybridization. Hydrogenated amorphous carbon (a-C:H) is mechanically very hard and hence is called diamondlike carbon (DLC) [3]. DLC also has a very high

Amorphous Semiconductors: Structural, Optical, and Electronic Properties, First Edition.
Kazuo Morigaki, Sándor Kugler, and Koichi Shimakawa.

potential for electronic applications and pn control is now possible with it [4]. As pure amorphous carbon (a-C) is the basic material, we must first discuss a-C. Carbon has two allotropes, with the graphite and diamond crystalline structures. Amorphous carbon does not have any crystalline structure. It contains a mixture of sp^2 and sp^3 hybrids and shows semiconducting behaviors; the major content is sp^2 bonds.

7.1.1 Basic Structure of a-C (sp^2 Hybrids)

Ethylene (C_2H_4) has a double bond between the carbon atoms. In this molecule, the wave functions of the carbon atoms are sp^2 hybrids, because one π bond is required for the double bond between the two carbon atoms. Only three σ bonds can be formed per carbon atom. In sp^2 hybridization, the 2s orbital is mixed with only two (p_x and p_y) of the three available 2p orbitals. These three hybrids are also orthonormalized states and are arranged in a planar structure. The mathematical forms of the hybrids are as follows:

$$1)\ sp^2 = \left(\frac{1}{3}\right)^{1/2} s + \left(\frac{2}{3}\right)^{1/2} p_x$$

$$2)\ sp^2 = \left(\frac{1}{3}\right)^{1/2} s - \left(\frac{1}{6}\right)^{1/2} p_x + \left(\frac{1}{2}\right)^{1/2} p_y$$

$$3)\ sp^2 = \left(\frac{1}{3}\right)^{1/2} s - \left(\frac{1}{6}\right)^{1/2} p_x - \left(\frac{1}{2}\right)^{1/2} p_y$$

In ethylene, the two carbon atoms form a σ bond by the overlapping of two sp^2 orbitals and each carbon atom forms two other covalent bonds with hydrogen by s–sp^2 overlap, all with 120° angles. The π bond between the carbon atoms perpendicular to the molecular plane is formed by $2p_z$–$2p_z$ overlap. The hydrogen–carbon bonds are all of equal strength and length (resonant bonds), which agrees with experimental data. In the honeycomb structure of the graphite crystal, such a local arrangement can be observed in the planar layers. The bond length is shorter by 0.01 nm than sp^3 hybrid bonds. The bond is stronger than in the diamond crystal because of the additional π bond. This local arrangement with sp^2 hybridization constructs a graphitelike amorphous carbon network. Distorted, nearly planar versions of this atomic arrangement can be observed in graphene and in nanotubes.

Finally, a linear arrangement, called sp hybridization, should be mentioned. The chemical bonding in acetylene (ethyne) (C_2H_2) consists of sp–sp overlap between the two carbon atoms, forming a σ bond, and two additional π bonds formed by p–p overlap. Each carbon also bonds to hydrogen in a σ bond with s–sp overlap, with 180° angles. The mathematical description of sp hybridization is as follows:

$$1)\ \mathrm{sp} = \left(\frac{1}{2}\right)^{1/2} \mathrm{s} + \left(\frac{1}{2}\right)^{1/2} \mathrm{p}_x$$

$$2)\ \mathrm{sp} = \left(\frac{1}{2}\right)^{1/2} \mathrm{s} - \left(\frac{1}{2}\right)^{1/2} \mathrm{p}_x$$

The chain version C_n–C_n is called carbyne. Carbyne chains are the strongest material known.

7.1.2 Preparation Techniques

Pure a-C films can be prepared by evaporation and by conventional RF sputtering techniques. For structural studies on a-C, for example neutron diffraction measurements, which require a large mass of film (~1 g), the evaporation of graphite is an appropriate technique [5]. As a very high temperature is required for graphite to evaporate, arc-evaporation of graphite rods under a pressure of 10^{-3} Pa is used. A large current between two mechanically contacted graphite rods induces evaporation of graphite by Joule heating. Note that C_{60} and carbon nanotubes were found by using this technique under a He atmosphere.

Diamondlike carbon can be prepared by a variety of methods, such as RF and DC plasma-enhanced chemical vapor deposition (PECVD), sputtering, ion beam deposition, and arc-evaporation [6]. Which technique should be chosen depends on the application (see Section 7.1.3). For electronic devices, including pn control, the PECVD technique is adopted, similarly to what is done in the preparation of a-Si:H films.

Similarly to DLC, a-SiC_x:H films can be prepared by various techniques. PECVD (DC, RF, or microwave) is the basic technique adopted for a-SiC_x:H, using decomposition of tetramethylsilane ($Si(CH_3)_4$, TMS). Mechanically hard films have been produced by decomposition of TMS using electron cyclotron resonance plasma flow in Ar [7].

7.1.3 Brief Review of Structural Studies on Amorphous Carbon

As already mentioned, a-C is the basic material for a-SiC:H and DLC, and hence we review the structural studies on a-C films that have been done. A pioneering electron diffraction measurement on an evaporated carbon film was reported more than half a century ago by a Japanese group [8]. They found that the film contained diamondlike and graphitelike crystalline regions. The regions had no mutual orientation and had a size of several angstroms.

Neutron diffraction measurements have been performed on amorphous carbon at low temperatures [5] in order to investigate the reasons for the anomalous electrical conduction observed below 50 K [9]. Using an arc vaporization method, a 1.0 g sample of pure amorphous C was prepared. The neutron diffraction experiment was carried out using the 7C2 spectrometer installed on the hot source at the Orphée reactor at Saclay. A momentum transfer range of 5–160 nm^{-1} was covered. In the measured pair correlation function, the positions of the first four main peaks were 0.142, 0.246, 0.283 (third-neighbor cross-ring distance), and 0.373 nm. In a perfect graphite crystal the first four nearest-neighbor distances are 0.142, 0.246, 0.284, and 0.364 nm. It seems likely that the sample contained a high proportion of threefold-coordinated carbon atoms; that is, it was a typical graphitelike material. Beyond 0.6 nm, an absence of atomic structure was found: $g(r)$ had no appreciable medium-range order. The $S(Q)$'s derived exhibited small differences due to variation of the temperature, but there was no significant difference between the radial distribution functions.

The structure of another a-C sample (20–30 mg, which is a very limited specimen mass for neutron diffraction), prepared by plasma-arc deposition, was determined by neutron diffraction using the twin-axis diffractometer D4 at the Institut Laue-Langevin, Grenoble [10]. The conclusion was exactly opposite to the above results: the structure factor and reduced radial distribution function were similar to those for a-Si and a-Ge, indicating a high proportion of tetrahedral bonding. The first three main interatomic distances were very close to the bond lengths in the diamond crystal (0.152–0.153 nm, 0.250–0.252 nm, and 0.296–0.296 nm). A fit to the data gave about 86% tetrahedral bonding.

Thick a-C films were prepared in a high-vacuum system with a base pressure of 10^{-7} Torr by RF sputtering on liquid-nitrogen-cooled Cu substrates [11]. To confirm the amorphous nature of the deposited films, Raman measurements were performed. A sample (0.8 g) was used for neutron diffraction measurements, performed at 300 K using the Special Environment Powder Diffractometer at the Intense Pulsed Neutron Source at Argonne National Laboratory, USA. The films were removed from the substrate by using dilute hydrochloric acid. The radial distribution function (RDF) was compared with theoretical models. The RDF results exhibited qualitative agreement with a number of basic models, indicating predominant threefold bonding. The absence of a specific peak at the graphite intrahexagon distance (third-neighbor peak) indicated that structural models with such intermediate-range correlations were not correct for this a-C. This distinguishes amorphous carbon from locally two-dimensionally ordered graphitelike materials.

7.1.4 Applications

DLC films are mechanically very hard, and hence these thin films are important in many industrial applications, such as surface coating of many devices and also of optical components [12–14]. DLC is therefore very important in the field of tribology. As DLC films contain uncontrollably high numbers of defect states (associated with low electron mobility), however, their applications in the field of electronics are limited. However, cold cathode field emitters with a low threshold field are very attractive [4].

a-SiC_x:H films may be also useful for their tribological properties, because their mechanical hardness is very high and they are chemically stable. In addition to these properties, a-SiC_x:H films can be expected to be used in electronic and optical devices, since p- and n-type a-SiC_x:H have been realized while maintaining a large optical band gap. The use of a boron-doped (p-type) layer as a window layer over the heterojunction in a-Si:H solar cells results in a significant improvement in efficiency [2, 15].

When high-temperature performance (in the environment) of electronic devices is required, transistors using a-C:H and a-SiC_x:H are promising materials, because they have a large band gap.

7.2 Amorphous Oxide Semiconductors

Among various amorphous oxides, amorphous In_2O_3–ZnO–Ga_2O_3 (IGZO) is the best-known amorphous oxide semiconductor (AOS) that has reached the market. Similarly to a-Chs, amorphous oxides have a long history. First, electronically conductive a-V_2O_5 was reported in 1954 [16]. This discovery was a breakthrough in glass science, because glasses had been believed to be insulators. Since then, electronic transport in glasses containing transition metals, for example the P_2O_5–V_2O_5 system, has been investigated extensively [17]. However, the carrier mobility is very small ($10^{-4}\,cm^2\,V^{-1}\,s^{-1}$), and hence no practical applications in electronics have been reported for amorphous oxides.

Recently, however, a new type of amorphous oxide, namely IGZO, was developed (see, for example, [18]). It is possible to prepare IGZO films at room temperature on flexible, optically transparent organic films (e.g., polyethylene terephthalate, PET), and hence large-area flexible thin-film transistors (TFTs) for liquid-crystal displays have been produced [19]. Following two excellent reviews [18, 19], the current status of IGZO will be briefly introduced in this section.

7.2.1 Preparation Techniques

Thin films of AOSs can be prepared easily by RF sputtering on conventional substrates in an O_2 + Ar atmosphere at room temperature (without heating of the substrate) [20]. IGZO films can also be prepared by pulsed laser deposition (PLD) in an appropriate atmosphere [21]. Free carriers (electrons) are produced by chemical doping, that is, alteration of the stoichiometry of the oxygen ions by controlling the oxygen pressure during the deposition processes or by ion implantation of appropriate cations such as H^+ and Ti^+. Such cations induce metallic transport behavior in AOSs [22–25]. Note that it is possible to prepare only n-type semiconductors by these preparation techniques.

For applications of AOSs, p-type material is very much required, and p-type material has actually been achieved with $CuAlO_2$ [26], which opens up a new field in oxide electronics.

7.2.2 Optical Properties

Figure 7.1 shows optical absorption spectra of high-quality (HQ) and low-quality (LQ) *metallic* IGZO films [19, 27], where "as" and "ann" denote as-deposited and annealed films, respectively. It is known that the conduction band minimum is composed of vacant s orbitals and the valence band maximum is composed of oxygen 2p orbitals. The principal (fundamental) optical transition occurs between these states. The optical band gap E_o (Tauc gap) is estimated to be ~3.2 eV for the HQ a-IZGO films (In : Ga : Zn = 1 : 1 : 1) and ~3.0 eV for the LQ films. Similarly to other amorphous semiconductors such as a-Si:H and a-Chs, an Urbach tail just below E_o is found (see Sections 5.5.1 and 6.4.1), with a tail energy of ~0.15 eV [28]. So-called subgap absorption around $E \sim 2.5$ eV is observed for both the HQ and LQ films, which may be due to deep gap states.

The optical band gap increases with the number of carriers n; this is due to the band-filling effect and is called the Burstein–Moss shift. This shift is given by [29]

$$\Delta E_g \propto \frac{n^{2/3}}{m^*} \tag{7.1}$$

where m^* is the effective mass of the carriers. From this shift, $m^* = 0.3$ has been deduced for a-IGZO [19, 30].

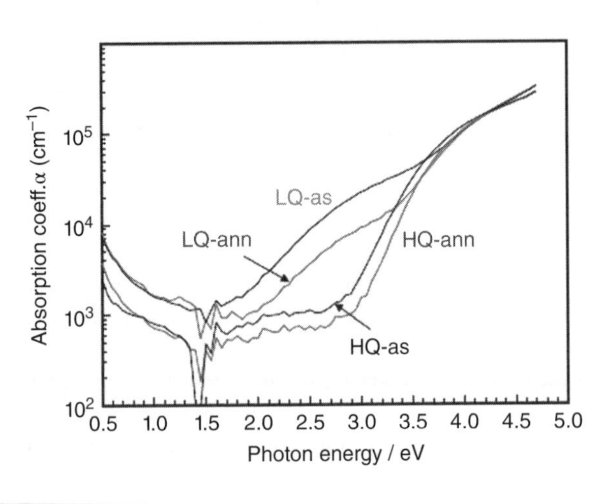

Figure 7.1 Optical absorption spectra of high-quality (HQ) and low-quality (LQ) metallic IGZO films: "as" and "ann" denote as-deposited and annealed states, respectively. *Source:* Kamiya *et al.* 2010 [19]. Used under https://creativecommons.org/licenses/by/3.0/.

Below $E \sim 1.5$ eV, the absorption increases with decreasing E, similarly to the behavior found originally in a-Cd_2GeO_4 films [31], which is attributed to free-carrier absorption (see Section 7.2.3) [23–25]. The free carrier relaxation time τ is estimated to be $\sim 3 \times 10^{-15}$ s in the HQ films, for example, supporting the picture of electronic transport which will be discussed in Section 7.2.3. The quantities τ and m^* are important physical parameters in electronic transport, and these will also be discussed in Section 7.2.3.

7.2.3 Electronic Properties

Figures 7.2(a) and (b) show typical examples of the density of free electrons deduced from Hall measurements and of the electronic conductivity, respectively, as a function of temperature in conducting (doped) HQ IZGO films [19]. Free electrons are provided by donors located ~0.1 eV from the conduction band edge [19]. A remarkable feature of Hall measurements on a-IGZO is that no sign anomaly exists, unlike the case for a-Si:H and a-Chs (Sections 5.4.3 and 6.3.2), suggesting that the carrier mean free path is long compared with that in a-Si:H and a-Chs.

While the number of electrons is independent of temperature when $n_{\mathrm{Hall}} > 10^{17}\,\mathrm{cm}^{-3}$, the conductivity σ_e depends weakly on temperature except when $n_{\mathrm{Hall}} \sim 10^{20}\,\mathrm{cm}^{-3}$. These results predict that the Hall mobility μ_{Hall} should be temperature dependent. Figures 7.3(a) and (b) show the Hall mobility as a function of carrier concentration and temperature, respectively, in a-IGZO [18]. As seen in Figure 7.3(b), μ_{Hall} is thermally activated in films with n_{Hall} less than

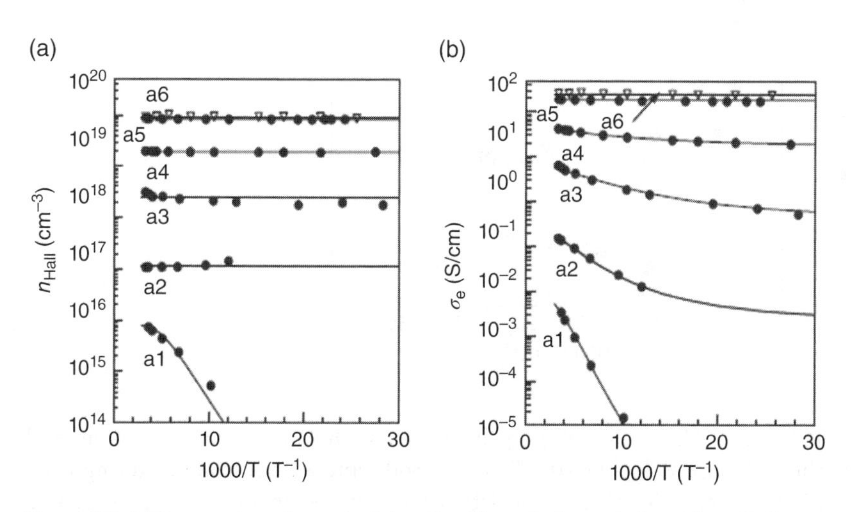

Figure 7.2 (a) Temperature-dependent density of free electrons deduced from Hall measurements, and (b) DC conductivity of HQ IGZO films. The symbols a1–a6 denote different doping levels. *Source:* Kamiya *et al.* 2010 [19]. Used under https://creativecommons.org/licenses/by/3.0/.

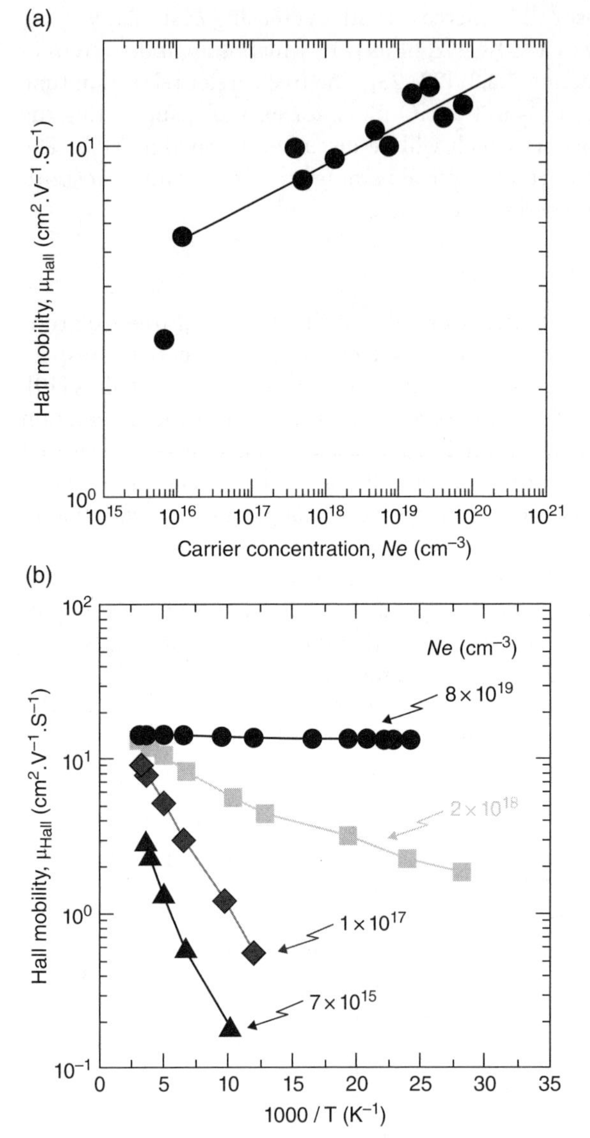

Figure 7.3 Hall mobility as a function of (a) carrier concentration and (b) temperature in a-IGZO films. *Source:* Hosono 2006 [18]. Reproduced with permission from Elsevier.

2×10^{19} cm^{-3}. The value of μ_{Hall} increases with carrier concentration and reaches ~10 cm^{2}V^{-1}s^{-1} at 10^{18-19} cm^{-3}, producing an electron scattering time $\tau \sim 1.7 \times 10^{-15}$ s. This value is consistent with that estimated from free-carrier absorption (Section 7.2.2).

Surprisingly, this trend is opposite to that for conventional semiconductors, in which carrier scattering by dopant ions dominates the mobility.

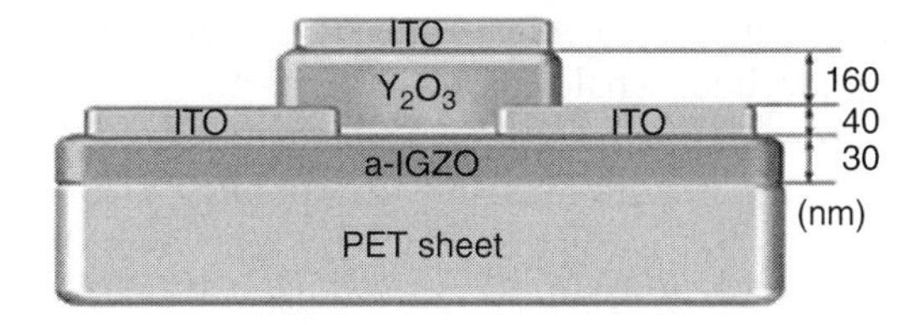

Figure 7.4 Schematic illustration of a primitive a-IGZO-based TFT structure. *Source:* Nomura *et al.* 2004 [34]. Reproduced with permission by Macmillan Publishers Ltd.

This behavior of μ_{Hall} can be interpreted by introducing a percolation transport model [32], in which the conduction band edge is assumed to fluctuate (long-range potential fluctuations) [33]. When the Fermi level E_{F} reaches the percolation threshold energy E_{pth} ($n_{\mathrm{Hall}} \sim 10^{20}\,\mathrm{cm}^{-3}$), completely metallic transport behavior should be observed. Note, however, that the transport is metallic even when the conductivity decreases with decreasing temperature (see samples a2–a5 in Figure 7.3(b)). Thus a metal–insulator transition is expected to occur at a critical number of free electrons $n_{\mathrm{c}} = n_{\mathrm{Hall}} \sim 1 \times 10^{17}\,\mathrm{cm}^{-3}$ [18, 19].

7.2.4 Applications

Flexible, transparent TFTs can be made using a-IGZO on PET sheet [34]. A primitive a-IGZO-based TFT structure is shown in Figure 7.4. The PLD technique was used for preparation of the a-IGZO layer. An advantage of using amorphous materials in flat panel displays (FPDs) is their uniformity and the absence of the grain boundary problems typical of polycrystalline FPDs, although the mobility is lower than that in polycrystalline Si ($\sim 100\,\mathrm{cm}^2\,\mathrm{V}^{-1}\,\mathrm{s}^{-1}$). Since then, a-IGZO has become a representative amorphous oxide semiconductor, and has been used in various FPDs, electronic papers, organic light-emitting-diode displays, and of course liquid crystal displays. For FPDs, which require very high uniformity, the largest size of 37 inches achieved in 2010 has been exceeded year-on-year. It should be stated once again that the main drawbacks of a-Si:H TFTs, that is, low mobility ($\sim 0.5\,\mathrm{cm}^2\,\mathrm{V}^{-1}\,\mathrm{s}^{-1}$) and instability, and the poor uniformity of polycrystalline Si TFTs have been overcome by using a-IGZO.

Nowadays, large-area, high-speed fabrication techniques are required for large-scale production of FPDs, and AOS TFTs are being prepared by RF/DC sputtering and RF magnetron sputtering [19]. We expect therefore that Si-based FPDs will be replaced by AOS FPDs in the near future.

7.3 Metal-Containing Amorphous Chalcogenides

Metal-containing chalcogenide glasses (and films) have potential applications in solid-state batteries, chemical sensors, and optoelectronic devices (in particular, optical memories, waveguides, gratings, microlenses, and so on).

Among the metals, Ag is the most attractive and popular partner for amorphous chalcogenides. In particular, its photoinduced dissolution and diffusion are scientifically and technologically important [35]. Thus, in this section, we will focus on amorphous Ag-containing chalcogenides (Ag-Chs). Fibers and waveguides are the most promising devices using metal-doped chalcogenides, but they do not use Ag, and hence we will discuss these devices in Section 8.2.4.

7.3.1 Preparation Techniques

Glasses containing Ag can be prepared by the classical melt-quenching method. Either homogeneous or phase-separated materials, or nanoparticles may be obtained by this technique. Two glass-forming regions are found in the Ag-Chs (e.g., Ag–As–S(Se) and Ag_2Se–Ga_2Se_3–$GeSe_2$). Phase-separated glasses often become homogeneous after thermal annealing at an appropriate temperature below the glass transition temperature [35].

Sol–gel methods and the use of Ag ion exchange from solutions are useful techniques for preparing glasses and thicker films [36]. For the preparation of such thicker films, spin-coating techniques, for example using *n*-butylamine solutions, may be helpful. Note, however, that the prepared films contain excess solvent and therefore removing it by vacuum heating may be required [37].

Thin films of Ag-containing glasses can be prepared by many of the usual techniques, that is, thermal evaporation, sputtering, chemical vapor deposition, and PLD. Photodoping of Ag into chalcogenides may be an alternative technique, where a bilayer of Ag (or an Ag compound) and a chalcogenide is illuminated with band gap light, and Ag then dissolves in the chalcogenide [38]. While the glass-forming regions of Ag-containing Chs are not large, a surprisingly large amount of Ag can be photodoped: for example, 57 at.% of Ag in GeS_3 glass is possible [39]. As photodoping of Ag is physically and technologically of interest, we will thus discuss this phenomenon in Section 7.3.3.

7.3.2 Structure of Ag-Chs and Related Physical Properties

The Ag is introduced as an Ag^+ ion, and its charge should be compensated by a onefold-coordinated chalcogen C_1^-. The Ag^+ is expected to be coordinated by two, three, or four bonds using d orbitals of the Ag and lone pairs of the chalcogen. In fact, we have found experimentally that the coordination numbers are in the range 3–3.5 [40]. For higher contents of Ag, the coordination number becomes lower, which can be interpreted by assuming a lack of accessible chalcogens in the vicinity of the Ag^+. In some cases, introduction of Ag into As–S(Se) or Ge–S(Se) produces homopolar As–As or Ge–Ge bonds. It is well known that there is no change in the near- or intermediate-range order for a low Ag content, for example less than 1 at.% of Ag in the Ag_2S–As_2S_3 system, whereas the ionic transport properties show a significant change [41].

In the context of percolation theory, a large change in physical properties, such as the ionic transport, cannot be expected in the range below 1 at.% of Ag. Hence a model of allowed volume, in which an Ag^+ ion occupies a space greater than its atomic volume, has been proposed to explain the compositional dependence of the conductivity. The details of the ionic transport properties of Ag-Chs have been discussed in Section 6.3.2. Above 1 at.% of Ag, the optical band gap decreases significantly with Ag content, and in this range a change in near- or intermediate-range structure is reported [42].

7.3.3 Photodoping

Photodoping is an effect where, when a film or Ag compound on an a-Ch is illuminated with band gap light, Ag is dissolved in the a-Ch film or bulk material [35, 38, 43]. The best material composition and configuration for this are an Ag film on an a-As_2S_3 film (~1 μm thick), with illumination from the As_2S_3 film side. However, when the illumination is done from the front Ag side, photodoping also occurs. This is an example of photoenhanced dissolution and diffusion of a metal (silver) into an a-Ch. The characteristic features of photodoping are the following. (i) A large amount of Ag, up to around 60 at.%, can be dissolved. (ii) The concentration profile of the Ag during diffusion is a sharp steplike diffusion front as shown in Figure 7.5, which does not follow Fick's law (which predicts the usual known as Gaussian diffusion profile). Note that the illumination was done from the substrate (rear) side. (iii) Diffusion is much enhanced when the chalcogenide is deposited on an electrically conducting substrate [44].

Although the mechanism is still not understood well, the photodoping mechanism is known to consist of several key steps. Let us consider the case

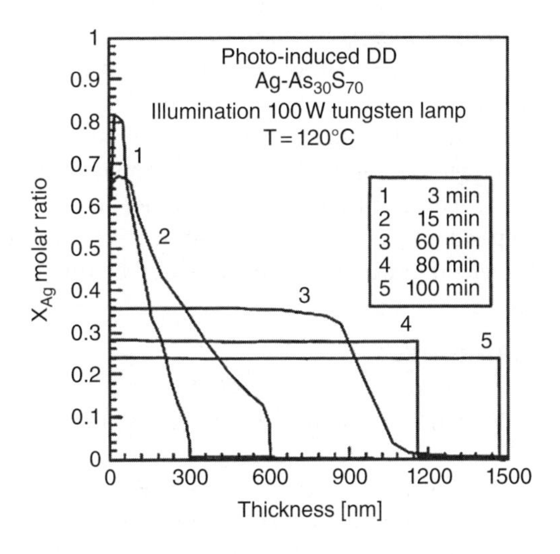

Figure 7.5 Concentration profile of photoenhanced Ag diffusion in Ag–$As_{30}S_{70}$ system. *Source:* Frumar and Wagner 2003 [35]. Reproduced with permission from Elsevier.

of Ag/As_2S_3. First, ionization of Ag occurs at the interface between silver and As_2S_3:

$$Ag + h\ (= Ag - e) \rightarrow Ag^+ \tag{7.2}$$

where holes (h) and electrons (e) are produced by photoillumination. On the other hand, Ag can react with sulfur:

$$\begin{aligned} &Ag + S \rightarrow Ag^+ + S^-, \qquad 2Ag + S \rightarrow 2Ag^+S^{2-} \qquad \text{or} \\ &4Ag + 2As_2S_3 \rightarrow As_4S_4 + 2Ag_2S \end{aligned} \tag{7.3}$$

Note that an increase in As_4S_4 structural units is found in thin-film Raman spectra [35].

Secondly, photoillumination produces a buildup of electric potential caused by separation of electrons and holes. The electrons reside near the illuminated surface (As_2S_3) (negative potential) and holes diffuse to the countersurface (Ag) (positive). Then the electric field accelerates the drifting of Ag^+. The composition of the doped region is $AgAsS_2$, corresponding to the good glass-forming region [45]. This implies that the photodoping occurs between compositions of minimal free energy, from As–S to $AgAsS_2$ [46]: the sharp edge may be the result of the border between As–S and $AgAsS_2$. As we have already discussed in Section 6.3.2, Ag-containing chalcogenides have higher diffusion coefficient of Ag^+ than that in the nondoped region, and hence Ag^+ ions tend at first to becomes distributed uniformly in $AgAsS_2$. Invasion of Ag^+ into a new area then follows after creating $AgAsS_2$. This may be the reason why a sharp doped edge is formed.

7.3.4 Applications

There are many potential applications of Ag-doped chalcogenides, for example solid electrolytes for batteries, electrochemical sensors, electrical memories, and holography. However, commercially, there are other competitive materials and hence Ag-Chs are still potentially only *expected* in the market [35]. Among the potential applications are a new type of diffraction gratings with nanometer-dimension lines, which can be formed by a focused laser beam in Ag-doped chalcogenides [47]. The width and height of the Ag-doped lines are highly dependent on the beam parameters, including the accelerating voltage. Excellent gratings can be fabricated by this technique.

References

1 Kugler, S. and Shimakawa, K. (2015) *Amorphous Semiconductors*, Cambridge University Press, Cambridge.

2 Demichelis, F., Pirri, C.F., Tresso, E., Della Mea, G., Rigato, V., and Rava, P. (1991) Physical properties of undoped hydrogenated amorphous silicon carbide. *Semicond. Sci. Technol.*, **6**, 1141–1146.

3 Filik, J., May, P.W., Pearce, S.R.J., Wild, R.K., and Hallam, K.R. (2003) XPS and laser Raman analysis of hydrogenated amorphous carbon films. *Diamond Relat. Mater.*, **12**, 974–978.
4 Milne, W.I. (2003) Electronic devices from diamond-like carbon. *Semicond. Sci. Technol.*, **18**, S81–S85.
5 Kugler, S., Shimakawa, K., Watanabe, T., Hayashi, K., Lazzlo, I., and Bellissent, R. (1993) The temperature dependence of the structure of amorphous carbon. *J. Non-Cryst. Solids*, **164–166**, 1143–1146.
6 Grill, A. (1993) Review of tribology of diamond-like carbon. *Wear*, **168**, 143–153.
7 Ito, H., Onitsuka, S., Gappa, R., Saitoh, H., Roacho, R., Pannell, K.H., Suzuki, T., Niibe, M., and Kanda, K. (2013) Fabrication of amorphous silicon carbide films from decomposition of tetrathylsilane using ECR plasma of Ar. *J. Phys.: Conf. Ser.*, **441**, 012039-1–6.
8 Kakinoki, J., Katada, K., Hanawa, T., and Ino, T. (1960) Electron diffraction study of evaporated carbon film. *Acta Cryst.*, **13**, 171–179.
9 Shimakawa, K., Hayashi, K., Kameyama, T., Watanabe, T., and Morigaki, K. (1991) Anomalous electrical conduction in graphite-vaporized films. *Philos. Mag. Lett.*, **64**, 375–378.
10 Gaskell, P.H., Saeed, A., Chiux, P., and McKenzie, D.R. (1991) Neutron-scattering studies of the structure of highly tetrahedral amorphous diamondlike carbon. *Phys. Rev. Lett.*, **67**, 1286–1289.
11 Li, F. and Lannin, J.S. (1990) Radial distribution function of amorphous carbon. *Phys. Rev. Lett.*, **65**, 1905–1908.
12 Toth, S., Veres, M., Fule, and Koos, M. (2006) Influence of layer thickness on the photoluminescence and Raman scattering of a-C:H prepared from benzene. *Diamond Relat. Mater.*, **15**, 967–971.
13 Veres, M., Toth, S., and Koos, M. (2008) New aspects of Raman scattering in carbon-based amorphous materials. *Diamond Relat. Mater.*, **17**, 1692–1696.
14 Yoshitake, T., Nakagawa, Y., Nagano, A., Ohtani, R., Setoya, H., Kobayashi, E., Sumitani, K., Agawa, Y., and Nagayama, K. (2010) Structural and physical characteristics of ultrananocrystalline diamond/hydrogenated amorphous carbon composite films deposited using a coaxial arc plasma gun. *Japan. J. Appl. Phys.*, **49**, 015503-1–4.
15 Levitas, V.I., Ma, Y., Selvi, E., Wu, J., and Patten, J.A. (2012) High-density amorphous phase of silicon carbide obtained under large plastic shear and high pressure. *Phys. Rev. B*, **85**, 054114-1–5.
16 Denton, E.P., Rawson, H., and Stanworth, J.E. (1954) Vanadate glasses. *Nature*, **173**, 1030–1032.
17 Shimakawa, K. (1989) On the mechanism of dc and ac transport in transition-metal oxide glasses. *Philos. Mag. B*, **60**, 377–389.
18 Hosono, H. (2006) Ionic amorphous oxide semiconductors: Material design, carrier transport, and device application. *J. Non-Cryst. Solids*, **352**, 851–858.

19 Kamiya, T., Nomura, K., and Hosono, H. (2010) Present status of amorphous In–Ga–Zn–O thin-film transistors. *Sci. Technol. Adv. Mater.*, **11**, 044305-1–23.

20 Orita, M., Ohta, H., Hirano, M., Narushima, S., and Hosono, H. (2001) Amorphous transparent conductive oxide $InGaO_3(ZnO)_m$ ($m < 4$). *Philos. Mag. B*, **81**, 501–515.

21 Takagi, A., Nomura, K., Ohta, H., Yanagi, H., Kamiya, T., Hirano, M., and Hosono, H. (2005) Carrier transport and electronic structure in amorphous oxide semiconductor, a-$InGaZnO_4$. *Thin Solid Films*, **486**, 38–41.

22 Hosono, H., Kikuchi, N., Ueda, N., Kawazoe, H. (1996) Working hypothesis to explore novel wide band gap electrically conducting amorphous oxides and examples. *J. Non-Cryst. Solids*, **198–200**, 165–169.

23 Shimakawa, K., Narushima, S., Hosono, H., and Kawazoe, H. (1999) Electronic transport in degenerate amorphous oxide semiconductors. *Philos. Mag. Lett.*, **79**, 755–761.

24 Narushima, S., Hosono, H., Jisun, J., Yoko, T., and Shimakawa, K. (2000) Electronic transport and optical properties of proton-implanted amorphous 2CdO–GeO_2 films. *J. Non-Cryst. Solids*, **274**, 313–318.

25 Singh, J. and Shimakawa, K. (2003) *Advances in Amorphous Semiconductors*, Taylor & Francis, London and New York.

26 Kawazoe, H., Yasukawa, M., Hyodou, H., Kurita, M., Yanagi, H., and Hosono, H. (1997) p-type electrical conduction in transparent thin films of $CuAlO_2$. *Nature*, **389**, 939–942.

27 Kamiya, T., Nomura, K., and Hosono, H. (2009) Electronic structure of the amorphous oxide semiconductor a-$InGaZnO_{4-x}$: Tauc–Lorentz optical model and origins of subgap states. *Phys. Stat. Sol. (a)*, **206**, 860–867.

28 Kamiya, T., Nomura, K., Hirano, M., and Hosono, H. (2008) Electronic structure of oxygen deficited amorphous oxide semiconductor a-$InGaZnO_{4-x}$: Optical analysis and first-principle calculations. *Phys. Stat. Sol. (c)*, **5**, 3098–3100.

29 Burstein, E. (1954) Anomalous optical absorption limit in InSb. *Phys. Rev.*, **93**, 632.

30 Kamiya, T., Nomura, K., and Hosono, H. (2009) Origins of high mobility and low operation voltage of amorphous oxide TFTs: Electronic structure, electron transport, defects and doping. *J. Disp. Technol.*, **5**, 273–288.

31 Hosono, H., Yasukawa, M., and Kawazoe, H. (1996) Novel oxide amorphous semiconductors: Transparent conducting amorphous oxides. *J. Non-Cryst. Solids*, **203**, 334–344.

32 Kamiya, T., Nomura, K., and Hosono, H. (2010) Origin of definite Hall voltage and positive slope in mobility–donor density relation in disordered oxide semiconductors. *Appl. Phys. Lett.*, **96**, 122103-1–3.

33 Shimakawa, K. and Ganjoo, A. (2002) AC photoconductivity of hydrogenated amorphous silicon: Influence of long-range potential fluctuations. *Phys. Rev. B*, **65**, 165213-1–5.

34 Nomura, K., Ohta, H., Takagi, A., Kamiya, T., Hirano, M., and Hosono, H. (2004) Room-temperature fabrication of transparent flexible thin-film transistors using amorphous oxide semiconductors. *Nature*, **432**, 488–492.
35 Frumar, M. and Wagner, T. (2003) Ag doped chalcogenide glasses and their applications. *Curr. Opin. Solid State Mater. Sci.*, **7**, 117–126.
36 Paje, S.E., Garcia, M.A., Villegas, M.A., and Llopis, J. (2000) Optical properties of silver ion-exchanged antimony doped glass. *J. Non-Cryst. Solids*, **266–269**, 128–136.
37 Wagner, T., Kohoutek, T., Vlcek, Mir., Vlcek, Mil., Munzar, M., and Frumar, M. (2002) Physico-chemical properties and structure of the $Ag_x(As_{0.35}S_{0.67})_{100-x}$ films prepared by spin-coating technique. 13th International Symposium on Non-Oxide and New Glasses, Part I, pp. 157–160.
38 Kolobov, A.V. and Elliott, S.R. (1991) Photodoping of amorphous chalcogenides by metals. *Adv. Phys.*, **40**, 625–684.
39 Wagner, T. and Frumar, M. (1990) Photoenhanced dissolution and diffusion of Ag in As–S layers. *J. Non-Cryst. Solids*, **116**, 269–276.
40 Fritzshe, H. (1999) The chemical bonding of silver photodissolved in chalcogenide glasses, in *Homage Book: A. Andrish* (ed. M. Popescu), INOE and INFM Publishing House, Bucharest, pp. 53–59.
41 Ribes, M., Bychkov, E., and Pradel, A. (2001) Ion transport in chalcogenide glasses; dynamics and structural studies. *J. Optoelectron. Adv. Mater.*, **3**, 665–674.
42 Wagner, T., Kasap, S.O., Vlcek, Mil., Frumar, M., Nesladek, P., and Vlcek, Mir. (2001) The preparation of the $Ag_x(As_{0.33}S_{0.67})_{100-x}$ amorphous films by optically-induced solid state reaction and the films properties. *Appl. Surface Sci.*, **175&176**, 117–122.
43 Wagner, T. and Frumar, M. (2003) Optically-induced diffusion and dissolution of metals in amorphous semiconductors, in *Photo-induced Metastability in Amorphous Semiconductors* (ed. A.V. Kolobov), Wiley-VCH, Berlin, pp. 196–216.
44 Wagner, T., Frumar, M., and Suskova, V. (1991) Photoenhanced dissolution and lateral diffusion of Ag in amorphous As–S layers. *J. Non-Cryst. Solids*, **128**, 197–207.
45 Owen, A.E., Firth, A.P., and Ewen, P.J.S. (1985) Photo-induced structural and physico-chemical changes in amorphous chalcogenide semiconductors. *Philos. Mag. B*, **52**, 347–362.
46 Tanaka, K. and Shimakawa, K. (2011) *Amorphous Chalcogenide Semiconductors and Related Materials*, Springer, New York.
47 Mietzsch, K. and Fitzgerald, A.G. (2000) Electron-beam-induced patterning of thin film arsenic-based chalcogenides. *Appl. Surf. Sci.*, **162&163**, 464–468.

8
Applications

In this chapter, the important devices using amorphous semiconductors that have already appeared on the market are introduced. One of the most successful device applications was electrophotography (or xerography) using amorphous selenium (a-Se) and hydrogenated amorphous silicon (a-Si:H). Everyone knows the term "xerography," since copying documents is very useful in our everyday life. Organic materials are now used in low-cost xerographic machines, and hence a-Se is less widely known today. Among many successful devices, photovoltaics (PVs) (i.e., solar cells) and thin-film transistors (TFTs) are being made from a-Si:H. Phase-change memory devices (the best-known type being DVDs), direct X-ray image sensors, and high-gain avalanche devices are being made from a-Chs. Therefore, as in the previous chapters, we will discuss the principles sustaining these devices for two categories of material systems, that is, a-Si:H and a-Chs, since the device action is highly dependent on the physical properties of the material. All of the devices we will now discuss were born in the "amorphous semiconductor society," and now these successful devices are creating new "academic societies" independently. This means that the role of the "amorphous semiconductor society" is very important.

8.1 Devices Using a-Si:H

8.1.1 Photovoltaics

Alternative energy sources, as a replacement for oil, are always required for many reasons, for example a lack of sources of oil and environmental problems, which we do not need to say any more about here. PVs made from various materials are, of course, useful. The markets has a high demand for large-area PVs that have a high conversion efficiency while maintaining cost-effectiveness. The first a-Si:H solar cells (on a laboratory scale) were fabricated at RCA

Amorphous Semiconductors: Structural, Optical, and Electronic Properties, First Edition.
Kazuo Morigaki, Sándor Kugler, and Koichi Shimakawa.

Laboratories [1]. Subsequently, Kuwano's group at Sanyo in Japan shipped PV devices to the market, although initially PVs prepared from a-Si:H were used only as power sources for small electronic devices such as watches and calculators. Now, megawatt-scale PV power stations have been constructed over almost all the world.

Among several demands made on PVs, the most important issue is the conversion efficiency η from light to electric power. The commonly used structures are p–i–n-type heterojunction structures, rather than the classic n–p structures used in crystalline Si (c-Si) solar cells to obtain an appropriate η. Note that free carriers (electrons and holes) are produced in the i-layer. The n–p type of cell is a carrier *diffusion* type and the p–i–n cell is a carrier *drift* type, in which the internal potential difference in the i-layer is effectively used, since the so-called $\mu\tau$ product of the photocarriers in a-Si:H is small. This accelerates the photocreated carriers with reduced carrier recombination, and hence the collection of electrons and holes at opposite electrodes is more effectively performed. This is why the p–i–n structure is adopted in a-Si:H and related PV cells.

As the optical absorption coefficient of a-Si:H is high in the visible spectral range, that is, in the range of the solar spectrum, a very thin film less than 1 μm thick is enough to achieve the same absorption as a thicker layer of c-Si. The commonly used a-Si:H-based PV cell configurations are tandem ones (dual and triple junctions), in which more than 10% efficiency is achieved in commercial large-area PV devices [2]. In the bulk i-layer, further improvement of the electronic transport properties is not easy, and hence light confinement by designing textured surfaces and by using plasmon effects with nanometals are being investigated [3].

As already stated in Section 5.5, the midgap states, for example D^0, in a-Si:H act as carrier recombination centers, which have effects on η. Unfortunately, in a-Si:H, photoillumination produces new D^0 states, which reduces the carrier lifetime, leading to a reduction in η.

8.1.2 Thin-Film Transistors

TFTs (in the form of field effect transistors, consisting of a gate, source, and drain) using a-Si:H were first reported on a laboratory scale [4, 5]. Two of the most important factors for TFTs are a high on/off current ratio and a small gate voltage, which were achieved by using a-Si:H with a-SiN$_x$ as an insulating interface. These characteristics make the devices suitable as switching transistors for liquid crystal displays (TFT-LCDs). Logic circuits (active matrix arrays, AMAs) using a-Si:H were then developed [6, 7]. The use of the Brown tube as a display device was completely replaced by the TFT-LCD system. After many improvements to TFT-LCDs, there is now a huge market for flat panel displays that produce clear images, from mobile telephones and personal computers to large display screens (over 1 m).

It should be noted that there is a high demand for flexible, optically transparent TFTs in next-generation flat panel displays. Transparent conductive oxides have been developed for this purpose (see, for example, [8]). Transparent TFTs have been developed using ionic oxides such as a-$InGaZnO_4$ (a-IGZO). The electron mobility of a-IZGO is larger than that of a-Si:H and the stability of TFTs made from this material is excellent. Samsung's group in Korea have developed a-IGZO TFT-LCD displays on a commercial basis. The high quality and stability of a-IGZO TFT-LCDs may dominate the "display world" in the near future for large-area displays. All of this has been covered in Section 7.2.

8.2 Devices Using a-Chs

8.2.1 Phase-Change Materials

"Phase-change" here means that there are changes in optical and electrical properties between the amorphous and crystalline states of a material. Phase-change materials (PCMs) are promising materials for data storage (memory) and electronic switching devices and are already used in rewritable optical data storage media (DVD, "digital versatile disc"). A great effort to commercialize nonvolatile electronic memory is continuing. Some unique properties of PCMs, which are very much different from those of conventional a-Chs, are the basis of the following applications. We therefore begin with a discussion of the fundamental physical properties of PCMs, before introducing the actual devices.

For PCMs, the following properties are required: (i) a high-speed phase transition (nanosecond timescale), (ii) a large change in optical and/or electrical properties, (iii) a large cycle number, and (iv) high physical and chemical stability [9]. The most successful PCMs which satisfy the above requirements are pseudobinary alloys along the GeTe–Sb_2Te_3 tie line shown in Figure 8.1 [9]. These cannot be obtained by the melt-quenching method, and hence can only be made in film form by sputtering techniques and so on. Among the PCMs along this line, $Ge_2Sb_2Te_5$ (often represented as "GST225") is a well-known commercialized DVD material, which has been developed over several years.

Let us see first the overall features of the real part of the conductivity (energy loss) of GST225 in the RF–THz–near infrared frequency range. Figure 8.2 shows the angular-frequency-dependent conductivity of GST225 films in both the crystalline and the amorphous state [10]. The crystalline GST225 (fcc structure) was obtained by thermal annealing of as-deposited amorphous films at 200 °C.

The circles and crosses represent the experimental data for the real part of the conductivity in the crystalline and amorphous states. The frequency-independent losses observed at lower frequencies (RF) in both the crystalline

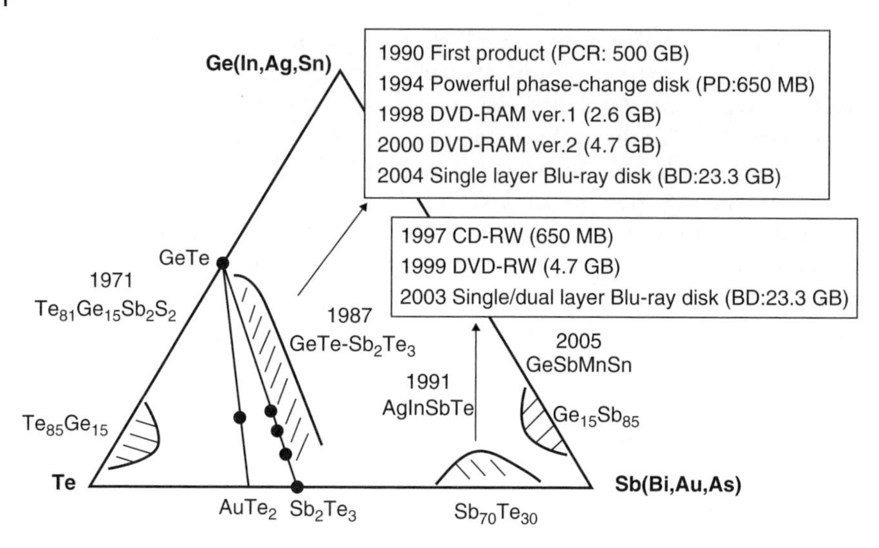

Figure 8.1 Ternary phase diagram for different phase-change alloys. *Source:* Matsumura and Hayama 1980 [6]. Reproduced with permission by Macmillan Publishers Ltd.

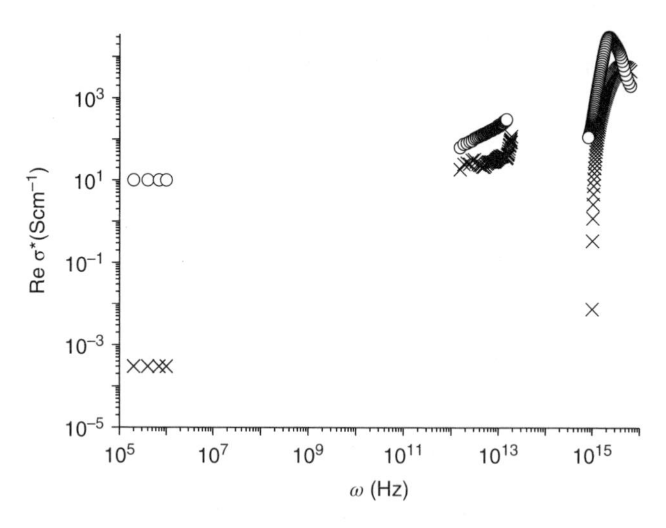

Figure 8.2 Angular-frequency-dependent conductivity (real part) of both crystalline (circles) and amorphous (crosses) states of GST225 films.

and the amorphous state give the DC conductivity, which is dominated by free carriers (holes). The large difference between the crystalline and amorphous states is useful in electrical phase-change memories. At $\omega \sim 10^{13}$ Hz, the conductivity increases with frequency in both the crystalline and the amorphous state.

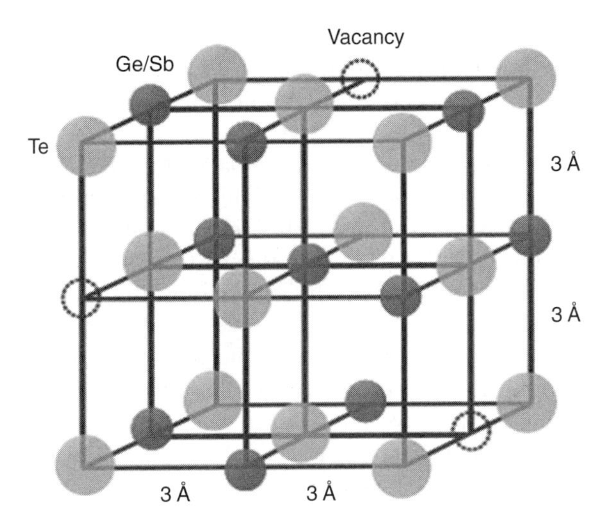

Figure 8.3 Schematic image of the crystal structure of the rock-salt-like structure of GST. *Source:* Matsumura and Hayama 1980 [6]. Reproduced with permission by Macmillan Publishers Ltd.

Phonons dominate the loss in this frequency range, as in standard a-Chs. In the crystalline state, however, a large amount of free carriers is expected to dominate the loss in the THz range, and the behavior is different from the prediction of the Drude law [11]. Loss peaks are observed in the near infrared and visible frequencies ($\omega \sim 10^{15}$ Hz). These are dominated by interband electronic transitions. The peak position (energy) for the crystalline state is lower than that for the amorphous state, indicating that the average band gap (or bond gap) in the crystalline state is smaller than that in the amorphous state. This kind of behavior is common to all GSTs [12].

The differences in electronic and optical properties between crystalline and amorphous GSTs may be due to their structure. The structure of the crystalline GSTs is clear, while the structure of the amorphous GST family is not well understood [9]. Figure 8.3 shows a schematic image of the crystal structure (rock-salt type) of GSTs [9], in which Te atoms occupy one sublattice, and random occupation of the other sublattice by Ge and Sb atoms with vacancies is reported. This large density of vacancies acts as acceptors and hence crystalline GSTs may behave as degenerate semiconductors. Note that only p orbitals are used for chemical bonding and the typical nearest-neighbor distance is 0.3 nm, which is larger than that (~0.26 nm) in the amorphous state. In amorphous GSTs, some hybridization of s and p orbitals can be induced and hence Ge may have fourfold coordination, which induces loss of long-range order.

Following this structural picture, an interesting proposal has been made that *resonance bonds* exist in crystalline GSTs, and that resonance bonds dominate the optical properties [12]. Resonance bonds require long-range structural order, and hence no resonance bonds exist in amorphous GSTs. This may provide a large contrast in optical properties between crystalline and amorphous GSTs. Details will be discussed in the following section.

Before closing the present section, we shall briefly summarize the operation of optical (DVD) and electronic (phase-change random access memory, PCRAM) applications of GSTs. The phase changes between the amorphous and crystalline states require thermal energy; for the change from the amorphous to the crystalline state (referred to as *set*), the material must be heated up to the crystallization temperature T_c, and for the inverse change from the crystalline to the amorphous state (*reset*), it needs to be heated to above the melting temperature T_m. To perform the *set* process, a long, low-intensity laser pulse is used with DVDs; for the *reset* process, a short, high-intensity laser pulse is used. In actual devices, a 650 nm wavelength laser is used with DVD-RAMs (4.7 GB) and a 405 nm laser with Blu-ray rewritable DVDs (50 GB).

8.2.1.1 Optical Memories (DVDs)

The open circles and the crosses in Figure 8.4 show the reflectance R versus photon energy E for a-GST225 and its crystalline (fcc) phase. R was obtained from ellipsometric measurements, that is, by deducing the refractive index n^* $(= n + ik)$ [13]. The big difference in R at 1.91 eV ($\lambda = 650$ nm) and 3.06 eV ($\lambda = 650$ nm) between the two phases is useful for the performance of DVDs. To get more precise insight, it is of interest to discuss the optical dielectric constant $\varepsilon^* = \varepsilon_1 + i\varepsilon_2$ in this energy range, where ε_1 and ε_2 are the real and imaginary parts, respectively, since R is directly related to ε^* (and n^*). The open circles

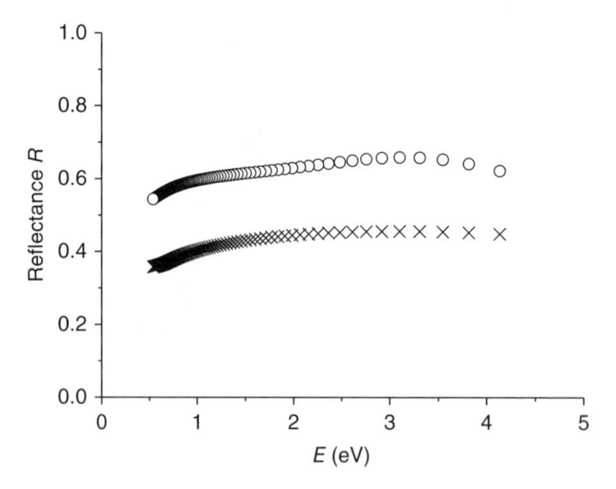

Figure 8.4 Reflectance R versus photon energy E for a-GST225 (crosses) and its crystalline (fcc) phase (circles), deduced from ellipsometric measurements.

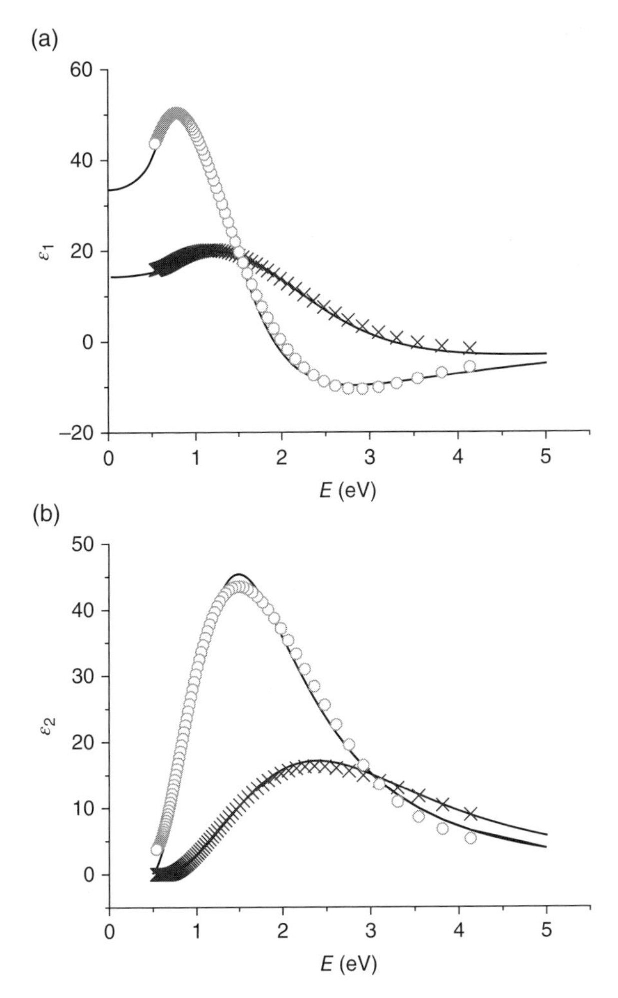

Figure 8.5 Optical dielectric constants (a) ε_1 and (b) ε_2 of crystalline (circles) and amorphous (crosses) phases of GST225, deduced from ellipsometry. *Source:* Shimakawa *et al.* 2015 [15]. Reproduced with permission from AIP Publishing LLC.

and the crosses in Figure 8.5(a) show ε_1 for the crystalline and amorphous phases, respectively, of GST225, and Figure 8.5(b) shows ε_2 similarly. The solid lines are theoretical predictions, which will be briefly discussed later. The phase change produces a significant change in ε_1 and ε_2. The position of the peak in ε_2 shifts to lower energy when the change from the amorphous to the crystalline phase occurs, with an increasing magnitude of the dielectric constant.

The solid lines show predictions from the so-called Tauc–Lorentz model [13, 14], in which both the classical Lorentz resonance (optical transitions, quantum mechanically) and the optical absorption near the band edge (the Tauc gap) are taken into account in the calculation. In the limit of $\omega \to 0$ ($E \to 0$), $\varepsilon_1(0)$ is given by [15]

$$\varepsilon_1(0) = 1 + f_m \left(\frac{\hbar\omega_p}{E_g} \right)^2 \tag{8.1}$$

where f_m (~1) is the oscillator strength, ω_p is the plasma frequency of the valence electrons ($= Ne^2/m\varepsilon_0$, where N is the number of valence electrons), and E_g is the average band gap. Equation (8.1) with $f_m = 1$ is identical to the Penn gap rule [16]. The energy at which ε_2 shows a peak corresponds to the *average* band gap (or bond gap) in a semiconductor [17]. The expression "$\varepsilon_1(0)$" here looks the static dielectric constant. However, we are now discussing electronic transitions in a higher-frequency range as compared with phonon frequencies, and hence this should be the high-frequency dielectric constant ε_∞ in the usual notation, instead of $\varepsilon_1(0)$.

Figure 8.6 shows $\varepsilon_\infty - 1$ versus $1/E_g^2$ for the GST family [15], together with values for Sb. As shown by the solid lines, the experimental data can be scaled by $1/E_g^2$, suggesting that $\varepsilon_1(0)$ (now ε_∞) is determined only by E_g. From the straight lines, $(\hbar\omega_p)^2 = 66$ is deduced for GST and $(\hbar\omega_p)^2 = 108$ for Sb, which give $N = 4.6 \times 10^{22}\,\text{cm}^{-3}$ for GST and $7.4 \times 10^{22}\,\text{cm}^{-3}$ for Sb. These values suggest that only two electrons per atom take part in the optical transition contributing to ε_∞ [15]. It should be noted that nothing happens in the transition matrix element (or oscillator strength) and the plasma frequency during the phase

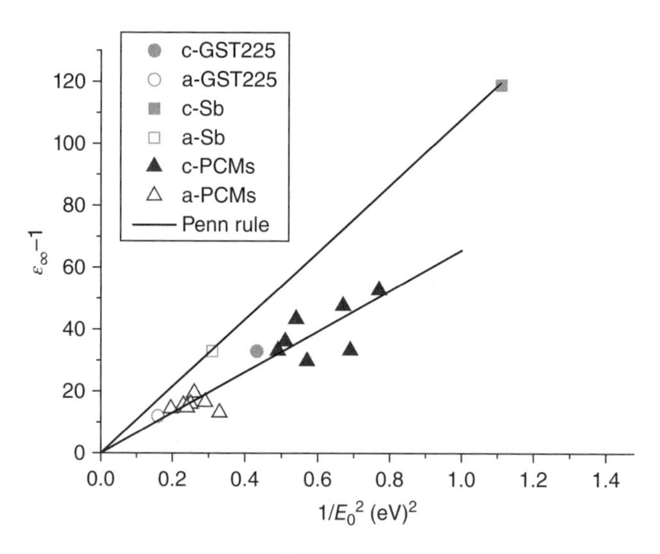

Figure 8.6 $\varepsilon_\infty - 1$ versus $1/E_g^2$ for the GST family, together with values for Sb. *Source:* Shimakawa *et al.* 2015 [15]. Reproduced with permission from AIP Publishing LLC.

change, which provides a very much different view from the proposal of resonance bonds in the GST family [12, 18]. It is emphasized here that the concept of resonance bonds is not necessary to explain the very high values of ε_∞ in the crystalline phase of GSTs.

Resonance bonding would not be a local or partial effect but a collective phenomenon throughout the crystalline phase. Thus *bonding alignment* (coherence or in-phase character of the wave functions) in three-dimensional space would therefore be required in addition to positional order, that is, an alternating lineup of the bonding and antibonding sites (the back lobes of the same p orbitals) would be a prerequisite for resonance to occur [19, 20]. In fact, however, the bonding and back-lobe antibonding sites are not always arranged coherently in space to form a rock-salt-like structure using p orbitals. The phase-flip between bonding and back-lobe sites may not have coherence in space, preventing the occurrence of resonance. In addition, the randomly distributed vacancies in GSTs would also disturb this coherence (or resonance). Once again, high values of ε_∞ are not a fingerprint of the existence of resonance bonds.

8.2.1.2 Electrical Memories (Phase-Change Random Access Memory)

PCRAM uses repeated *electrical* switching of two phases associated with a large change (~4 orders of magnitude) in resistance, where short nanosecond-scale electrical pulses are used to set, reset, and read the memory [9, 21, 22]. Information is stored and read by measuring the resistance of a large array of PCRAM cells in matrix form. PCRAM can compete in speed (switching and reading time) and in scalability (memory size) with dynamic access memory (DRAM), as well as with the established optical memory technology (DVD). Note that PCRAM is a nonvolatile type of memory, but DRAM is volatile.

Figure 8.7 shows the basic switching actions, in which the typical current–voltage characteristics are shown with current ranges. When the applied voltage is increased while the material is in the high-resistance amorphous phase (we call this the *reset* state), switching occurs, with an increase in current (voltage snapback). At threshold, carrier multiplication by impact ionization and/or Joule heating occurs, and when we keep the voltage in the *set* current zone the material is transformed into the crystalline phase (the *set* state). This means that the low-resistance state is memorized even after the voltage is removed. Deleting the memory proceeds as follows: when the applied voltage (*reset* pulse) exceeds a certain level (corresponding to the *reset* current zone), the crystalline phase melts, and rapid cooling just after the one-shot pulse is stopped produces the high-resistance amorphous state again. The melting point is about 600 °C for GSTs, as already mentioned. The structure of the PCRAM cell is simple [21] and the manufacturing process is easy, and hence the cost-effectiveness might be better than that of Si technologies.

In 2008, PCRAM devices were shipped onto the market. However, PCRAM still does not have a major position among computer devices. One of the

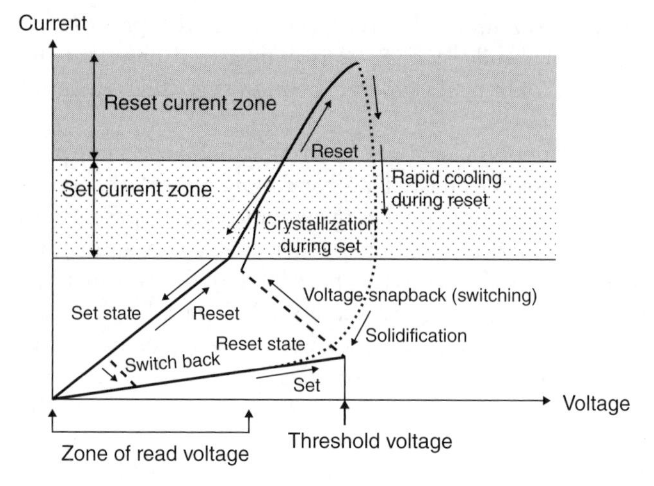

Figure 8.7 Schematic current–voltage characteristics of a threshold switching device and a PCRAM using a phase-change material. *Source:* Terao *et al.* 2009 [21]. Reproduced with permission of The Japan Society of Applied Physics.

reasons for this is the resistance drift of GSTs in the amorphous phase after resetting [23]. The drift of the resistance $R(t)$, an increase in the resistance of the amorphous phase, is described by

$$R(t) = R(t_0)(t/t_0)^{\alpha} \tag{8.2}$$

where t_0 is an arbitrarily chosen zero of time and α is an empirical parameter in the range from 0.03 to 0.1 depending on the device and the initial conditions of the reset.

This issue of resistance drift is still a matter of debate [24]. Among various models, stress relaxation [23, 25] and the defect-related issues [25] have been suggested. All of these models suggest an increase in the "mobility gap," causing an increase in resistance and hence an increase in the switching threshold voltage V_{th}. This increase in V_{th}, which is a device instability, is the matter of most concern in the PCRAM field at present.

A new application of electrical switching memory, using intermediate resistance states (multilevel states) that are controlled by set pulses, also has a high potential for a so-called brain computer [22, 26]. By utilizing continuous transitions between multilevel states in an analog manner, this type of memory system may mimic the behavior of synapses. Resistance drift therefore also disturbs the potential use of a brain computer, when the drift in resistance is larger than the resistance difference between levels.

8.2.2 Direct X-ray Image Sensors for Medical Use

Thick a-Se films have been selected as a candidate for X-ray photoconductors for medical use [27]. This means that X-rays produce enough free electrons and holes in a-Se for this purpose. The charge carriers are collected (using local electronic charge as a memory) and detected in the following way. To fabricate the device, a-Se is coated, usually by evaporation, onto an AMA made from a-Si:H and is connected to storage capacitors for collection of the charges [28]. A metal electrode is deposited on top of the a-Se to apply a biasing voltage. The X-ray-generated carriers in the a-Se travel along the electric field and reach a storage capacitor equipped with a TFT. One set consisting of a TFT and a storage capacitor is called a pixel. Each pixel electrode carries an amount of charge that is proportional to the amount of incident X-ray radiation. Images are obtained by scanning the charge through the AMA.

This type of detector produces excellent X-ray images. A typical example of an X-ray image of a phantom hand obtained from a flat panel detector is shown in Figure 8.8. The resolution is determined by the pixel size of around 100 μm. It should be mentioned that there are other X-ray image sensors, using single-crystal materials such as GaAs and CdTe. The main advantages of using a-Se

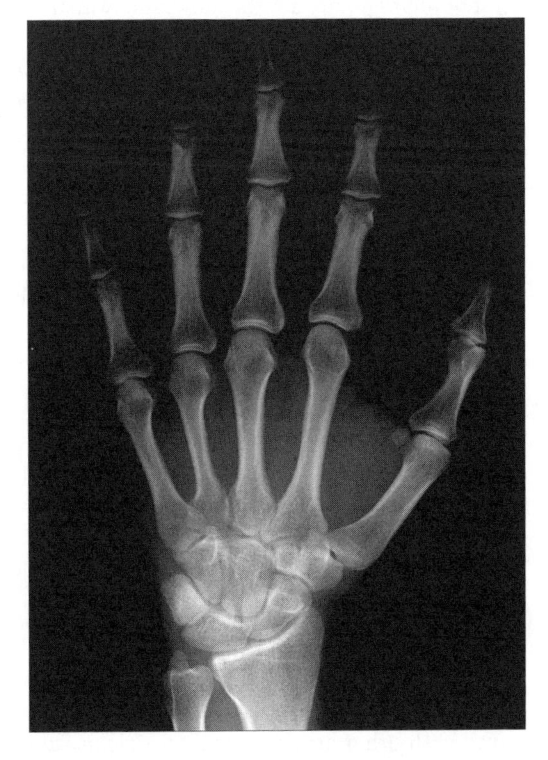

Figure 8.8 X-ray image of a phantom hand obtained from a flat panel image detector using a-Se. *Source:* Kasap *et al.* 2011 [28]. Reproduced with permission.

are easier preparation of large, uniform areas without damaging the underlying AMA substrate, although thicker films are required for sufficient X-ray absorption. For example, for mammography at 20 keV, the required thickness is about 100 μm and for chest radiology at 70 keV, it is about 2 mm, which is, however, not a difficult task. Cesium iodide (CsI) doped with Tl has been used as a scintillator to convert X-ray quanta into visible photon emission on an AMA made of a-Si:H [29]. This is called an *indirect* method, and its cost-effectiveness and sensitivity are almost the same as those of a-Se direct X-ray sensors. Thus, severe competition is continuing in the market.

8.2.3 High-Gain Avalanche Rushing Amorphous Semiconductor Vidicon

Avalanche photomultiplication has been found in a-Se [30]. A device utilizing the avalanche effect in a-Se has been developed by NHK (Japan Broadcasting Corporation) [31]. The name HARP (high-gain avalanche rushing amorphous photoconductor) has been given to vidicon tubes for TV cameras. The HARP vidicon TV camera has more than 100 times the sensitivity of a CCD camera. Color pictures of rainbows at Mount Fuji in Japan and the Iguazu Falls in Brazil under *moonlight* are a well-known demonstration of this TV camera. This success is due to high photosensitivity to visible light and the occurrence of an avalanche effect with free holes under a high electric field. It is of interest to discuss the avalanche effect in amorphous semiconductors, since in general the carrier mean free path is expected to be very short compared with crystalline Si, for example. It might be expected, therefore, that it would not be feasible to produce an avalanche effect in an amorphous semiconductor.

8.2.3.1 Avalanche Effect in Amorphous Semiconductors

Carrier multiplication due to an avalanche effect in a-Se can be observed in either a sandwich or a vidicon structure. In both cases, carrier injection is suppressed by insulating layers. Under avalanche breakdown conditions, the electrical current density increases by up to two orders of magnitude [30, 31]. The most important parameter for characterizing the avalanche effect is the impact ionization factor γ, which is related to the current density J by

$$J = A\exp(\gamma d) \tag{8.3}$$

where A is a constant and d is the thickness. The data symbols in Figure 8.9 show the electric-field dependence of γ obtained by several research groups. The solid line shows the calculated result [32] obtained using the lucky-drift model, which was originally proposed for crystalline solids [33]. In the lucky-drift model applied to a-Se, the existence of less phonon scattering, as compared with other amorphous materials, is suggested to be the most important

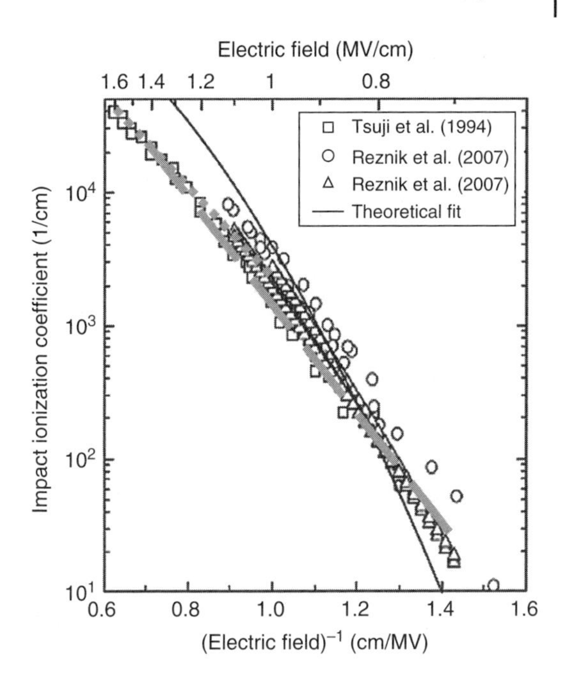

Figure 8.9 Electric-field dependence of impact ionization factor γ obtained by several research groups. *Source:* Tanaka 2014 [35]. Reproduced with permission.

factor for the occurrence of photomultiplication of holes [32]. That is why avalanche breakdown occurs only in a-Se. Note that the similar effect has been found in a-Si:H [34]. However, it is not clear whether or not an avalanche effect actually occurs in a-Si:H [32].

The lucky-drift model for avalanche breakdown appears to work well for a-Se. However, the carrier mean free path between elastic collisions of free holes is deduced to be around 0.4 nm, which is comparable to the interatomic distance [35]. The small mean free path of free carriers violates the fundamental criteria for Boltzmann transport [36]. Hence it should be concluded that the lucky-drift model, applied to a-Se, is still incomplete. Note that the lucky-drift model predicts much longer mean free paths of free electrons in crystalline materials, for example 8 nm for elastic scattering in Si. Similarly to impact ionization in doped crystalline semiconductors [37], the subgap states (localized holes in the valence band tails) may play a role in avalanche breakdown; that is, the initial step of carrier generation originates from the excitation of localized holes [35].

8.2.3.2 Device Structures

The HARP tube is a traditional vacuum-type vidicon camera tube, in which weak light illumination produces a focused optical image on a-Se, which is converted to a very highly conductive state (resulting in more charge) through avalanche carrier multiplication. The conductive pattern is transformed to a current signal by scanning with an electron beam. We need a high electric field, for example 2 kV between the electrodes, to induce avalanche multiplication, and a glass tube. Instead of using a glass tube, a combination of a-Se avalanche multiplication with an active-matrix electron emitter array (like the AMA in the previous section) has been proposed in order to reduce the size of the device (to 5 cm diameter and 1 cm thickness) [38].

8.2.4 Optical Fibers and Waveguides

The development of silica fibers is the most important cornerstone of optical communication. At present, the transmission loss through silica optical fibers is ~0.2 dB/km (optical signals are transmitted over ~80 km). A single fiber transmits signals at, for example, 32 different wavelengths simultaneously. This system is referred to as *wavelength division multiplexing.* The signals of different wavelength are combined and resolved by dispersive elements (fiber Bragg gratings), in which all of the signals are optically amplified using stimulated emission in an Er^{3+}-doped fiber amplifier (EDFA), which will be discussed later.

As chalcogenides transmit infrared light very well, optical fibers made from a-Chs can be useful in particular infrared regions. Note that silica fibers are not useful in the infrared. Thus, chalcogenide fibers are employed for biological and medical use [39–41]. For example, 10–100 W of infrared light ($\lambda = 5$ and 10.6 μm) emitted from a CO or CO_2 laser can be transmitted with a relatively high loss ~12 dB/km in As_2S_3. However, for medical use, this may be sufficient, because a long fiber is not required, unlike the case for communication purposes. A problem may arise with using As_2S_3 for medical applications, however, since it contains As. The search for other chalcogenide fibers is continuing.

Owing to the disorder and flexibility in chalcogenides, incorporating metallic elements such as rare earth ions and transition metals into chalcogenides is easier [42, 43], allowing one to produce active devices such as fiber amplifiers (EDFAs). In a fiber amplifier, rare earth ions function as stimulated-emission centers [44]. There are many useful energy levels for this purpose in, for example, Er^{3+} and Pr^{3+}. Er^{3+} is useful for amplification of 1.5 μm light (0.83 eV) when it is excited by 0.8 μm light (1.55 eV).

Finally, we should mention waveguide devices, which are important components in optical communications. It is possible to fabricate a rare earth ion amplifier in an *optical* integrated circuit. A waveguide with a high refractive index of ~2.5 is suitable for confining light propagation, and a three-dimensional $Ge_{22}As_{20}Se_{58}$ waveguide buried in As_2S_3 glass, with 0.04 dB/cm loss at a wavelength of 9.3 μm, has been made [45].

Ultrafast all-optical waveguide switches using a-Chs are promising. Since the optical nonlinearity of a-Chs is relatively large, that is, the refractive index

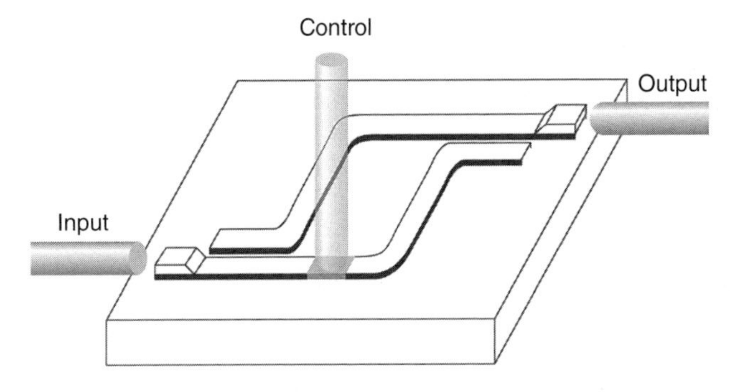

Figure 8.10 An all-optical waveguide switch using a Mach–Zehnder interferometer. *Source:* Tanaka and Shimakawa 2011 [46]. Reproduced with permission from Springer Science and Business Media.

depends strongly on the light intensity, we can modify the optical path length of an arm in a waveguide interferometer as shown in Figure 8.10. Control beam pulses can control the switching on and off of the output signal at high speed (on the order of picoseconds). This device may replace the present commercially used optical switches, which use a Pockels cell and direct laser modulation [46].

References

1 Carlson, D.E. and Wronski, C.R. (1976) Amorphous silicon solar cell. *Appl. Phys. Lett.*, **28**, 671–673.

2 Carlson, D.E., Arya, R.R., Benett, M., *et al.* (1996) Commercialization of multijunction amorphous silicon modules. *Conf. Record 25th IEEE*, 1023–1028.

3 Atwater, H.A. and Polman, A. (2010) Plasmonics for improved photovoltaic devices. *Nature Mater.*, **9**, 205–213.

4 Powell, M.J. (1984) Material properties controlling the performance of amorphous silicon thin film transistors. *Mater. Res. Soc. Symp. Proc.*, **33**, 259–274.

5 Le Comber, P.G. (1989) Present and future applications of amorphous silicon and its alloys. *J. Non-Cryst. Solids*, **115**, 1–13.

6 Matsumura, M. and Hayama, H. (1980) Amorphous-silicon integrated circuit. *Proc. IEEE*, **68**, 1349–1352.

7 Snell, A.J., Mackenzie, K.D., Spear, W.E., Le Comber, P.G., and Hughes, A.J. (1981) Application of amorphous silicon field effect transistors in integrated circuits. *Appl. Phys. A*, **26**, 83–86.

8 Kamiya, T., Nomura, K., and Hosono, H. (2010) Present status of amorphous In–Ga–Zn–O thin-film transistors. *Sci. Technol. Adv. Mater.*, **11**, 044305-23.

9 Wuttig, M. and Yamada, N. (2007) Phase-change materials for rewritable data storage. *Nature Mater.*, **6**, 824–832.
10 Shimakawa, K., Kadlec, F., Kadlec, C., Prikryl, J., Wagner, T., Frumar, M., and Kasap, S. (2016) Effects of grain boundaries on THz conductivity in the crystalline states of GeSbTe phase-change materials: Correlation with dc loss. *J. Appl. Phys.*, submitted.
11 Shimakawa, K., Wagner, T., Frumar, M., Kadlec, F., Kadlec, C., and Kasap, S. (2013) Terahertz and direct current losses and the origin of non-Drude terahertz conductivity in the crystalline states of phase change materials. *J. Appl. Phys.*, **114**, 233105-7.
12 Shportko, K., Kremers, S., Woda, M., Lencer, D., Robertson, J., and Wuttig, M. (2008) Resonant bonding in crystalline phase-change materials. *Nature Mater.*, **7**, 653–658.
13 Orava, J., Wagner, T., Sik, J., Prikyl, J., Frumar, M., and Benes, L. (2008) Optical properties and phase change transition in $Ge_2Sb_2Te_5$ flash evaporated thin films studied by temperature dependent spectroscopic ellipsometry. *J. Appl. Phys.*, **104**, 043523-5.
14 Jellison, G.E. Jr. and Modine, F.A. (1996) Parameterization of the optical functions of amorphous materials in the interband region. *Appl. Phys. Lett.*, **69**, 371–373. Erratum (1996) *Appl. Phys. Lett.*, **69**, 2137.
15 Shimakawa, K., Strizik, L., Wagner, T., and Frumar, M. (2015) Penn gap rule in phase-change memory materials: No clear evidence for resonance bonds. *APL Mater.*, 041801-5.
16 Penn, D.R. (1962) Wave-number-dependent dielectric function of semiconductors. *Phys. Rev.*, **128**, 2093–2097.
17 J.C. Phillips (1973) *Bonds and Bands in Semiconductors*, Academic Press, New York and London.
18 Huang, B. and Robertson, J. (2010) Bonding origin of optical contrast in phase-change memory materials. *Phys. Rev. B*, **81**, 081204 (R)-4.
19 Kolobov, A.V., Krbal, M., Fons, P., Tominaga, J., and Uruga, T. (2011) Distortion-triggered loss of long-range order in solids with bonding energy hierarchy. *Nature Chem.*, **3**, 311–316.
20 Kolobov, A.V., Fons, P., Tominaga, J., and Hase, M. (2014) Excitation-assisted disordering of GeTe and related solids with resonant bonding. *J. Phys. Chem.*, **118**, 10248–10253.
21 Terao, M., Morikawa, T., and Ohta, T. (2009) Electrical phase change memory: State of the art. *Japan. J. Appl. Phys.*, **48**, 080001-14.
22 Raoux, S., Xiong, F., Wuttig, M., and Pop, E. (2014) Phase change materials and phase change memory. *MRS Bull.*, **39**, 703–710.
23 Karpov, I.V., Mitra, M., Kau, D., Spandini, G., Kryukov, Y., and Kaprov, G. (2007) Fundamental drift of parameters in chalcogenide phase change memory. *J. Appl. Phys.*, **102**, 124503-6.

24 Mitra, M., Jung, Y., Gianola, D.S., and Agarwal, R. (2010) Extremely low drift of resistance and threshold voltage in amorphous phase change nanowire devices. *Appl. Phys. Lett.*, **96**, 222111-1–3.

25 Ielmini, D., Lacaita, A.L., and Mantegazza, D. (2007) Recovery and drift dynamics of resistance and threshold voltages in phase-change memories. *IEEE Trans. Electron Devices*, **54**, 308–314.

26 Suri, M., Bichler, O., Querlioz, D., Traor, B., Cueto, O., Perniola, L., Sousa, V., Vuillaume, D., Gamrat, C., and DeSalvo, B. (2012) Physical aspects of low power synapses based on phase change memory devices. *J. Appl. Phys.*, **112**, 054904–10.

27 Rawlands, J.A. and Kasap, S.O. (1997) Amorphous semiconductors usher in digital X-ray imaging. *Phys. Today*, **50**, 24–30.

28 Kasap, S., Frey, J.B., Belev, G., Tousignant, O., Mani, H., Greenspan, J., Laperriere, L., Buben, O., Reznik, A., DeCrescenzo, G., Karim, K.S., and Rowland, J.A. (2011) Amorphous and polycrystalline photoconductors for direct conversion flat panel x-ray image sensors. *Sensors*, **11**, 5112–5157.

29 Hohei, M., Arques, M., Chabbal, J., Chaussat, C., Ducourant, T., Hahm, G., Horbascheck, H., Shulz, R., and Spahn, M. (1998) Amorphous silicon X-ray detectors. *J. Non-Cryst. Solids*, **227–230**, 1300–1305.

30 Juska, G., Arlauskas, K., and Montrimas, E. (1987) Features of carriers at very high electric fields in a-Se and a-Si:H. *J. Non-Cryst. Solids*, **97–98**, 559–561.

31 Tanioka, K. (2007) The ultrasensitive TV pickup tube from conception to recent development. *J. Mater. Sci.*, **18**, S321–S325.

32 Rubel, O., Potvin, A., and Laughton, D. (2011) Generalized lucky-drift model for impact ionization in semiconductors with disorder, *J. Phys.: Condens. Matter*, **23**, 055802-7.

33 Riedley, B.K. (1983) Lucky-drift mechanism for impact ionization in semiconductors. *J. Phys. C: Solid State Phys.*, **16**, 3373–3388.

34 Akiyama, M., Hanada, M., Sawada, K., and Ishida, M. (2003) Multiplication characteristics of a-Si:H p-i-n photodiode film in high electric field. *Japan. J. Appl. Phys.*, **42**, 2345–2348.

35 Tanaka, K. (2014) Avalanche breakdown in amorphous selenium (a-Se) and related materials: Brief review, critique, and proposal. *J. Optoelectron. Adv. Mater.*, **16**, 243–251.

36 Mott, N.F. (1993) *Conduction in Non-Crystalline Materials*, 2nd edn, Oxford University Press, Oxford.

37 Bringuier, E. (1994) High-field transport statistics and impact excitation in semiconductors, *Phys. Rev. B*, **49**, 7974–7989.

38 Negishi, N., Matsuba, Y., Tanaka, R., Nakada, K. Sakemura, T., Okuda, Y., Watanabe, A., Yoshikawa, T., and Ogasawara, K. (2007) Development of a high-resolution active-matrix electron emitter array for application to high-sensitivity image sensing. *J. Vac. Sci. Technol.*, **25**, 661–665.

39 Nishii, J. and Yamashita, T. (1998) Chalcogenide glass-based fibers, in *Infrared Fiber Optics* (eds J.S. Sanghera and I.D. Aggarwal), CRC Press, Boca Raton, FL, pp. 143–184.

40 Snopatin, R.E., Churbanov, M.F., Pushikin, A.A., Gerasimenko, V.V., Dianov, E.M., and Plotnichenko, V.G. (2009) High purity arsenic-sulfide glasses and fibers with minimum attenuation of 12 dB/km. *J. Optoelectron. Adv. Mater. Rapid Commun.*, **3**, 669–671.

41 Bureau, B., Maurugeon, S., Charpentier, F., Adam, J.-L., Boussard-Pledel, C., and Zhang, X. (2009) Chalcogenide glass fibers for infrared sensing and space optics. *Fiber Integrated Opt.*, **28**, 65–80.

42 Morishita, Y. and Tanaka, K. (2003) Microscopic structures in Co-doped SiO_2–GeO_2 glasses and fibers. *Japan. J. Appl. Phys.*, **42**, 7456–7460.

43 Tver'yanovichi, Yu.S. and Tverjanovich, A. (2004) Rare-earth doped chalcogenide glass, in *Semiconducting Chalcogenide Glass III* (eds R. Fairman and B. Ushkov), Elsevier, Amsterdam, pp. 169–207.

44 Desurvire, E. (1994) *Erbium-Doped Fiber Amplifiers: Principles and Applications*, John Wiley & Sons, Inc., New York.

45 Coulombier, Q., Zhang, S.Q., Zhang, X.H., Bureau, B., Lucas, J., Boussard-Pledel, C., Troles, J., Calvez, L., Ma, H., Maurugeon, S., and Guillevic, S. (2008) Planar waveguide obtained by burying a $Ge_{22}As_{20}Se_{38}$ fiber in As_2S_3 glass. *Appl. Opt.*, **47**, 5750–5752.

46 Tanaka, K. and Shimakawa, K. (2011) *Amorphous Chalcogenide Semiconductors and Related Materials*, Springer, New York.

Index

a

a-As_2S_3 films, 176, 177
Abell–Tersoff potentials, 23
AC conduction, 80–86, 206
  continuous-time random walk (CTRW), 83
  Debye-type dielectric relaxation, 82
  dispersive AC loss, 82
  Dyre's PPM approach, 84
  effective medium approximation (EMA), 83
  hopping carriers, 82
  hopping/tunneling of electronic/ionic charge, 82
  Maxwell–Wagner type effect, 82, 83
  Maxwell–Wagner type inhomogeneities, 85
  Maxwell–Wagner type relaxation, 85
  percolation path method (PPM), 84
  photoillumination, 86
AC transport, 83, 168–169
amorphous Ag-containing chalcogenides (Ag-Chs)
  applications, 242
  photodoping, 241–242
  preparation techniques, 240
  structure and physical properties, 240–241
amorphous carbon (a-C), 83
  applications, 234–235
  basic structure, 232–233
  preparation techniques, 233
  sp^2 and sp^3 hybrids, 232
  structural studies, 233–234
amorphous chalcogenides (a-Chs), 3
  applications, 157–158
  direct X-ray image sensors, 257–258
  electronic transport
    AC transport, 168–169
    DC transport, 165–167
    Hall effect, 167
    thermoelectric power, 167–168
  evaporation, 7–8
  fundamental optical absorption
    a-As_2S_3 films, 176, 177
    fractal concepts, 175
    Tauc relation, 175, 176
    valence band, 175
  glass formation
    crystallization, 162–165
    glass transition temperature, 160–162
    ternary alloy systems, 160
    thermal analysis/differential scanning calorimetry, 160

Amorphous Semiconductors: Structural, Optical, and Electronic Properties, First Edition.
Kazuo Morigaki, Sándor Kugler, and Koichi Shimakawa.

amorphous chalcogenides (a-Chs) (*cont'd*)
- high-gain avalanche rushing amorphous photoconductor
  - avalanche effect, 258–259
  - device structures, 259–260
- ionic transport
  - Dyre's random-walk process, 171, 172
  - equivalent electrical circuit, 171
  - impedance spectroscopy (IS), 170
  - power-law dependence, 172
  - silver concentration dependence, 172, 173, 175
- light-induced effects
  - defect creation kinetics, 207–210
  - electronic transport, 206–207
  - electron spin resonance, 200–202
  - optical absorption, 202–203
  - photoconductivity, 205–206
  - photodarkening, 210–213
  - photoinduced volume expansion (PVE), 213–215
  - photoluminescence, 203–204
- MQ method, 7
- nature of defects and defect spectroscopy
  - electronic transport, 199–200
  - electron-phonon coupling, 192
  - electron spin resonance, 196–197
  - electrophotography, 199
  - ESR signals, 192
  - IVAPs, 195
  - law of mass action, 195
  - n-type doping, 195
  - optical absorption, 197
  - primary photoconductivity, 197
  - p-type doping, 195
  - secondary photoconductivity, 197–198
- optical fibers and waveguides, 260–261
- phase-change materials (PCM)
  - angular-frequency-dependent conductivity, GST225 films, 249
  - crystalline and amorphous states, 250
  - electrical memories (PCRAM), 255–256
  - electronic transitions, 251
  - melt-quenching method, 249
  - nonvolatile electronic memory, 249
  - optical memories (DVDs), 252–255
  - physical properties, 249
  - resonance bonds, 251, 252
- photoconduction
  - primary photoconduction measurements, 183–185
  - secondary photoconduction measurements, 186–191
- photoluminescence (PL)
  - broadband excitation, 183
  - characteristic features, 180
  - charged defect concept, 181
  - continuous-wave (CW) light sources, 179
  - frequency-resolved photoluminescence, 182
  - frequency-resolved spectroscopy (FRS), 179
  - photoluminescence peak energy *vs.* photoluminescence excitation (PLE) peak energy, 181, 182
  - QFRS technique, 183
  - quadrature frequency-resolved spectroscopy (QFRS), 179
  - self-trapped exciton (STE), 180
- pulsed laser deposition, 8
- Urbach and weak absorption tails, 178–179

amorphous oxide semiconductor (AOS)
- electronic properties, 237–239
- optical properties, 236–237
- preparation techniques, 235–236

amorphous selenium (a-Se), 1, 3, 11, 13, 157
  charged defects, 193
  chemical bond argument, 192
  molecular dynamics simulation, 35–37
  neutron diffraction, 19–20
  reverse Monte Carlo (RMC) simulation, 30–31
amorphous semiconductors
  applications, 3
  avalanche effect, 258–259
  bonding structures
    in Column IV Elements, 44–45
    in Column VI Elements, 45–46
  charge distribution, 49–52
  and crystalline semiconductors, 49
  dangling bonds, 54–56
  density of states, 52–54
  doping, 57–58
  electronic structure, 46–47
  Fermi energy, 47–49
  structures of, 16
amorphous silicon (a-Si)
  MC simulation, 25–26
  molecular dynamics simulation, 34
  neutron diffraction, 17–18
  reverse Monte Carlo (RMC) simulation, 28–30
avalanche effect, 258–259

b

band conduction, 75–76
band tails
  LESR measurements
    decay of, 63
    ENDOR measurements, 62
    isotropic and anisotropic hyperfine interaction constants, 64
    ODENDOR measurements, 63
    photomodulated infrared (IR) measurements, 65
    self trapped hole centre, 64, 65
    time resolved ODMR measurements, 62
bonding defects, 159

c

chalcogenide glasses. *see* amorphous chalcogenides (a-Chs)
charged-dangling bond (CDB) model, 192
compositional disorder, 1, 178
computer simulations
  a-Si Model, MC simulation, 25–26
  atomic interactions
    Abell-Tersoff potentials, 23
    defect free diamond-type crystals, 21
    empirical potentials, 21
    Keating potential, 22
    quantum mechanical energy calculations, 24–25
    Stillinger-Weber empirical potential, 22, 23
  Car and Parrinello Method, 37
  molecular dynamics simulation
    a-Se model, 35–37
    a-Si:H model, 34–35
    a-Si model, 34
    ATOMDEP, 33
    film growth, 33
    rapid-cooling technique, 33
  Monte Carlo-type methods, 20–21
  reverse Monte Carlo (RMC) simulation
    a-Se model, 30–31
    a-Si model, 28–30
    constraints, 27–28
    RMC algorithm, 26–27
    structure factor and radial distribution function, 26
continuous-time random walk (CTRW), 83
continuous-wave (CW) light sources, 179
coordination defect, 191

d

dangling bonds, 54–56, 74
DC conduction
  band conduction, 75–76
  hopping conduction
    nearest-neighbor hopping, 76–77
    variable range hopping, 78–80
DC transport, 165–167
diamondlike carbon (DLC), 231, 233
digital versatile disc (DVD), 157
direct X-ray image sensors, 257–258
doping effect, 89–92
  doping efficiency, 89, 90
  n-type and p-type doping, 89
  valence electrons, 91
Dyre's PPM approach, 84

e

effective medium approximation (EMA), 83
electrically detected magnetic resonance (EDMR), 74
electronic density of states (DOS), 159
electronic transport
  AC transport, 168–169
  DC transport, 165–167
    Boltzmann transport theory, 167
    Meyer–Neldel rule, 165, 166
    multiexcitation entropy, 166
  Hall effect, 167
  thermoelectric power, 167–168
electron magnetic resonance
  pulsed electron–nuclear double resonance
    Davies method, 118
    ENDOR efficiency, 119
    ENDOR measurement, 117, 118
    hydrogen-related dangling bonds, 120–121
    microwave power dependences, 121
    three-pulse sequence, 119
  rabi oscillations
    dangling bond ODMR signal, 128
    direct and indirect observations, 124
    Fourier transform, 125
    free induction decay, 124
    magnetic field dependence, 127
    PL and photoconductivity (PC), 124
    threedimensional contour plot, TN signal, 127
  spin echo phenomena and ESEEM, 112–117
    electron-nucleus interaction, 114
    motion of magnetization vectors, 112
    pulse sequence, 113
    three-pulse ESEEM spectra, 116, 117
    three-pulse spin echo method, 113
    two-pulse echo decay and time-domain ESEEM spectrum, 115
electron-nuclear double resonance (ENDOR), 2
electron–phonon coupling, 192
electron spin resonance, 200–202
  light-induced effects, 200–202
  nature of defects and defect spectroscopy, 196–197
electron spin resonance (ESR), 2, 192
electrophotography, 157, 199, 247
energy space, 159
equivalent electrical circuit (EEC), 171

f

Fermi energy, 47–49
flat panel displays (FPDs), 239
frequency-resolved photoluminescence, 182
frequency resolved spectroscopy (FRS)
  geminate and nongeminate recombination, 99–100

localized triplet excitons, 100–101
principle, 96–99
frequency-resolved spectroscopy (FRS), 179

g
Ge–S(Se) systems, 196
glass-forming process, 12

h
Hall effect, 167
HWCVD technique, 6
hydrogenated amorphous silicon (a-Si:H)
applications, 3
band tails, 62–66
electrical properties
AC conduction, 80–86
DC conduction, 74–80
doping effect, 89–92
Hall effect, 87–88
thermoelectric power, 88–89
electron magnetic resonance, 112–128
HWCVD technique, 6
light-induced defect creation
dangling bond density, 139
hydrogen-related dangling bond, 135
hydrogen-related dangling bonds, 141
light-induced dangling bonds, 143
Monte Carlo computer simulation, 144
normal dangling bond, 135
rate equation model, 136
recombination energy, 135
values of parameters, 139
light-induced phenomena
conductivity and photoconductivity, 132–133
ESR, ODMR, 133–134
photoluminescene, 133
molecular dynamics simulation, 34–35
neutron diffraction, 18–19
photovoltaics, 247–248
plasma-enhanced chemical vapor deposition (PECVD), 5–6
recombination processes (*see* recombination processes)
spin-dependent properties
spin-dependent photoinduced absorption, 129–131
spin-dependent transport, 129
structural defects, 66–68
thin-film transistors, 248–249

i
impedance spectroscopy (IS), 170
In_2O_3-ZnO-Ga_2O_3 films, 235
International Union of Crystallography, 12

k
Keating potential, 22
KJMA approach, 164

l
law of mass action, 194
LESR measurements
decay of, 63
ENDOR measurements, 62
isotropic and anisotropic hyperfine interaction constants, 64
ODENDOR measurements, 63
photomodulated infrared (IR) measurements, 65
self trapped hole centre, 64, 65
time resolved ODMR measurements, 62
lowest unoccupied molecular orbital (LUMO), 215

m
Maxwell–Wagner type effect, 82, 83
melt-quenching method, 249
microwave PECVD (MW PECVD), 6
molecular dynamics simulation
  a-Se model, 35–37
  a-Si:H model, 34–35
  a-Si model, 34
  ATOMDEP, 33
  film growth, 33
  rapid-cooling technique, 33
Monte Carlo computer simulation, 144
Monte Carlo-type methods, 20–21

n
negative-U defects, 191
neutron diffraction
  amorphous selenium, 19–20
  amorphous silicon, 17–18
  hydrogenated amorphous silicon, 18–19

o
optical absorption, 202–203
optical properties
  frequency resolved spectroscopy (FRS), 96–101
  fundamental optical absorption, 92–94
  photoconductivity
    photocurrent, 101–102
    range III photocondction characteristics, 107–109
    range II potoconduction characteristics, 106–107
    range I potoconduction characteristics, 104–106
    recombination, 101
    trapping and recombination, 102
  photoluminescence, 96
  weak absorption, 94–96
optical spectroscopy
  infrared (IR) absorption, 13–15
  Raman scattering, 12–13

p
percolation path method (PPM), 84
phase-change materials (PCM)
  angular-frequency-dependent conductivity, GST225 films, 249
  crystalline and amorphous states, 250
  electrical memories (PCRAM), 255–256
  electronic transitions, 251
  melt-quenching method, 249
  nonvolatile electronic memory, 249
  optical memories (DVDs), 252–255
  physical properties, 249
  resonance bonds, 251, 252
phase-change random access memory (PRAM), 158
photoconductivity, 205–206
  primary photoconduction measurements, 183–185
  secondary photoconduction measurements, 186–191
photodarkening, 159, 210–213
  definition, 210
  short-time dynamics, 212, 213
  temperature dependence, 210
  transient change, 212
photodoping, 241–242
photoillumination, 242
photoinduced defect creation (PDC), 208
photoinduced volume expansion (PVE), 213–215
photoluminescence (PL), 203–204
  broadband excitation, 183
  characteristic features, 180
  charged defect concept, 181
  continuous-wave (CW) light sources, 179

frequency-resolved photoluminescence, 182
frequency-resolved spectroscopy (FRS), 179
QFRS technique, 183
quadrature frequency-resolved spectroscopy (QFRS), 179
self-trapped exciton (STE), 180
toluminescence peak energy *vs.* photoluminescence excitation (PLE) peak energy, 181, 182
photovoltaics, 247–248
plasma-arc deposition, 234
plasma-enhanced chemical vapor deposition (PECVD), 5–6
plasmon effects, 248
primary photoconductivity, 197
pulsed electron–nuclear double resonance
Davies method, 118
ENDOR efficiency, 119
ENDOR measurement, 117, 118
hydrogen-related dangling bonds, 120–121
microwave power dependences, 121
three-pulse sequence, 119
pulsed laser deposition (PLD), 8, 235

q

quadrature frequency-resolved spectroscopy (QFRS), 179, 183

r

Rabi oscillations
dangling bond ODMR signal, 128
direct and indirect observations, 124
Fourier transform, 125
free induction decay, 124
magnetic field dependence, 127
PL and photoconductivity (PC), 124
threedimensional contour plot, TN signal, 127
radial distribution function (RDF), 16, 18, 26, 36, 37, 234
recombination processes
nonradiative recombination, 70–72
radiative recombination
exciton recombination, 70
geminate electron–hole pair recombination, 68
nongeminate electron–hole pair recombination, 68–70
and recombination centers, in a-Si:H, 72–73
spin dependent recombination, 73–74
reverse Monte Carlo (RMC) simulation
a-Se model, 30–31
a-Si model, 28–30
constraints, 27–28
RMC algorithm, 26–27
structure factor and radial distribution function, 26

s

secondary photoconductivity, 197–198
self-trapped exciton (STE), 180
solar cells, 3, 11, 247, 248
Special Environment Powder Diffractometer, 234
spin echo phenomena and ESEEM, 112–117
electron-nucleus interaction, 114
motion of magnetization vectors, 112
pulse sequence, 113
three-pulse ESEEM spectra, 116, 117
three-pulse spin echo method, 113
two-pulse echo decay and time-domain ESEEM spectrum, 115
Staebler–Wronski effect, 131
Stillinger–Weber empirical potential, 22, 23

structural defects, 1
  ENDOR measurements, 67
  hydrogen related dangling bond, 68
  normal dangling bond, 67
  self rapped holes, 67
  vacancy hydrogen complex, 67

t

thermal deposition, 6
thermoelectric power, 167–168
thin-film transistors, (TFTs) 3, 235, 248–249
Twyman–Green interferometry, 213

v

valence-alternation pair (VAP), 193
very high frequency field PECVD, 6

x

xerography, 247